Tourism Employment in Nordic Countries

"A great read that offers critical insights into a variety of issues relating to tourism employment. Human interactions are at the core of the tourism experience, and the importance of the people working in the tourism and hospitality industry cannot be overstated. This book tackles the key issues of tourism employment in the Nordic countries and is a must read for tourism professionals and decision makers, as well as tourism and hospitality students and academics. Certainly, an important contribution to the field of tourism."

—Guðrún Þóra Gunnarsdóttir, *Director, The Icelandic Tourism Research Centre*

"*Tourism Employment in Nordic Countries* provides the first major overview of an agonizingly neglected topic in global tourism research. The editors collect a wide array of cases from all over the Nordic areas covering plentiful aspects of tourism employment and contributing important knowledge with significance far beyond the Nordic region. Although based on pre-Covid-19 experiences, the volume offers many insights that will help enable a recovery of the tourism industry into a righteous state."

—Dieter Müller, *Professor of Human Geography, Umeå University, Sweden*

Andreas Walmsley · Kajsa Åberg ·
Petra Blinnikka · Gunnar Thór Jóhannesson
Editors

Tourism Employment in Nordic Countries

Trends, Practices, and Opportunities

palgrave
macmillan

Editors
Andreas Walmsley
International Centre for Transformational Entrepreneurship
Coventry University
Coventry, UK

Petra Blinnikka
JAMK University of Applied Sciences
Jyväskylä, Finland

Kajsa Åberg
Region Västerbotten
Umeå, Sweden

Gunnar Thór Jóhannesson
Faculty of Life and Environmental Sciences
University of Iceland
Reykjavík, Iceland

ISBN 978-3-030-47812-4 ISBN 978-3-030-47813-1 (eBook)
https://doi.org/10.1007/978-3-030-47813-1

This Palgrave Macmillan imprint is published by the registered company Springer Nature Switzerland AG
The registered company address is: Gewerbestrasse 11, 6330 Cham, Switzerland

Contents

1 **Introduction** 1
Andreas Walmsley, Kajsa Åberg, Petra Blinnikka and Gunnar Thór Jóhannesson

2 **Guides on a Crossroad: Between Deregulation and Entrepreneurship** 19
Jane Widtfeldt Meged

3 **Tourism Employment and Education in a Danish Context** 37
Ida Marie Visbech Andersen

4 **Tourism Work: Public Management of the Tourism Workforce in Finland** 57
Anu Harju-Myllyaho, Maria Hakkarainen and Mari Vähäkuopus

5 **A Potential Treasure for Tourism: Crafts as Employment and a Cultural Experience Service in the Nordic North** 77
Outi Kugapi, Maria Huhmarniemi and Laura Laivamaa

6 **Hardworking, Adaptive, and Friendly: The Marketing of Volunteers in Iceland** 101
Jónína Einarsdóttir and Guðbjörg Linda Rafnsdóttir

7 **On the Move: Migrant Workers in Icelandic Hotels** 123
Margrét Wendt, Gunnar Thór Jóhannesson and Unnur Dís Skaptadóttir

8 **"You Need to Consider How It Looks in the Eyes of the Guest": The Work of Teenage Girls in Tourism in Iceland** 143
Anna Vilborg Einarsdóttir and Laufey Haraldsdóttir

9 **Migrant Workers in Tourism: Challenges of Unions and Workers in the Icelandic Tourism Boom** 169
Magnfríður Júlíusdóttir and Íris H. Halldórsdóttir

10 **Employee Motivation and Satisfaction Practices: A Case from Iceland** 195
Magnus Asgeirsson, Paulína Neshybová, Brynjar Thor Thorsteinsson and Ester Gústavsdóttir

11 **Managerial Practices of Co-creation and Psychosocial Work Outcomes** 213
Olga Gjerald and Trude Furunes

12 **Seasonal Workers as Innovation Triggers** 235
Birgitta Ericsson, Kjell Overvåg and Cecilia Möller

13 **Gateway, Fast Lane, or Early Exit? Tourism and Hospitality as a First Employer of Norwegian Youth** 257
Åse Helene Bakkevig Dagsland, Richard N. S. Robinson and Matthew L. Brenner

14 **Labour Mobility in the Tourism and Hospitality Sector in Sweden** 279
Mats Lundmark

15 **Downshifting Dutch Rural Tourism Entrepreneurs in Sweden: Challenges, Opportunities and Implications for the Swedish Welfare State** 303
Marco Eimermann, Charlotta Hedberg and Urban Lindgren

16 **Sustainable Tourism Employment, the Concept of Decent Work, and Sweden** 327
Tara Duncan, Anna Gudmundsson Hillman and Jörgen Elbe

17 **A Labour Regime Perspective on Workforce Formation in Nordic Tourism: Exploring National Tourism Policy and Strategy Documents** 349
Dorothee Bohn and Cecilia De Bernardi

18 **Battling the Past: Social, Economic, and Political Challenges to Indigenous Tourism Employment** 375
Ellen A. Ahlness

19 **Hospitality Through Hospitableness: Offering a Welcome to Migrants Through Employment in the Hospitality Industry** 401
Tone Therese Linge, Trude Furunes, Tom Baum and Tara Duncan

20 **Tourism Employment in Nordic Countries: Trends, Practices and Opportunities** 425
Andreas Walmsley, Kajsa Åberg, Petra Blinnikka and Gunnar Thór Jóhannesson

Index 443

Notes on Contributors

Kajsa Åberg Tourism strategist at the regional development organisation of Västerbotten in northern Sweden, engaged in strengthening the connections between academia, policymaking and practitioners. Current and prior endeavours address the view on knowledge in both operational and strategic spheres of tourism development and scrutiny of destination collaborations between private and public actors. Her doctoral thesis "Anyone could do that" Nordic perspectives on competence in tourism was presented at the Department of Geography and Economic History, Umeå University. Kajsa's background in tourism includes being an employee, employer, destination organisation board member and founder of a lighthouse museum. Associations in academia include the Nordic Society for Tourism and Hospitality Research, the Tourism Education Future Initiatives and the Nordic Tourism Workforce Research Group based in ITRC (Icelandic Tourism Research Centre), Reykjavik.

Ellen A. Ahlness is a Fellow with the Washington Institute for the Study of Inequality and Race at the University of Washington. She attended Telemark University College and Northern Arctic Federal University's international Ph.D. programme Russia in the Arctic:

Global and Local Contexts. She currently sits on the advisory board for the *Upping the Anti Journal* and assists with the Jackson School of International Studies' 2020 Arctic Task Force. Her work on circumpolar Indigenous affairs has been published in Arctic Yearbook, Ecologia Politica, Current Developments in Arctic Law and Managing Multicultural Scandinavia, among other journals.

Ida Marie Visbech Andersen is a Lecturer in Tourism Management at UCL, University College Denmark. She attended the University of Southern Denmark, where she received her Master's Degree in International Tourism and Leisure Management in 2012. She also holds a Graduate Diploma in International Tourism and Hospitality from Griffith University, Australia. Her main research interests are Sustainable Tourism Development, E-tourism and Smart Tourism Businesses.

Magnus Asgeirsson is an Adjunct at the University of Iceland, Department of Tourism and Geography. He holds a Master's Degree in business, from the University of Iceland and is currently working towards his Ph.D. in the field of service management. His main research focus is on service culture and service design, customer satisfaction, internal marketing and service quality.

Tom Baum is Professor of Tourism Employment at the University of Strathclyde, Glasgow, Scotland. Baum researches tourism workforce issues in a social and cultural context and has addressed these issues at a policy and practice level with a particular focus on Africa and Asia. Baum writes, teaches and consults extensively in this area.

Petra Blinnikka (Ph.D. student, M.Sc. [Econ], BHM) works as a senior lecturer in tourism and services management in School of Business, JAMK University of Applied Sciences Jyväskylä, Finland. She has worked within tourism development and education during the last eighteen years in hotel business, education and development, rural tourism, responsible tourism and event management. Her special interests are responsibility in tourism, tourism business development, event management and congress and meeting management. Petra has worked as a project manager in several tourism development projects, mainly focusing on responsibility and sustainability issues. She is a founder and coordinator of the International Centre for Responsible Tourism Finland and has been working within responsibility issues in tourism both in local, regional, national and international level. Most of her research and publications are about sustainable and responsible tourism and accessibility in tourism. At the moment she is a Ph.D. student at the University of Lapland focusing on the effects of strong tourism growth on the destination area and tourism policy, and to find solutions to prevent overtourism. Associations in academia include The Finnish Society for Tourism Research and Tourism Workforce Research Group based in ITRC (Icelandic Tourism Research Centre), Reykjavik. She is a chairman in Jyväskylä Congress Bureau Steering committee.

Dorothee Bohn is a Ph.D. candidate in Human Geography at Umeå University, Umeå, Sweden. Her dissertation focuses on the nexus between political economy and tourism development in the European Arctic. She is a member of the Formas funded research project: Climate Change and the Double Amplifications of Arctic Tourism: Challenges and Potential Solutions for Tourism and Sustainable Development in an Arctic Context. In addition to issues of tourism policy and planning, Dorothee is interested in exploring labour relations in the service sector and the intersection of economy and environment.

Matthew L. Brenner earned a Ph.D. from the University of Queensland in Australia. He currently serves as a sessional lecturer and research assistant at UQ. With an academic background in accounting, Matthew also completed a Master of Science in Hotel, Restaurant and Institutional Management from The Pennsylvania State University in the United States. While at Penn State, Matthew also was an instructor in the School of Hospitality Management. Prior to his time in academia, Matthew held numerous culinary and foodservice management positions in various hotels, resorts, restaurants and private clubs throughout Australia, Canada, and the United States.

Åse Helene Bakkevig Dagsland is Associate Professor at the University of Stavanger, Norwegian School of Hotel Management, with teaching and research in the field Leadership and organisations, and Human Resource Management. Her Ph.D. in Leadership investigated socialisation of young workforce to the hospitality industry, focusing specifically on the apprentices' situation. Her academic background is in pedagogics, with a higher/final degree from the University of Oslo in Educational Science, and with her expertise in teaching and supervision, she is a valued lecturer and course supervisor. Before joining faculty, she held several leader positions within the university college/university system, as rector, general manager and HR-manager.

Cecilia De Bernardi is currently a Doctoral Candidate at the Centre for Tourism and Leisure Research (CeTLeR) at Dalarna University in Sweden. She is also affiliated with the Multidimensional Tourism Institute at the University of Lapland in Rovaniemi, Finland. Her publications concentrate mostly on the conceptualisation of authenticity in indigenous tourism, especially regarding marketing. She has also published on philosophical issues in academic work and on tourism policy related to winter tourism in the Nordic countries. Cecilia has several research interests, both closely related to her discipline, tourism, but also just pertaining to the philosophical approach critical realism. She is now in the final stages of her doctoral research process.

Tara Duncan is a Senior Lecturer at Dalarna University, Sweden and Chair of ATLAS (Association for Tourism and Leisure Education and Research). Duncan's main area of research focuses on the intersections between tourism, work and mobilities with a focus on young budget travellers, academic mobility and sustainability within tourism and hospitality careers.

Marco Eimermann is Research Assistant Professor at the Department of Geography, and affiliated with the Arctic Research Centre (ARCUM) Umeå University (Sweden). His research interests regard lifestyle migration, downshifting practices and sustainable entrepreneurship in rural, remote and northern sparsely populated areas. Eimermann is involved in the Lifestyle Migration Hub, an expanding network of migration scholars studying social rather than economic reasons for voluntary mobility and migration across the globe. He has published peer-reviewed articles, book chapters and books about transient and strategically switching populations in Europe.

Anna Vilborg Einarsdóttir holds an M.Sc. degree in tourism from the University of Iceland. Her main work has been teaching and management in the Icelandic School system and for a short time Iceland's representant in the Nordic network for adult learning. Now Anna Vilborg is Assistant Professor at Holar University in Iceland where she has taken part in diverse research and conducted her own within the field of workforce in tourism with focus on girls, under 18-years old, work in the tourism industry as well as the role of tour guides in nature conservation.

Jónína Einarsdóttir is a Professor of Anthropology. She has a Ph.D. in Anthropology from Stockholm University, Sweden. Einarsdóttir has studied maternal grief, child death, infanticide, breastfeeding and child trafficking in Guinea-Bissau. In Iceland, her research focuses on ethical questions related to the treatment of extremely preterm infants and the implications of their birth on the daily life of their families, domestic violence and punishment of children, as well as the custom to send urban children to stay and work on farms during the summertime. Additional themes of research are fragile states, development and disaster aid, and health care systems.

Jörgen Elbe holds a Ph.D. from Uppsala University, Sweden, and is a Professor in Business Studies at Dalarna University, Sweden. He has a theoretical interest in economic sociology which he uses in empirical studies of tourism development and destination marketing, as well as studies in other areas. He has published several articles in international journals, books and book chapters. He currently holds the position of Pro-Vice-Chancellor of Dalarna University, Sweden.

Birgitta Ericsson Eastern Norway Research Institute, Inland Norway University of Applied Sciences, Norway. Research Professor, M.A. in Planning. Main research topics are tourism, tourism impacts, regional economics, regional development and innovation, as well as cultural and creative industries and festivals in a regional and economic context.

Trude Furunes is Professor in Leadership and Organisation at the Norwegian School of Hotel Management, University of Stavanger, Norway. She holds a Ph.D. in Leadership from the University of

Stavanger, focusing on leaders' attitudes towards older workers in hospitality and public service workplaces. Her main research interests are leadership, human resource management and diversity management in the hospitality industry. Furunes is the Chief Editor of *Scandinavian Journal of Hospitality and Tourism.*

Olga Gjerald is Associate Professor in Service Management at the Norwegian School of Hotel Management, University of Stavanger, Norway. She holds a Ph.D. in Leadership from the University of Stavanger, focusing on basic assumptions of service employees about guests, co-workers and competitors in relation to employee job outcomes. Her main research interests are service organisation and management of service employees, and consumer experiences in service and tourism management.

Ester Gústavsdóttir is the HR specialist at the Reykjavik University. She holds a Bachelor's Degree in Sociology and Media and a Master's Degree in Human Resource Management from the University of Iceland. She holds a Teaching Diploma from the University of Akureyri. She was appointed as one of the three Editors of the *Business Research Journal* for the business institution at the University of Iceland. She has also worked as a part-time lecturer at the University of Iceland in Strategy, HRM and Management in both bachelor's and master's level.

Maria Hakkarainen works as a Senior Lecturer in the University of Lapland/Multidimensional tourism institute. In her doctoral dissertation, completed in the autumn of 2017, she has studied the terms of developing tourism and tourism work in a small and remote village community. Her current research interests are in tourism participatory economies: tourism mobility ecosystems, tourism sharing economy, participatory research methods and project development. She has been involved in several national and regional level tourism development processes and projects. Currently she works as a project manager in Foreign Individual Travelers hospitality and Mobility Ecosystem, FIT ME! (Business Finland)—project.

Íris H. Halldórsdóttir is a researcher at the Icelandic Tourism Research Centre and adjunct in Tourism Studies at the University of Iceland. She has a background in business-related studies and tourism and has worked in the field of tourism for many years. Her interests are in the areas of tourism management and regional development.

Laufey Haraldsdóttir holds a Master of Arts degree in Ethnology from the University of Iceland, and Cand. Med. Vet. degree in Veterinary Medicine from the Norwegian University of Life Sciences in Norway. Laufey is an Assistant Professor in the Department of Rural Tourism at the Hólar University in North Iceland since 2007, and Head of the Department since 2015. Last few year's Laufey's main task has been administrative work within academia. Her research is on food, culture and rural tourism, and spans topics such as place and placemaking, locality, identity, interpretation and New Nordic Cuisine.

Anu Harju-Myllyaho works as the Head of expertise group (Responsibility in Business and Services) at the Lapland University of Applied Sciences, Multidimensional tourism institute. She has a Master's Degree in Hospitality Management (2012) and Social sciences (2018). Currently she is working on her Ph.D. at the University of Lapland. Her research interests concern inclusion in tourism, tourism employment and futures studies.

Charlotta Hedberg is Associate Professor in Human Geography at the Department of Geography, Umeå University (Sweden). Her research focusses on migration and spatial mobility, particularly international labour migration, but also with an interest in internal and lifestyle-related mobilities and regional development. Hedberg is the author of 14 scientifically published papers, and a number of book chapters and policy reports.

Anna Gudmundsson Hillman is a Ph.D. Candidate at the Gothenburg School of Economics and Commercial Law. She has 20 years of work experience with The European Tourism Research Institute (ETOUR) and is former Programme Director Tourism and Destination Development at Mid Sweden University. She currently holds a position as Programme Director Personnel and Work Life at Dalarna University, Sweden.

Maria Huhmarniemi is Doctor of Art and an artist and a teacher at the University of Lapland, Faculty of Art and Design. She studies socially engaged art, arts-based methods and education for sustainability through art. She has developed transdisciplinary collaboration of artists and researchers. She has exhibited in various exhibitions curated international art events and published research articles.

Gunnar Thór Jóhannesson is Professor at the Department of Life and Environmental Sciences at the University of Iceland. His recent research has been on destination dynamics and placemaking with a focus on the entanglement of nature and culture. This has involved studies on entrepreneurship, innovation and policymaking. He has published his research in various books and journals. Most recently, he co-edited a volume titled: *Co-creating Tourism Research: Towards Collaborative Ways of Knowing*, published with Routledge in 2018. Associations in academia include the Nordic Society for Tourism and Hospitality Research, AIRTH—Alliance for Innovators and Researchers in Tourism and Hospitality, and the Nordic Tourism Workforce Research Group based at the Icelandic Tourism Research Centre, Reykjavík.

Magnfríður Júlíusdóttir is Associate Professor in Geography at the University of Iceland. She was a doctoral student at the Lund University, Sweden, in the 1990s. Her research interests are mobility of people from an intersectional approach, focusing on gender, ethnicity and class, in the context of both global and regional development.

Outi Kugapi is a doctoral candidate in tourism research at the University of Lapland, Multidimensional Tourism Institute. She knits and purls together her Ph.D., woollen accessories and two projects concerning cultural sensitivity (ARCTISEN, NPA) and handicraft tourism (Handmade in Lapland, ESF). Her primary research interests are: craft-based tourism, tourist experiences and souvenirs, and she has published a few research papers around these issues.

Laura Laivamaa is a doctoral candidate at the University of Lapland, Faculty of Art and Design (MA, MA). Her background is in industrial design, service design and interior & textile design. She is currently working with issues related to handcrafts and creative tourism

(Handmade in Lapland, ESF), and with her artistic research working with ceramics and textiles. In addition, she has been designing linen and woollen interior textiles for Finnish textile company Lapuan Kankurit, in which her research interest for the concept of Arctic design is shown in a commercial context.

Urban Lindgren is Professor of Human Geography at the Department of Geography, Umeå University (Sweden). He has published within the fields of economic geography, population geography, and more recently in the intersection of health and place. His core research activities have contributed to the development of theories on the links between labour mobility, knowledge spillovers and firm performance. Lindgren has been a driving force behind the establishment of the Umeå SIMSAM Lab, which is an interdisciplinary research infrastructure that has made individual-level register data available to the research community. He has authored more than 50 peer-reviewed journal articles and book chapters, and 50 reports.

Tone Therese Linge is Associate Professor at the Norwegian School of Hotel Management, University of Stavanger, Norway. She holds a Ph.D. in Media and Communication from the University of Oslo, focusing on communication competence in multicultural hotel workplaces. Her main research interests are diversity management, human resource management, organisational and intercultural communication in the hospitality industry. Before joining the Norwegian School of Hotel Management, Linge worked several years as an advisor for multicultural work in the municipal sector in Norway.

Mats Lundmark is since 2003 Professor of Human Geography at Örebro University, Department of Social and Political Sciences. Lundmark is a member of the multidisciplinary Centre for Urban and Regional Studies (CUReS) at Örebro University. He received his Ph.D. at Uppsala University in 1989, where he held a position as Senior Lecturer between 1990 and 2003. His research interests are mainly in the field of economic geography. His publications deal with local and regional development, location analysis, regional policy and changing labour markets. Recently he has been a research coordinator of a project on labour mobility and career paths in the hospitality and tourism sector, financed by BFUF (the R&D Fund of the Swedish Tourism & Hospitality Industry). He is also doing research on schoolteacher's mobility in the labour market.

Jane Widtfeldt Meged is Associate Professor at Roskilde University, Denmark. Jane Meged's research interests are on guided tours, network innovation, the sharing economy, sustainability and precarious working life. She has published several articles and book chapters on the sharing economy as networked innovation, but also on new labour and precarious working life applying a critical perspective on the tourism economy.

Cecilia Möller Karlstad University, Sweden. Ph.D. in Human Geography. Main research topics include geographies of gender, work and mobility, and the intersections with tourism studies. Her research on cross-border regions includes a range of different mobilities such as commuting for work, cross-border shopping and tourism.

Paulína Neshybová holds a B.A. and an M.A. degree in History of Art from Comenius University in Bratislava, Slovakia and an M.A. degree in Tourism from the University of Iceland in Reykjavík. After moving to Iceland in 2016, Paulína spent the following years working in the hotel industry, where she discovered the need to focus her academic path on various aspects of hotel and HR management issues in the Icelandic context.

Kjell Overvåg Inland Norway University of Applied Sciences, Norway. Professor in tourism, Ph.D. in Human Geography. Working on tourism, recreation and leisure linked to regional development, governance, urban–rural relations, mobility and place development. Including topics like second homes, national parks, mountain areas, ski resorts and planning.

Guðbjörg Linda Rafnsdóttir is a Professor of Sociology and Pro-Rector of science at the University of Iceland. She has a Ph.D. in sociology from Lund University, Sweden. Rafnsdóttir has experience in sociological and interdisciplinary research mainly on working life and organisational issues. She has, for instance, studied occupational health and well-being, workplace violence, gendered labour, work–family balance and virtual work. Her research interest also includes social stratification, mobility, global work and new trends in working life. In addition to a research project on volunteers, she participates in multinational projects on gender balance and diversity in research, innovation and business leadership.

Richard N. S. Robinson managed foodservice operations in the private club, heritage facility and hotel sectors, before joining The University of Queensland in 2005. He completed his Ph.D. in 2011, investigating the reasons chefs quit, leading to his current research on the mediatisation of cookery and chefs' mental health & well-being. He teaches hospitality and tourism management and professional development and his expertise and scholarship in teaching & learning is recognised by awards, advisory and programme reviewer appointments in Australia and internationally. He holds a UQ Research Development Fellowship, to investigate hospitality employment opportunities for disadvantaged persons.

Unnur Dís Skaptadóttir is a Professor of Anthropology, Faculty of Sociology, Anthropology and Folkloristics, University of Iceland, Reykjavík, Iceland. She earned her Ph.D. from CUNY Graduate Centre, New York, USA. Her research interests are in the areas of mobility, migration, transnationalism and gender. She has conducted research among various groups of migrants in Iceland. She has published widely on these topics in peer-reviewed journals including *Migration Studies*, *Journal of Immigrant & Refugee Studies*, *Nordic Journal of Migration Research* and *Inclusive Society*. She is the co-editor of *Undoing Homogeneity in the Nordic Region: Migration, Difference and politics of Solidarity* (Routledge).

Brynjar Thor Thorsteinsson is an Assistant Professor at Bifrost University and is working towards his Ph.D. degree at the University of Iceland. Brynjar received his master's at Copenhagen Business School in International Business Studies. His interests are marketing, digital transformation, services and branding in various contexts. Brynjar is teaching relevant subjects and his research has been within the tourism sector focusing on market segmentation, responsible tourism and destination image.

Mari Vähäkuopus is a Senior Lecturer at the Lapland University of Applied Sciences, (Responsibility in Business and services) Rovaniemi. She has several years of professional experience in project planning and management tasks in the field of tourism development, cross-border co-operation and organisational development. Vähäkuopus holds a Master's Degree in Business and Administration (Leadership and Management) from the University of Vaasa. She is currently working in research and development projects related to employee experience in tourism.

Andreas Walmsley is Associate Professor for Entrepreneurship with the International Centre for Transformational Entrepreneurship at Coventry University. His interests in entrepreneurship aside, he also researches issues relating to employment in tourism, especially those that focus on youth employment and responsible employment. In 2015 he published *Youth Employment in Tourism and Hospitality: A Critical Review* (Goodfellow Publishers) and in 2019 *Entrepreneurship in Tourism* (Routledge). He currently co-leads the Tourism Workforce Research Group as part of the Icelandic Tourism Research Centre of which he is an Associate Member.

Margrét Wendt is a Research Assistant in the Department of Geography and Tourism at the University of Iceland. She holds a Bachelor's and Master's Degree in Tourism Research from the University of Iceland. She also has considerable experience in working in the tourism industry.

List of Figures

Fig. 1.1	Growth in international arrivals in Nordic countries 2007–2017 (*Source* OECD)	8
Fig. 4.1	Tourism stakeholders (Kyyrä 2019)	62
Fig. 5.1	Aspects of craft-based tourism development in Finnish Lapland	85
Fig. 8.1	Percentages of employees in tourism by age in 2016–2017	144
Fig. 8.2	Percentage of 13–17 year old girls in the tourism and hospitality industry 2008–2018 (*Source* Statistics Iceland 2019d)	146
Fig. 8.3	Percentage of girls 13–17 years old in diverse occupations within the tourism industry 2008–2018 (*Source* Statistics Iceland 2019e)	147
Fig. 9.1	Proportion of foreign employees in tourism subsectors, 2017 (*Source* Of underlying data: Statistics Iceland, n.d.c)	172
Fig. 9.2	Employees by foreign or Icelandic citizenship and region in 2017 (*Source* Statistics Iceland. *Map* Íris H. Halldórsdóttir & Benjamin David Hennig)	173
Fig. 12.1	Seasonal workers' assessment of the most decisive factor for involvement in innovation activities. Implemented actions. Percent ($N=67$)	250

Fig. 15.1 Hagfors and Munkfors in county Värmland, Sweden. Cartography: Martin Hedlund 310
Fig. 16.1 Dominant themes and actions 338
Fig. 16.2 Detailed analysis of themes I—an example of conflicting findings 339
Fig. 16.3 Detailed analysis of themes II—an example of conflicting findings 340
Fig. 16.4 Detailed analysis of themes III 341

List of Tables

Table 1.1	Features of employment in Nordic countries	5
Table 3.1	Total employment derived from tourism in 2017 distributed into regions	39
Table 6.1	Workaway and HelpX photo analysis	106
Table 8.1	Determinants of job satisfaction. Lillo-Bañuls et al. (2018)	151
Table 14.1	Selected subsectors of the hospitality and tourism industry according to SNI 2007. Total employment in 2008 and 2014 in the four counties	287
Table 14.2	Individual and work-related characteristics in 2014 in the hospitality and tourism industry and in the labour market as a whole	288
Table 14.3	Logistic regressions on persons changing workplace between 2008 and 2011	292
Table 15.1	Migrant entrepreneurs included in the study	314
Table 16.1	Publication year	337
Table 16.2	Top 10 publishing journals	337

Table 17.1 Selected Nordic tourism strategies and policy documents 357
Table 17.2 Tourism labour regimes created in Nordic tourism policy and strategy documents 359
Table 18.1 Challenges to Indigenous peoples in the tourism industry, by themes in literature 382

1

Introduction

Andreas Walmsley, Kajsa Åberg, Petra Blinnikka and Gunnar Thór Jóhannesson

Tourism's job creation potential is unequivocal. According to World Travel and Tourism Council (2017) estimates, no fewer than one in ten jobs is now generated by tourism (directly and indirectly). As tourism continues its inexorable rise, international tourist arrivals growing from

A. Walmsley (✉)
International Centre for Transformational Entrepreneurship, Coventry University, Coventry, UK
e-mail: ad3412@coventry.ac.uk

K. Åberg
Region Västerbotten, Umeå, Sweden

P. Blinnikka
JAMK University of Applied Sciences, Jyväskylä, Finland
e-mail: petra.blinnikka@jamk.fi

G. T. Jóhannesson
Faculty of Life and Environmental Sciences, University of Iceland, Reykjavík, Iceland
e-mail: gtj@hi.is

A. Walmsley et al. (eds.), *Tourism Employment in Nordic Countries*,
https://doi.org/10.1007/978-3-030-47813-1_1

278 million in 1950 to 1235 million in 2016 (UNWTO 2017) with a forecast of further growth to 1.8 billion by 2030 (UNWTO 2011), so will it continue to offer employment opportunities to increasing numbers of people globally.

A case can quite easily be made then why policymakers might be interested in tourism and its development with regard specifically to the generation of jobs: despite recent rapid advances associated with the Fourth Industrial Revolution and artificial intelligence (Bort 2017; Ford 2015; Frey and Osborne 2017; World Economic Forum 2016), it is and will in the foreseeable future remain a labour-intensive industry. It is commonly regarded as a low skills sector, and therefore one that can provide work to those who have not entered the so-called knowledge economy (frequently it is looked down upon because of this low skills status). Tourism also frequently takes place in remote, often peripheral (not just geographically, socially and economically) regions and as such one of few sectors that may offer work, and work that it is difficult, if not impossible, to outsource (Denmark's Tourism Strategy makes specific reference to 'jobs that cannot be relocated to other countries' The Danish Government 2014, p. 11).

But, employment in tourism is not one-dimensional, its features may be interpreted in different ways (which makes it such an interesting topic of study). We have already alluded to the purported low skills nature of many jobs in tourism. Based on wages as a measure of skills then undoubtedly this is true; tourism is a low wage sector. The extent to which it is truly low skilled or whether this reflects bias towards soft skills is another matter (Burns 1997). As is typically the case, the devil is in the detail, for example Åberg and Müller (2018) describe how in Sweden the characteristics of the workforce in tourism differ between urban and rural locations with low-skilled, young tourism workers predominantly found in the urban regions, whereas in more rural and peripheral regions it employs the relatively higher educated. Or we could highlight Underthun and Jordhus-Lier's (2018) study of hotel workers in Oslo where 'working tourists' mix with less privileged expatriate workers, often undertaking the same functional roles.

Extending the discussion beyond wages and skills, more broadly there continues to be an ongoing debate around the nature of work and working conditions in tourism (Winchenback et al. 2019; Baum 2018; Walmsley et al. 2018; Robinson et al. 2019) with some emphasis being placed on the International Labour Office's Decent Work agenda with the notion of Decent Work featuring prominently in the UN Sustainable Development Goals (Goal 8: Decent Work and Economic Growth). More generally, there has been a shift in policymaking circles from a focus solely on quantity to a focus also on quality of jobs. The question is no longer solely 'how many?' but 'what kind?' (e.g. Taylor 2017). In a changing world of work, it is instructive to inspect the extent to which policy prerogatives filter their way down to practice, and to explore in some detail the nature, characteristics and meaning of work in tourism, here from a Nordic perspective.

The Nordic Context

Commenting on the end of the Cold War Francis Fukuyama famously suggested we had reached the 'end of history' (Fukuyama 1989) with the 'unabashed victory of economic and political liberalism' (Fukuyama 1989, p. 1). Today, his words ring a little hollow in an era of increased political division. That there still exists a dominant politico-economic paradigm is no longer a given; in Coen and Roberts' (2012, p. 5) words: 'The easy consensus on policy which typified the last years of the age of liberalization…has collapsed' (Coen and Roberts 2012, p. 5). Within Europe, but also further afield (e.g. in the United States) recent history has seen a move from centrist politics. Voices that question the dominant Anglo-American (frequently referred to as neo-liberal) form of capitalism are now regularly heard (e.g. Manolescu 2011; Chang 2011; Küng 2010; Mason 2015; Collier 2018; Picketty 2014). The point is, rather than a sure-footed, steady march towards homogenisation of politico-economic systems in the Anglo-Saxon mould, we are today witnessing a renewed interest in variations of the capitalist model.

Upon this backdrop *Tourism Employment in Nordic Countries* seeks to explore and make sense of facets of employment as they relate specifically to what are commonly regarded as Nordic Countries (generally said to comprise Denmark, Finland, Iceland, Norway, Sweden and associated territories; The Faroe Islands, Greenland and Åland islands). Viewed through a politico-economic lens, Nordic countries share what is often referred to as the 'Nordic Model' comprising features such as:

- a comprehensive welfare state financed by taxes on labour (Kolm and Tonin 2015)
- more equitable income distribution (OECD 2017)
- high spending on childcare (Kolm and Tonin 2015)
- emphasis on a social democratic element, or the element of coordinated market economy, as different from a pure, or liberalist one (Gustavsen 2007)
- lifelong learning policies (Jochem 2011)
- the value of equality (Kvist and Greve 2011).

While Nordic countries share similarities in terms of the structure of their politico-economic systems, especially their welfare and labour market institutions (Høgedahl and Kongshøj 2017) frequently contrasted with Liberal or Conservative models (Kvist and Greve 2011), the status of Nordic Countries as a distinct analytical category extends also to shared socio-cultural values and norms, derived from a shared historical development (in part), climate and ecosystems. The Nordic as a backdrop thus serves less as a geographical limitation than as a frame for addressing issues that arise in a society characterised by institutional structures torn between free market capitalism, with added pressures of globalisation, and what could be described as socialist elements. Thus, the strong regulatory power of the governments in the Nordic countries has resulted in established employment security and regulated working environment. However, the other side of the coin is a relatively heavy administrative as well as economic burden for the employer.

Nonetheless, while we argue it makes sense to view these Nordic countries as sharing characteristics, and in acknowledging 'predominantly similar labour market conditions from a global perspective' (Kjeldstad and Nymoen 2012, p. 208), both in terms of social, economic, cultural and political sense, as well as in relation to tourism demand and supply, they are far from homogeneous. This sameness and difference is evidenced in Table 1.1.

Table 1.1 Features of employment in Nordic countries

	Harmonised unemployment rate first quarter 2019	Income inequality, Gini coefficient (0 = complete equality, 1 = complete inequality), 2015	Gender equality gap (% of gap closed) 2018$^{>}$	Earnings quality (2010)	Tax on personal income (% of GDP) 2016	Benefits in unemployment (share of previous income. After 1 year, % of previous in-work income, 2016)	Trade union density (%) 2013
Denmark	5.3	0.263	77.8	28.26	24.7	85	66.8
Finland	6.7	0.26	82.2	20.42	13.2	59	66.3
Iceland	3	0.255	85	19.26	13.6	52	80.5
Norway	3.7	0.272	83.2	26.32	10.7	68	51.9
Sweden	6.3	0.278	82.2	19.01	13.1	65	67.4$^{\wedge}$
OECD	5.3	0.33	75.8$^{\wedge}$	16.65*	8.24	44	27.3

Based on latest available data (all data sourced June 2019 from OECD). *Average across OECD where data available. ^2012. ^Western Europe. $^{>}$Data from WEF 2018. For further details please contact the authors
Source OECD

From Table 1.1 we can identify some typical characteristics of Nordic labour markets: Unemployment is generally low, mirroring average OECD figures. There is less inequality, including gender inequality in particular where the Nordic countries score very high (Iceland tops the table here, and globally, as the most gender equal country in terms of pay levels). Social protection, here represented by benefits in unemployment, are higher across the board than the OECD average as is taxation (we can see in particular that Denmark is a high tax-high welfare society). Finally, trade union membership is a key feature of Nordic countries, with trade union density far higher in Nordic countries than the OECD average. These are features Kvist and Greve (2011) describe as pertaining to the Nordic welfare model, specifically: comprehensive state responsibility, universal coverage, individualism, high employment, equality of opportunities and results, high quality (public) services, high generosity and social dialogue (among others).

Tourism in Nordic countries shares many characteristics that will be explored in further detail in the book and only receive a brief outline here. National and regional authorities in the Nordic countries have for some time seen tourism as a convenient tool for socioeconomic development (Hall et al. 2009). A heavy feature of tourism in all Nordic countries is nature-based tourism. With the exception of Denmark (133 persons per km^2), Nordic countries are sparsely populated. Sweden ranks 193rd most densely populated country with 22 persons per km^2, Norway is 203rd, Finland 204th and Iceland with only 3 people per km^2 is 226th. By way of comparison, excluding The Vatican, San Marino and Malta, the Netherlands with 408 persons per km^2 is the most densely populated European nation (World Population Review 2019).

Other issues affecting Nordic countries, according to the State of the Nordic Region 2020 report (Grunfelder et al. 2020), include urbanisation, decreasing fertility rates and increasing life expectancy which will shape the regions and societies of Nordic countries and will present a challenge to the Nordic welfare model in the future. Ageing of the population is likely to impact policymaking in relation to migrant workers. Although the ageing of the population structure is a common feature in Nordic Countries, there are also differences within the countries. There are more older people than children in Denmark, Finland, Sweden and Åland, while in Norway there are nearly equal numbers of older people and children. It is important to note that all the Nordic countries are characterised by a strong centre-periphery dynamic. The administrative centres as well as most of the population resides in the southern parts of the respective countries while the northern parts are much more sparsely populated, traditionally dependent on primary industries such as forestry, fisheries and agriculture as well as reindeer herding in the case of Norway, Sweden and Finland. In Iceland, more than 80% of the population lives in the vicinity of the capital Reykjavík. Thus, an abundant natural heritage, from lakes, forests, the Fjord coastline, glaciers and tundra landscapes are a primary attraction for tourists, especially in the peripheral areas of the Nordic countries. There are certainly various other draws for tourists in and around the larger cities in the Nordic countries with food, architecture and vivid cultural experiences as key attractions for tourists. Cities seek to promote themselves as destinations in their own right offering buzzing urban experiences with particular flair, such as the 'Copenhagen lifestyle' and 'season of hygge' in the case of Copenhagen (Visit Denmark 2020).

In relation to international tourism, while all Nordic countries have seen growth, Iceland in particular, but also Denmark, have seen a rapid expansion of inbound tourism in recent years (see Fig. 1.1).[1] The reasons for this vary but certainly present an interesting case given that the tourist offer in both countries is quite distinct (while Denmark attracts many holidaymakers from Northern Europe who visit the Danish coast, Iceland's appeal is more global and tends to attract people to its dramatic landscapes). Both Copenhagen and Reykjavík are destinations in their own right, with the latter recently starting to promote itself as such, seeking to create a position of more-than-a-gateway to natural attractions (Reykjavíkurborg 2019). We will now pay a closer look at tourism development, the position of tourism as a sector as well as key points with regard to tourism employment in each of the Nordic countries.

[1]We cross-checked data with data from the World Tourism Organisation for Denmark given the remarkable growth in inbound tourism between 2004 and 2006 where this growth was confirmed. We presume part of this growth may be down to measurement issues. We also note that care should be taken when comparing data across nations given data collection methods—as a guide though these figures are instructive.

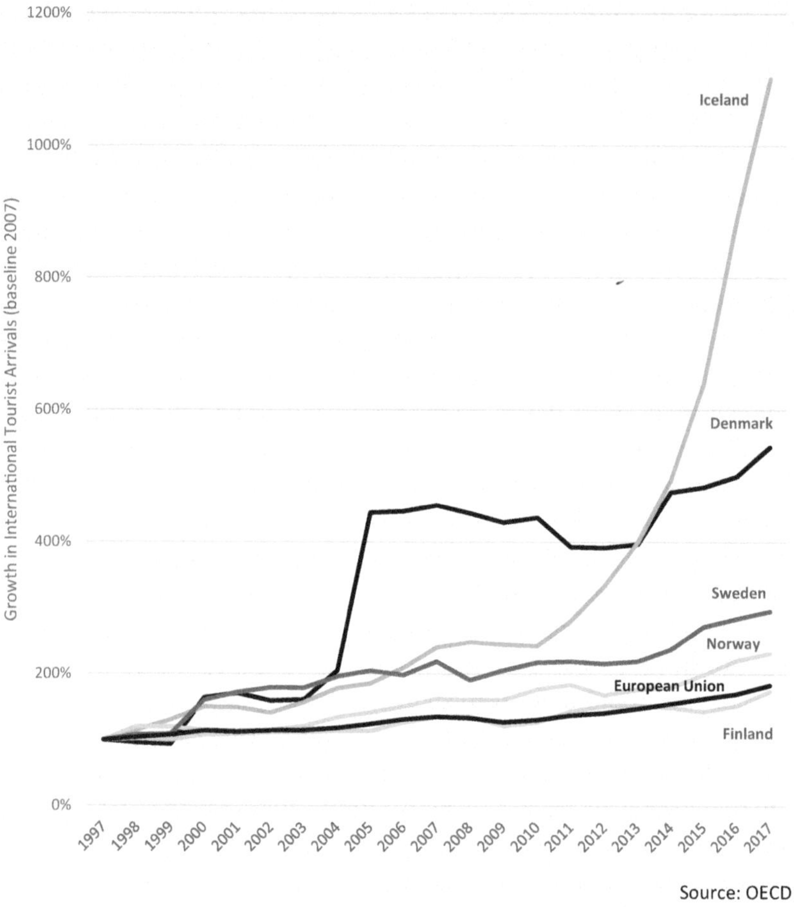

Fig. 1.1 Growth in international arrivals in Nordic countries 2007–2017 (*Source* OECD)

Iceland

Modern tourism started to emerge in Iceland after the mid-twentieth century when Icelandic airlines started to offer flights to both Europe and North America. Tourism growth was however rather modest, and tourism was not regarded as an important economic sector by the central authorities vis-a-vis the traditional backbone of the economy, fisheries and heavy industries (Jóhannesson et al. 2010). This changed abruptly in the wake of the global financial crisis in 2008 that hit Iceland hard. Tourism got on the radar of the central authorities, desperately needing foreign currency and their focus was to get as many tourists to the country as possible (Jóhannesson and Huijbens 2010). Currency devaluation and severe competition in international aviation helped to make Iceland, for a while, a cheaper more accessible destination than before. Moreover, in 2010 the volcano Eyjafjallajökull erupted, which caused havoc in European airspace but simultaneously made Iceland highly visible and, for many, an interesting place to visit (Benediktsson et al. 2011; Heimisdóttir et al. 2019). In 2010, registered tourist arrivals were 488 thousand and tourism provided 3.4% of GDP. In 2018 tourist arrivals reached 2.3 million and tourism's share in GDP was 8.6% (Ferðamálastofa 2020). The tourism sector has during the last ten years become the single most important provider of foreign currency, the largest provider of jobs (16% of total number of employees in the labour market in 2018) and a recognised fourth pillar of the economy in national political discourse (see also Chapter 9).

Tourism in Iceland has been driven by a few large companies and a multitude of small and micro-sized companies, many of which operate on lifestyle motives. The rapid growth seems to have levelled off a little in 2019, and while this causes concerns for some it also provides an opportunity to get a better grip on tourism development and its challenges. One of the major challenges is the seasonality of tourism. While it has in practice disappeared in the capital area with a fast growth in winter tourism, it still remains a big obstacle in other parts of the country due to more difficult access and different travel patterns during winter (Rantala et al. 2019). Currently policies are being developed by the central authorities with the general aim to make tourism in Iceland a leading force for sustainable development (Stjórnarráð Íslands 2019).

Norway

Tourism contributes 4.2% of Norway's GDP and is responsible for one in fifteen jobs with 30% of total tourism consumption relating to foreign visitors (Innovation Norway 2019). Mirroring growth in global tourism, in 2017 Norway witnessed its seventh year of consecutive growth in international arrivals (Innovation Norway 2019). This is not to suggest tourism in Norway is without its challenges; there exists a high degree of seasonality (May–August) and spatially most of it takes place in 'Fjord Norway' and 'Northern Norway' putting pressure on destinations in these locations.

Like other Nordic countries, Norway has a very low unemployment rate (see Table 1.1) and a high labour force participation rate. One of the reasons for this is the so-called 'tripartism' (a broad agreement among unions, employers and government to maintain a high level of coordination in wage bargaining (Nilsen 2018). Some of the key positive characteristics of the Norwegian labour market according to Nilsen (2018) are a downward sloping trend of part-time work among women, relatively small and stable wage inequality, and low levels of unemployment, even among youth. These positives are set in contrast to high levels of worker absenteeism putting pressure on the welfare state, as well as a highly gender-segregated labour market with regard to sector and occupation. As per other Nordic countries, the majority of workers are covered by collective agreements (approximately 80%; Nilsen 2018).

Sweden

Sweden is on a level with Finland, below both Norway and Iceland and slightly above Denmark when measuring tourism's share of GNP. In terms of employment, tourism is calculated to encompass 3.4% of the working population, a group that has grown by 32% since the year 2000, to be compared with the general growth in employment which was 19% during that period. In addition, the annual export value (consumption made by international visitors in Sweden) has surpassed the prior dominating sectors of iron and steel as well as cars. The main bulk of tourism consumption and visitor numbers in Sweden are however found in domestic tourism. Approximately 73% of all commercial guest nights are undertaken by Swedes (Swedish Agency for Economic and Regional Growth 2019).

In 2017, the government presented a thorough governmental investigation regarding the potential in tourism and how it can best be exploited—in a sustainable way (Ministry of Enterprise and Innovation

2017). Government involvement may be explained by the perceived potential rather than the actual impact of the industry with strategies at national and regional levels aiming to create economic activities in less populated, rural areas struck by restructuring (Lundmark 2006). The effects of the tourism sector are primarily found in the few metropolitan regions together with already developed tourism destinations.

The relatively high taxes placed on labour and corporations finance the comprehensive welfare state and inclusive growth strategy once termed *The Swedish Model* (Ministry of Finance 2017). The potential gains in numbers of jobs are thus somewhat counteracted by high costs related to employment. In 2019, the employer must pay approximately 31% of each employee's gross pay as general payroll tax including a fee to the national insurance. During 2007 and 2016, this was lowered for employees in the age group of 19–27 and a report published in 2018 showed that the initiative led to an estimated 16,400 new jobs and that the effect was most evident in firms with a large proportion of young employees (Hui Research 2018), a well-known characteristic of the tourism sector.

Finland

Recent employment data indicate unemployment stood at almost 7% in Finland (Statistics Finland 2019) which is slightly higher than in other Nordic countries. The tourism cluster employs over 140,000 people, which comprises 5.5% of the working population. Total tourism demand was €15 billion in 2017 contributing approximately 2.6% to Finland's GDP (Ministry of Economic Affairs and Employment of Finland 2018).

Finland has four major tourism regions that are unique even on a global scale: the Helsinki region, Lapland, the Finnish lake district, and the coast and archipelago (Business Finland 2020). Currently Finland has the lowest number of international arrivals among the Nordic Countries but tourism growth in the last few years has been strong and, as a result, tourism has become more important to the Finnish economy. In 2017, there were 15.2 million overnight stays of domestic tourists and 6.7 million overnight stays of foreign tourists in Finland (Ministry of Economic Affairs and Employment of Finland 2018).

In 2019 international overnight stays increased by 2.9% to a record 7 million (Business Finland 2020). Russia is still the most important market in terms of international arrivals. The other largest groups of tourists were Estonians and Swedes (Jänkälä 2019). The biggest growth was reported from Chinese markets (+35.3%) (Ministry of Economic Affairs and Employment of Finland 2018).

Finland's attractiveness as a travel destination continues to grow. Finland is aiming to become the most 'sustainably growing' tourist destination among the Nordic countries. The drive to grow tourism comes from the desire to create jobs and wealth via services more generally. On one hand, improving accessibility is essential for the Finnish tourism industry. On the other hand, it is acknowledged that tourism growth must be sustainable (Jänkälä 2019).

Denmark

In 2014 the Danish Government launched its Tourism Strategy recognising the growth potential and importance of tourism to the Danish economy. The strategy made direct reference to the job creation potential of tourism (according to the report tourism is responsible for 4.3% of jobs in Denmark), highlighting in particular tourism employment's 'cohesive effect on Danish society' in that 'it also creates jobs for people with short-cycle education and for a great many non-ethnic Danes' (The Danish Government 2014, p. 1). In fact, in common with many higher wage countries, such as the Nordic countries, the enlargement of the EU has seen an increase in the number of migrants working in Denmark and concomitant concerns around precarious work (Rasmussen et al. 2016; Simkunas and Thomsen 2018).

The Danish labour market shares many of the characteristics of those of its northern neighbours. A key feature of the Danish labour market are collective agreements between employer and employee associations, including trade unions. Thus, legislation still plays a role in determining labour market characteristics and working conditions, but only alongside the collective agreements (for example, there is no minimum wage in Denmark, as wages are determined at a sector level by these agreements).

In 2012, no fewer than 84% of Danish employees were covered by collective agreements (Danish Employers Federation, DA 2013).

Alongside labour market characteristics should be considered social security and social protection legislation as the two go hand-in-hand, particularly in relation to social security in case of redundancy. Thus, in Denmark most unemployed are entitled to unemployment insurance benefits (arbejdsløshedsdagpenge) if they have paid into the insurance scheme for at least one year. However, the ability to receive unemployment benefits was reduced from four to two years as part of a labour market reform (Dagpengereform) which reflects ongoing pressure among EU countries to cut social security expenditure, and more broadly policy debates surrounding the need to liberalise (deregulate) labour markets to remain internationally competitive (see also Høgedahl and Kongshøj 2017). As Rasmussen et al. (2016) have claimed, many including trade unions complained that this was reducing the security element of Denmark's 'Flexicurity' model, a system that provides flexibility to employers while also providing security to employees. As a consequence, there were instances of trade unions demanding longer redundancy notice periods, and higher levels of compensation (Klindt 2014 cited in Rasmussen et al. 2016). In quite a damning analysis of changes in Denmark's welfare system Knudsen and Lind (2018) even suggest it should now be called 'flexinsecurity' because employment policies are now geared to achieving employer-determined flexibility by fear and insecurity instead of social security networks. And yet, Kvist and Greve (2011) despite recognising ongoing changes to the Nordic welfare model still believe it has not changed fundamentally.

Overview of Book

Structuring the book around key themes, although the initial intention, proved not to be possible, at least not in any compelling way. Thus, distinctions between a micro and a macro focus, between the individual worker featuring centre stage or managerial prescriptions, between policy orientation or organisational needs, the role of migrant labour, self-employment, decent work and youth employment all provided themes that could have warranted their own sections and a method of structuring the text. The difficulty though is that all chapters covered a number of what we regard as key and recurrent themes, further demonstrating the interconnectedness and complexity inherent in issues relating to tourism employment (Ladkin 2011). This was amply brought home to us when we tried to create a table identifying key themes by chapter—it was simply not possible to reach a convincing consensus in this regard. We therefore adopted the approach of segmenting the chapters by country, with three chapters at the end of the volume covering more than one country. We also refer to the chapter abstracts which will give the reader a useful overview of the book's contents. The concluding chapter seeks to summarise the key issues, challenges and opportunities, to provide the reader with an overview of the contributions as a whole and offers a stepping stone for future work on tourism employment in Nordic countries. We trust the reader will find much of interest here, and indeed may serve to strengthen or awaken further interest in exploring an intriguing and intricate aspect of tourism.

References

Åberg, Kajsa Grandics, and Dieter K. Müller. 2018. The development of geographical differences in education levels within the Swedish tourism industry. *Tourism Geographies* 20 (1): 67–84.

Baum, Tom. 2018. Sustainable human resource management as a driver in tourism policy and planning: A serious sin of omission? *Journal of Sustainable Tourism* 26 (6): 873–889.

Benediktsson, K., K.A. Lund, and E. Huijbens. 2011. Inspired by eruptions? Eyjafjallajökull and Icelandic tourism. *Mobilities* 6 (1): 77–84.

Bort, Ryan. 2017. Will AI take over? Artificial intelligence will best humans at everything by 2060, experts say. *Newsweek*. Accessed on July 11, 2017.

Burns, Peter. 1997. Hard-skills, soft-skills: Undervaluing hospitality's 'service with a smile'. *Progress in Tourism and Hospitality Research* 3 (3): 239–248.

Business Finland. 2020. Seitsemän miljoonan kansainvälisen yöpymisen raja rikki matkailussa, January 31. https://www.businessfinland.fi/ajankohtaista/uutiset/tiedotteet/2020/seitseman-miljoonan-kansainvalisen-yopymisen-raja-rikki-matkailussa/.

Chang, Ha-Joon. 2011. *23 Things They Don't Tell You About Capitalism*. London: Penguin Books.
Coen, David, and Alisdair Roberts. 2012. A new age of uncertainty. *Governance* 25 (1): 5–9.
Collier, Paul. 2018. *The Future of Capitalism: Facing the New Anxieties*. London: Allen Lane.
DA. 2013. *Forskellige ansættelsesformer på det danske arbejdsmarked, rapport*. Copenhagen: DA.
Ferðamálastofa. 2020. Mælaborð ferðaþjónustunnar: Hagstærðir.
Ford, M. 2015. *The Rise of the Robots: Technology and the Threat of Mass Unemployment*. London: Oneworld.
Frey, Carl Benedikt, and Michael A. Osborne. 2017. The future of employment: How susceptible are jobs to computerisation? *Techological Forecasting and Social Change* 114: 254–280.
Fukuyama, Francis. 1989. The end of history. *The National Interest* 16: 3–18.
Grunfelder, Julien, Gustaf Norlén, Linda Randall, and Nora Sánche. 2020. *State of the Nordic Region*. Copenhagen, DK: Nordic Council of Ministers, February 4. https://pub.norden.org/nord2020-001/nord2020-001.pdf.
Gustavsen, Bjørn. 2007. Work organization and 'the Scandinavian model'. *Economic and Industrial Democracy* 28 (4): 650–671.
Hall, C.M., D. Müller, and J. Saarinen. 2009. *Nordic Tourism: Issues and Cases*. Bristol: Channel View Publications.
Heimisdóttir, Þ., A.D. Sæþórsdóttir, and G. Gísladóttir. 2019. The sublime attraction of active volcanoes: An exploration of tourists' experiences during a long-distance hiking route in Iceland. *Tourist Studies* 19 (2): 258–275.
Høgedahl, Laust, and Kristian Kongshøj. 2017. New trajectories of unionization in the Nordic Ghent countries: Changing labour market and welfare institutions. *European Journal of Industrial Relations* 23 (4): 365–380.
Hui Research. 2018. *Kan sänkta arbetskraftskostnader bidra till ökad sysselsättning? En utvärdering av de sänkta arbetsgivaravgifterna för unga*. Stockholm: Handelns Forskningsinstitut.
Innovation Norway. 2019. *Key Figures for Norwegian Tourism 2017*. Oslo: Innovation Norway. https://assets.simpleviewcms.com/simpleview/image/upload/v1/clients/norway/Key_Figures_2017_pages_9b3f82d5_43f4_4fe9_968c_7a85a36704b2_ed2d345f-e93b-410c-9fee-6c349da8eff0.pdf.
Jänkälä, Susanna. 2019. Kevään 2019 toimialojen näkymät: Matkailuala. Valtioneuvosto Stadsradet. http://julkaisut.valtioneuvosto.fi/handle/10024/161641. Accessed on October 23, 2019.

Jochem, Sven. 2011. Nordic employment policies—Change and continuity before and during the financial crisis. *Social Policy & Administration* 45 (2): 131–145.

Jóhannesson, G.T., and E. Huijbens. 2010. Tourism in times of crisis: Exploring the discourse of tourism development in Iceland. *Current Issues in Tourism* 13 (5): 419–434.

Jóhannesson, G.T., E. Huijbens, and R. Sharpley. 2010. Icelandic tourism: Past directions—Future challenges. *Tourism Geographies* 12 (2): 278–310.

Kjeldstad, Randi, and Erik Nymoen. 2012. Underemployment in a gender-segregated labour market. *Economic and Industrial Democracy* 33 (2): 207–224.

Knudsen, Herman, and Jens Lind. 2018. The "Danish Models" of labour market regulations and their status after recent changes. *New Zealand Journal of Employment Relations* 43 (1): 83–98.

Kolm, Ann-Sofie, and Mirco Tonin. 2015. Benefits conditional on work and the Nordic model. *Journal of Public Economics* 127: 115–126.

Küng, Hans. 2010. *Anständig wirtschaften: Warum Ökonomie Moral braucht.* München: Piper Verlag GmbH.

Kvist, Jon, and Bent Greve. 2011. Has the Nordic welfare model been transformed? *Social Policy & Administration* 45 (2): 146–160.

Ladkin, Adele. 2011. Exploring tourism labour. *Annals of Tourism Research* 38 (3): 1135–1155.

Lundmark, Linda. 2006. *Restructuring and Employment Change in Sparsely Populated Areas: Examples from Northern Sweden and Finland.* Umeå, Sweden: Umeå University, Department of Social and Economic Geography.

Manolescu, Elena. 2011. Socio-economic models during the period of crisis. *Theoretical and Applied Economics* 18 (9): 173–180.

Mason, Paul. 2015. *Postcapitalism: A Guide to Our Future.* London: Allen Lane.

Ministry of Economic Affairs and Employment of Finland. 2018. *Tourism Account Report 2018.* Helsinki, Finland: Ministry of Economic Affairs and Employment.

Ministry of Enterprise and Innovation. 2017. *Ett land att besöka.* Stockholm, Sweden: Statens Offentliga Utredningar.

Ministry of Finance. 2017. *The Swedish Model.* Government Offices of Sweden, Digital Asset. www.government.se. Accessed on August 8, 2019.

Nilsen, Oivind. 2018. *The Labor Market in Norway, 2000–2016.* IZA World of Labor, May. https://wol.iza.org/uploads/articles/424/pdfs/the-labor-market-in-norway.pdf?v=1.

OECD. 2017. *Understanding the Socio-Economic Divide in Europe.* Organisation for Economic Cooperation and Development. Centre for Opportunity and Equality, January 26. https://www.oecd.org/els/soc/cope-divide-europe-2017-background-report.pdf.

Picketty, T. 2014. *Capital in the Twenty-First Century.* Cambridge: Harvard University Press.

Rantala, O., S. de la Barre, B. Granås, G.T. Jóhannesson, D.K. Müller, J. Saarinen, and M. Niskala. 2019. *Arctic Tourism in Times of Change: Seasonality.* Copenhagen: Nordic Council of Ministers.

Rasmussen, Stine, Bjarke Refslund, Ole H. Sørensen, and Trine Larsen. 2016. *Reducing Precarious Work in Europe Through Social Dialogue: The Case of Denmark.* Aalborg University. http://www.research.mbs.ac.uk/ewerc/Our-research/Current-projects/.

Reykjavíkurborg. 2019. Áfangastaðurinn Reykjavík. Accessed on February 1, 2020.

Robinson, Richard, Antje Martins, David Solnet, and Tom Baum. 2019. Sustaining precarity: Critically examining tourism and employment. *Journal of Sustainable Tourism.* https://doi.org/10.1080/09669582.2018.1538230.

Simkunas, Doris Pljevaljcic, and Trine Lund Thomsen. 2018. Precarious work? Migrants' narratives of coping with working conditions in the Danish labour market. *Central and Eastern European Migration Review* 7 (2): 35–51.

Statistics Finland. 2019. Labour Force Survey 2019. https://www.stat.fi/til/tyti/2019/12/tyti_2019_12_2020-01-24_tie_001_en.html.

Stjórnarráð Íslands. 2019. Leiðandi í sjálfbærri þróun: Íslensk ferðaþjónusta til 2030. https://www.stjornarradid.is/lisalib/getfile.aspx?itemid=0bef70be-e109-11e9-944d-005056bc4d74.

Swedish Agency for Economic and Regional Growth. 2019. *Fakta om svensk turism 2018/Swedish tourism in numbers 2018.* Tillväxtverket/Swedish Agency for Economic and Regional Growth Report 0282, Stockholm, Sweden.

Taylor, Matthew. 2017. *Good Work: The Taylor Review of Modern Working Practices.* Department for Business, Energy and Industrial Strategy, July 11. https://www.gov.uk/government/publications/good-work-the-taylor-review-of-modern-working-practices.

The Danish Government. 2014. *Denmark at Work: Plan for Tourism Growth in Danish Tourism.* Copenhagen: The Danish Government. https://eng.em.dk/media/10590/summary-plan-for-growth-in-danish-tourism.pdf.

Underthun, Anders, and David Christopher Jordhus-Lier. 2018. Liminality at work in Norwegian hotels. *Tourism Geographies* 20 (1): 11–28.

UNWTO. 2011. *Tourism Towards 2030.* Madrid: UN World Tourism Organisation.

UNWTO. 2017. *UNWTO Tourism Highlights 2017 Edition.* Madrid: UNWTO.

Visit Denmark. 2020. VisitCopenhagen. https://www.visitcopenhagen.com/. Accessed on February 1, 2020.

Walmsley, Andreas, Shobana Partington, Rebecca Armstrong, and Harold Goodwin. 2018. Reactions to the national living wage in hospitality. *Employee Relations* 41 (1): 253–268. https://doi.org/10.1108/ER-02-2018-0044.

Winchenback, Anke, Paul Hanna, and Graham Miller. 2019. Rethinking decent work: The value of dignity in tourism employment. *Journal of Sustainable Tourism*. https://doi.org/10.1080/09669582.2019.1566346.

World Economic Forum. 2016. *The Future of Jobs: Employment, Skills and Workforce Strategy for the Fourth Industrial Revolution.* Geneva: World Economic Forum.

World Population Review. 2019. World Countries by Population Density 2019. http://worldpopulationreview.com/countries/countries-by-density/. Accessed on July 3, 2019.

World Travel and Tourism Council. 2017. Tourism Supports 1 in 10 Jobs, Outpacing Global Economy for 6th Consecutive Year, WTTC. WTTC. Last Modified March 20, 2017. https://www.wttc.org/about/media-centre/press-releases/press-releases/2017/tourism-supports-1-in-10-jobs-outpacing-global-economy-for-6th-consecutive-year-wttc/. Accessed on April 4, 2019.

2

Guides on a Crossroad: Between Deregulation and Entrepreneurship

Jane Widtfeldt Meged

Introduction

Tourism related jobs, occupation and employment are often precarious, low-paid and labour-intensive (Veijola 2010, p. 84), and most are subjected to seasonal fluctuations while regulated by a very liberal market mechanism. However, for several decades, the certified guides in Copenhagen have kept a unique position in the market of guided tours, because their union The Association of Authorized Tourist Guides (AATG) managed to negotiate a collective agreement signed and respected by most of the major employers in the tourism industry. The agreement secured decent tariffs, supplements, booking and cancellation conditions, whereby employers only booked non-certified guides in extreme situations. In recent years, however, the market for guided tours has changed radically in parts to the detriment of the certified guides, and in 2018, the collective agreement was not renewed. As of today, the

J. W. Meged (✉)
Department of People and Technology, Roskilde University,
Roskilde, Denmark
e-mail: janewm@ruc.dk

A. Walmsley et al. (eds.), *Tourism Employment in Nordic Countries*,
https://doi.org/10.1007/978-3-030-47813-1_2

certified guides work according to different tariffs decided and produced by each incoming agency. The tariffs are presented to the guides individually as a "take it or leave it", thus, completely circumventing their union. This loss of collective bargaining rights and in many cases cut back in tariffs may in part be explained by the rise of the so-called sharing economy and the influx of guide lights fuelled by the fast growth in tourism.

From 2008 to 2018 overnight stays in Copenhagen rose by 80% from 5.5 to 9.9 m (Visitdenmark 2018), and in the same period the number of cruise passengers rose 55% from 560,000 (CruiseCopenhagen 2017) to 868,000 (Bomberg 2018). The growth appears to continue, and the DMO Wonderful Copenhagen forecasts the number of tourists to Copenhagen will double by 2030 from 8.8 to 16 m (Baumgarten 2018). In this growth market, the demand for guides has by far exceeded the supply, but instead of strengthening the position of the certified guides as a valuable and scarce resource, the growth paved the way for masses of migrant tourism workers, who with a crash course become the so-called "guide-lights". The unlicenced "guide-lights" receive a minimal training provided by the industry, and they work far below the tariffs of the certified guides. Very fast, guide lights are replacing the certified guides on the market of standard guided tours, and they have become a crucial economic competitive parameter, particularly in the cruise industry notoriously known for its' low salaries (Weaver 2005; Becker 2014).

Since 2011, Copenhagen has also witnessed the arrival of several free guided walking tour companies, which in reality are not free, but are tips-only based tours, that by the end of the day generate a substantial income to the guides and the free walking tour companies. They began by addressing young, independent low budget traveller, in a peer-to-peer economy, but they have had a rapid uptake, and now they guide hundreds of thousands of tourists, thus capturing a large part of the mainstream market of guided tours (Meged and Zillinger 2018).

The above development reflects well the effects of the Fourth Industrial Revolution, which is defined as a total transformation in the nature of income-generating work (Gans 2016). The driving forces behind this transformation are robotization, informatization, artificial intelligence and extreme focus on cost-efficiency due to growing globalization and individualization (Savage 2015; Wilson 2017 cited in Johannessen 2019, p. 3). As a consequence we see a growing precariat, with less salaried workers, and still more people working in precarious job functions in sharp competition from digital actors, robotics, but also other humans, and hence with little security (Standing 2014). So far, live guides appear to thrive well next to digital guides. However, the nature of their work is transformed due to digitalization and cost efficiency in the sharing economy, as well as the growing globalization of the workforce. First, the chapter sets out to explore how the certified guides on an individual level face the transformation of the Fourth Industrial Revolution, and through the lens of Job Crafting theory, we seek to understand how guides have agency and ability to shape their own job in order to stay on top of the competition. Secondly, the chapter will address the organizational level by looking at the position and agency of The Association of Authorized Tourist Guides (AATG) as it unfolds in the Fourth Industrial Revolution. Will individualization, fragmentation and loose networks replace earlier stable working communities as platforms of solidarity, thus reducing the powers of bureaucracies and hierarchies as predicted by Johannessen (2019)? Thirdly, the chapter addresses the role and agency of The Tourist Guide Diploma Programme at Roskilde University. How does it align with the ethos of the Fourth Industrial Revolution as producers of knowledge entrepreneurs, who educate frontline experience designers (Johannessen 2019)? The chapter concludes by discussing the future prospects of certified guides, and identifies where we lack knowledge in order to propose a future research agenda.

"Jobcrafting" Guides Adjusting to Change

"Certified guides are exemplary, well-educated, self-employed, casual labourers in a liberal and competitive job market" (Meged 2017, p. 374), and although the increased competition from guide lights and free guided tours is quite substantial, the certified guides have a preparedness to address the challenges. Like all workers, certified guides have an agency to craft scope, content and relations in their work, making themselves employable, flexible and resilient to change. Berg et al. (2007, frontpage) explain:

> Job crafting captures the active changes employees make to their own job designs in ways that can bring about numerous positive outcomes, including engagement, job satisfaction, resilience, and thriving.

A study from 2017 by Meged shows how certified guides are driven by intrinsic values, when they cognitively craft the perception of their job. They are doing the job for its' own sake, where they are less focused on financial reward and career advancement and more on fulfilment of meaningful work and personal growth. Rather than accrediting themselves for having accomplished a quite demanding education at Roskilde University, they view themselves as "natural born" guides, who engage in self-actualization. They describe themselves as "people-persons", outgoing, able to lead and improvise in chaos, and with good communication skills, and the guide job is just the right arena to exercise these skills (Meged 2017), as a consequence guides have a strong motivation to face competition and stay in the market despite deteriorating pay and working conditions.

The certified guides have always spent considerable time, money and energy solving unpaid tasks associated with the job, in order to make themselves employable. Besides, conducting the actual tours, the guides spend hours in preparation for customized tours, they assume extra tasks, they engage in continuing professional development and they rearrange their calendars in order to take on last minute jobs. They exert wage flexibility, temporal flexibility and functional flexibility, and externalize the costs beyond the actual tour for the individual employer (Meged 2017). These are inherent features in the guide profession, and the guides exert them vigorously as a consequence of recent changes as described above.

The Tourist Guide Diploma Programme at Roskilde University always invites the newly qualified to a dialogue meeting after their first season as guides. This allows for fresh insights and keeps the programme updated on changes in the market. Ten to fifteen years ago most new graduates would explain how they mainly guided standard tours such as three-hour city sightseeing by bus; walking tours to the city centre with groups from the cruise ships and as newcomers they were rarely offered the more complex and longer tours. This has changed completely, now the certified guides, newly qualified as well as experienced, are booked to conduct the complex tours, and an increasing number of customized tours. As one guide puts it after her first season "I don't think I have conducted two tours alike this summer". This is a radical shift, where guide lights and the tips-only tour companies capture the market of standard tours, while the certified guides get the demanding specialized tours. On the one hand this provides an inspiring and rewarding working life aligning with the guides' image of self-actualization, but on the other hand it poses a more demanding, stressful and time-consuming working life, in terms of still more preparation, and unpredictability not least from individual customers on private tours.

One result of the above development is a growing differentiation between the certified guides themselves. Many certified guides embrace the new challenges, but some miss the security of the standard tours, that constitute bread and butter income, while they retain that guiding standard tours also is satisfying as no group and no day is ever alike (Larsen and Meged 2013; Meged 2017). The guides who work for the major incoming agencies are from 2019 forced to accept lower payment. The incoming agencies have each produced their list of tariffs, where they have kept the hourly payment unchanged from the previous year, but they have cancelled central pay supplements, thus in reality lowering the income. The guides consent on an individual level, thus circumventing AATG. Most guides have agreed to work on the new terms, as they otherwise, will be deprived a source of employment.

In parallel, we find the entrepreneurial guide, who develops and markets his or her own tours, and still more guides create a personal website giving them direct access to the customers. This has fuelled a surge in the small niche business, where the certified guides develop innovative tour products, while they collaborate in a network. The entrepreneurial guides fit nicely into Johannessen's (2019) category of knowledge entrepreneurs in the Fourth Industrial Revolution, as people with a higher education, who cannot get or do not wish for permanent employment. Johannessen explains, "However, only a few can make a living out of working as a knowledge entrepreneur. Therefore, they alternate between their own business and temporary contract jobs in large companies" (2019 p. 9), which pinpoints the situation for most entrepreneurial guides.

Guy Standing (2014a, b cited in Johannessen 2019) has identified a new class, whom he calls the precariat, which has emerged through globalization, liberalization and increasing robotization and digitization, i.e. the Fourth Industrial Revolution. However the situation is not the same for all actors in the precariat and Johannessen (2019) proposes a typology of four, divided by two parameters; degree of autonomy and degree of satisfaction. The first type is the underemployed, who once had stable salaried jobs, and often also good education, but now are employed in fragmented short-term contractual work with much less security and benefits. They have a high degree of autonomy in their work but a low degree of satisfaction. The second type is the underpaid, who undertake work that is below their qualifications, e.g. young people, who enter the job market for the first time. The underpaid have both a low degree of autonomy and a low degree of satisfaction. Both categories are frustrated as they wish for and expected a much more secure and stable working life.

The third type is the vagabond worker. They are often migrants, who occupy the jobs lowest in the hierarchies, where they are subject to a low degree of autonomy in their work functions, and are low-paid. Nonetheless, they are amongst the more satisfied in the precariat, because they have few other alternatives, and they are happy just to get a foothold in a foreign labour market. This is where we find many tips-only guides and guide lights. Tips-only guides are to a large degree recruited amongst migrant workers (Meged and Zillinger 2018), and according to Johannessen (2019, p. 4), they are satisfied, because their work allows them to get a foothold in a new country. Meged and Christensen (2017) explain how tips-only tour guides feel emotionally attached to their job, exactly because they become a part of a social community of unskilled tourism workers in Copenhagen. We do not have any research on guide lights so far, but they are also mainly recruited amongst migrant workers, and if they are as satisfied as are the tips-only guides, they pose tough competition to the certified guides.

The fourth type is the knowledge entrepreneur, the well-educated person, who choose to be freelancers, and thus place herself in the precariat. Knowledge entrepreneurs have a high degree of autonomy and show a high degree of satisfaction, and this is where we find many of the certified guides. Certified guides are driven by intrinsic motivations, and they do not take the job for money or to pursue a fully fledged career, but rather for fulfilment of meaningful work and personal growth (Meged 2017). Their jobs were always seasonal, and even with the collective agreement, they were hired one tour at a time by a host of different employers. Certified guides have always acted as "the company of one" or free agents managing their own career and flow of income (Johannessen 2019, p. 60), and as ardent job crafters they adapt well to changes.

The guides' trade union AATG, however, has not been able to recognize and integrate the entrepreneurial aspects of the guides' work life in their structures. AATG have seriously discussed if entrepreneurial guides could be members at all, because they risked working below the tariffs in their own companies, e.g. on open tours, where the number of participants and thus income would be unknown beforehand. This clearly demonstrates that unions operate in a paradigm shaped in the wake of the third industrial revolution, with a clear dichotomy between employer and employees with set roles, tasks and economic structures.

The Association of Authorized Tourist Guide—Declining Powers in the Fourth Industrial Revolution

> The Association of Authorised Tourist Guides was established in 1933 and is the oldest tourist guide association in the world. Today, we have 286 members speaking 27 different languages. (Guides.dk 2019)

AATG was born as a professional association representing the interests of the certified guides, and they have a history of very high membership rates. They developed professional standards in collaboration with the tourism industry, and they engaged actively to place their education at

Roskilde University by the end of the 1980s. They also negotiated a list of tariffs, which most employers respected; however, it was not until the end of 1990s when the AATG became a regular union.

Through the unique Nordic system of collective bargaining, AATG together with the major employers then organized in Danish Incoming Fraction (DIF), managed to agree on four years' collective agreements with fair and relatively high tariffs, allowances, and booking and cancellation terms. The guide profession was never protected by law in Denmark, like in southern Europe, however, AATG and DIF had an agreement, where the latter would only use non-certified or as we call them "guide lights" if the certified guides were unavailable and they had agreed with AATG in advance. This system reflects the ethos of the Nordic flexicurity model, which allows for flexibility on behalf of the employers but still secures the employees. For almost thirty years, the certified guides in Copenhagen have been in a sweet spot, where the guide jobs were secured during the season at decent tariffs, but this is changing now.

The rapid growth in tourism in Copenhagen not least in the cruise industry, has led to a demand for certified guides way beyond the supply in certain periods. When the incoming agencies, as handling agents of land arrangements, initially asked AATG for permission to use guide lights, the latter was often reluctant, and along with the Tourist Guide Diploma Programme at Roskilde University, they rejected the very idea of creating a formal crash course for guide lights. They feared it would undermine the professional standards and their hard-fought for labour rights. As tourism grew, the Incoming agencies left the collective agreement one by one and some started to organize yearly in-house training for numerous guide lights right before the start of the season. Many guide lights are migrant workers, like the young well-educated Southern Europeans, who suffer from high rates of youth unemployment in their home countries. They are offered crash courses, and in return, they must sign a contract where they have to remain with the company for the season, working for a salary much below the official tariffs. The guide lights are not only conquering the market of standard tours, but they take on a formal look with a dress code, and a badge to match the certified guides. In a facebook discussion on 27th 2019, the certified guides complain that some guide lights have not been properly trained in guide ethics and code of behaviour, and the guide lights act at the pier as if it is the certified guides who are the intruders and not vice versa.

The incoming agencies, who have honoured the collective agreement for longest, are the ones who regard certified guides as knowledge entrepreneurs providing them with a qualitative advantage. These agencies address the upscale market, using certified guides to profile themselves. However, even the most loyal employers have now caved into price competition. The model of collective bargaining has as of today fallen apart and AATG has not been able to renegotiate an agreement for the coming four years, and it is quite possible, there never will be a collective agreement again, which will be a mortal blow to AATG as a trade union.

AATG has neither been able to counteract the surge of tips-only based walking tour companies in Copenhagen. Akin to the platform economic posterchild Airbnb, Sandemans New Europe offered the first free walking tour company in Berlin in 2003, and since it has had a rapid uptake now offers tours all over Europe and North America. The concept is easy, free guided tours or rather tips-only, where the tourists are urged to pay tips according to their likings. The tips constitute the income for the guides, and the platform owner, as they receive a certain amount per tourist called marketing fee. The phenomenon arrived in Copenhagen in 2011, and in the beginning, the ethos was peer-to-peer guiding, where young people guided young people in an informal, friendly, party-like, atmosphere (Meged and Zillinger 2018). However, free guided tours had a rapid uptake in the mainstream market for guided tours and in 2017, and the independent Copenhagen Free Walking Tours estimated they alone would guide around 100.000 tourists (Meged and Zillinger 2018). The free tour companies address a new type of tourist, one who may never have attended guided tours otherwise, but undoubtedly, they also take shares from the segments of tourists, who normally attend guided tours, and who find the new product satisfying and cheap. The platform economies are disruptive in their nature, and the free guided tour companies have put AATG and the certified guides under pressure in the market of standard tours (Meged and Zillinger 2018).

Unlike taxi driving, Danish law does not protect the profession of the certified guides, and while the trade associations of taxi drivers have successfully managed to ban Uber as an illegal and anti-competitive practice in Denmark (nyheder.tv2.dk 2017), AATG has had little luck doing the same in the guided tours market. The free guided tour companies assemble hundreds and hundreds of tourists for departure right in front of the City Hall every day, but Liberal mayor of occupation and integration in Copenhagen welcomes the sharing or platform economy in tourism as a progressive way to create business, revenue and low entry jobs not least to migrant workers (55 Minutter 2019). The profound changes taking place on the labour market these years are embraced by Liberals but regretted by others, such as the unions, who struggle to define their role and room for manoeuvre.

> In the fourth industrial revolution work is characterized by increased individualization, fragmentation, mobility, survival syndrome, narcissism, Teflon relationships and weak ties. The personal crisis resulting from such a transition to a new work logic will lead to less emphasis on collective solutions and a greater emphasis on individual solutions pointing to a power shift where the old structures and hierarchies loose power. (Johannessen 2019, p. 83)

So far, the certified guides union AATG has enjoyed a very high rate of membership, but the question remains, how does it respond to the current crisis? AATG's inability to integrate their own entrepreneurial guides, the breakdown of the collective agreements and finally not being able to defend the profession in the face of the fast-rising numbers of guide lights and tips-only guides has put the AATG under severe strain. The guides are questioning their membership, but until now, most have remained loyal hoping for inside reforms.

AATG has not always been a union, and it constitutes the only official forum for certified guides, why the majority of guides still hope the AATG will be able to embrace the changes and transform into a modern association, that helps to shape the future roles of the certified guides. AATG will have to bridge the span where guides as knowledge entrepreneurs are both employees and micro-entrepreneurs, they have to find solutions to the divide between emerging masses of new types of guides and the "old" classical certified guide, and they must address the platform economy that poses challenges, but also offers new opportunities. On the plus side, the market of guided tours is expanding due to the general growth in tourism, and as tourism matures the demand for customized tours is also on the rise, and here certified guides specialize and carve out niches, which ultimately alters their working conditions. The question is how can AATG sustain the guides in this transition and just as important is the question whether their education, the Tourist Guide Diploma Programme, can adjust to these changes?

The Tourist Guide Diploma Programme—Towards Flexibilization, Entrepreneurship and Innovation

Education is at the heart of The Fourth Industrial Revolution, and to stay on top of a still more precarious labour market, the certified guides must continuously develop their skills as knowledge entrepreneurs. This implies less emphasis on knowledge reproduction and interpretation and more on innovation and experience design. In the Fourth Industrial Revolution all organizations "must focus on experience design, which are the processes and the coupling of customers and workers who do what the organisations is designed to do i.e. the people in the frontline" (Johannessen 2019, p. 41). The guides are exactly such frontline workers, and if they also act as entrepreneurs, they are in a privileged position to innovate not only new tours and tourism experiences (Holm and Kaae 2017), but also innovate new business models (Meged and Zillinger 2018).

The Tourist Guide Diploma Programme (TGDP) is not a democratic institution, but it has developed in close dialogue with AATG and the tourism industry, and as an adult education programme with a life-long learning perspective, it is forced to evolve over time and adjust to changes.

In 2012 by law, TGDP had to reorganize a hitherto holistic closely knitted education into several separate modules. The aim was to create flexibility for the individual students, so they could better tailor their own education. It also prompted TGDP to offer new and different modules, like “Innovative Tour and Experience Design”, which has coloured new generations of certified guides to see themselves as innovators forging their own professional niches.

Entrepreneurship and innovation are also at the core of the research project “Innovation and development of Urban Ecotourism” run by TGDP Roskilde University in collaboration with Copenhagen University and Norrøn Architects from 2017 to 2019. The project aims to create a niche of actors developing urban ecotourism. The project takes place in Amager Nature Park adjacent to Copenhagen and The National Park Skjoldungernes Land where Roskilde city lies within the park borders. In 2018, 21 guide-entrepreneurs took part in the project as innovators along with a host of actors in- and outside the two parks. They developed 15 new tours or events building on the principles of ecological, social and economic sustainability, while moulding their own role as normative guides “educating the travellers”. In 2019, the principles of the project were integrated in the curriculum of TGDP, and new modules like “Sustainable Tourism and Tour Design” and “Entrepreneurship within Tour and Experience Design” were created. The latter ran successfully in 2019 with a mix of Tourist Guide Diploma students, certified guides and entrepreneurs from the tourism industry.

Despite TGDP’s efforts to adjust to the development, the sharp competition not least from private unauthorized providers puts the education under pressure. From having a monopoly like position, educating the guides to a relative protected labour market 25 years ago, TGDP now tends to be an expensive and exclusive education sought by people, who already are well-prepared knowledge workers. The typical tourist guide student is a mature adult, who already holds a Master or Bachelor degree and has a lot of professional experience. Over the years, the classes have become smaller, and TGDP now appears to educate a niche segment of guides—while the masses of guides for the standard market receive in-house training by incoming agencies or free walking tour companies. As stated before, guiding is not a protected profession, and anybody can, in theory, work as a guide in Copenhagen. In fact we can witness already a host of guide types like segway guides, canal tour guides, private people who act as local guides, homeless guides, etc., who are either self-taught or trained by organizations or private businesses.

As a university degree in the adult educational system, TGDP like AATG must decide if it can and will open up to other types of guides, without lowering standards. A recent change in the curriculum makes it possible to attend designated modules on, e.g., bus training on guided standard tours, making it cheaper and more flexible to obtain formal qualifications without having to complete the entire programme. This option was unthinkable only ten years ago, and is yet another proof of how TGDP follows the ethos of the Fourth Industrial Revolution exerting flexibility, individualization, customization and innovation.

Discussion

According to the UN's Sustainable Development Goal 8 tourism must "promote sustained inclusive and sustainable economic growth, employment and decent work for all" (World Tourism Organization and United Nations Development Programme 2017, p. 17). The question is how you define a decent and sustainable working life in a Danish context. In Scandinavia employees enjoy extensive rights and benefits, as they are protected by a universal welfare system based on relatively stable and decent salaried jobs, high level of taxation and an effective social security system. In this regard, many of the new jobs created in the tourism sector are potentially disruptive to the traditional welfare state as they circumvent old hierarchies and some fly under the radar like in the sharing economy (Meged and Zillinger 2018).

However, some tourism workers fight for their labour rights applying tried and tested methods with the help from their unions. In summer 2018, the numerous Strömma canal tour and bus guides sent out a strike notice to the Swedish owned company Strömma, who operates in Copenhagen. The bus and canal tour guides receive a rudimentary in-house training, and then they work for a minimal salary, for very long hours with little or no breaks. The general working conditions are objectionable, and the entire personnel in Strömma sympathized with the guides and sent in strike notices in solidarity (Arbejderen 2018). Eventually, Strömma averted the strike, when they signed a new collective agreement, making minimal necessary improvements. The tariffs for Strömma guides are to this day far below that of the certified guides'.

Another example close to the guides are the drivers of tourist busses in Copenhagen. In Summer 2019, they engaged in an action against "pirate" busses and drivers from Eastern Europe and the Baltic countries accusing them of economic and social dumping (Eriksen 2019). In an attempt to get political attention, the drivers' and their union 3F blocked the entrance to the parking lot at Amalienborg Castle, a main attraction in Copenhagen. A Polish driver, driving a Lithuanian bus in Copenhagen, expressed his sympathy, and said that he ought to be paid like the Danish drivers. Guy Standing (2014) calls the growing precariat a dangerous class, and he predicts they eventually will come together to fight for their rights. The above action, however, only resulted in modest media attention and to the author's knowledge nothing has come out of it so far.

As knowledge workers and specifically knowledge entrepreneurs, the certified guides are in a somewhat different position than the bus drivers and the Strömma guides. Johannessen (2019) foresees that in future work will be characterized by extreme specialisation and cascades of innovation that builds on a competitive educational system, a competitive infra-and info structures, a competitive research system and a competitive entrepreneurship system. The certified guides in Copenhagen hold a degree from a university that has research at its core and that has started to integrate innovation, entrepreneurship and sustainability in the curriculum. Furthermore, the certified guides are ardent job crafters doing the job for its own sake, and they are used to work as a "company of one". Disregarding their recent loss of labour rights, the certified guides appear to be well prepared to face the challenges of future work. However, according to Johannessen (2019, p. 70) success depends on peoples' willingness to take risks and as entrepreneurs to see opportunities for profit where others see problems and limitations. This puts hard demands on the certified guides, and it is important future research explores how they manage the increasing competition and precarity, and whether and to what extent there exists a growing inequality between the certified themselves.

While we have some research on the working life of the certified guides (Meged 2017) and the tips only based guides (Meged and Christensen 2017; Meged and Zillinger 2018) in Copenhagen, we know little of the fast-growing number of guide lights and the host of other types of guides that is cropping up these years. We know that many of them are migrant workers, and research that is more comprehensive is needed to understand their working conditions as well as their impact on the Danish labour market.

Conclusion

The current situation of the certified guides' working life provides an excellent showcase of the paradigmatic shift from the Third to the Fourth Industrial Revolution. Up until 2019, the certified guides had a privileged position on the market of guided tours, as they were protected by a favourable collective agreement, but this year it was not renewed. Growth in tourism, the technological development and sharp price competition has become a leeway to numerous guide lights, and tips-only based guides, who have captured large parts of the market of standard tours at much lower prices. However, the certified guides are resilient, and as highly motivated knowledge workers and increasingly as knowledge entrepreneurs they craft new niches and do customized tours, while they balance between their own business and work for the incoming agencies.

The Association of Authorized Tourist Guides has lost most of its' bargaining power and hence its' position as a union, and is now struggling to find its' role in times of transition. Knowledge and education are at the heart of the Fourth Industrial Revolution, and The Guide Diploma Programme has managed to adjust to the new ethos including research, innovation and entrepreneurship into the formal education.

While we have some research on the working life of certified guides and tips-only based guides, we know little of the guide lights and other types of guides, where many are migrant workers, and more comprehensive research is needed to understand the transformation that is taking place in the Danish labour market these years.

References

Arbejderen. 2018. Nu kommer strejken - Ansatte står sammen i kampen mod kanal tours. *Arbejderen.* https://arbejderen.dk/fagligt/ansatte-st%C3%A5r-sammen-i-kampen-mod-canal-tours. Accessed on August 8, 2019.

Baumgarten, Henrik. 2018. København forudser fordobling i antallet af turister. *Standby Danmark*, October 14. https://standby.dk/koebenhavn-forudser-fordobling-af-antal-turister/. Accessed on May 27, 2019.

Becker, Elizabeth. 2014. Destination Nowhere: The Dark Side of the Cruise Industry. *The Saturday Evening Post*, April 17. https://www.saturdayeveningpost.com/2014/04/the-dark-side-of-the-cruise-ship-industry/. Accessed on June 3, 2019.

Berg, Justin M., Jane E. Dutton, and Amy Wrzesniewski. 2007. *What is job crafting and why does it matter?* Michigan: Ross School of Business. Retrieved from http://positiveorgs.bus.umich.edu/wpcontent/uploads/What-is-Job-Crafting-and-Why-Does-it-Matter1.pdf.

Bomberg, Leif. 2018. 30 mio. På krydstogt næste år. *Trends and travel.dk*, December 21. http://trendsandtravel.dk/30-mio-paa-krydstogt-naeste-aar/. Accessed on May 27, 2019.

CruiseCopenhagen. 2017. CruiseCopenhagen Årsberetning 2017. https://www.visitcopenhagen.com/sites/default/files/asp/visitcopenhagen/CruiseCopenhagen/PDF/arsberetning_2017.pdf. Accessed on June 15, 2019.

Eriksen, Asger H. 2019. 3f og arbejdsgivere i fælles aktion mod piratbusser. *Fagbladet 3F*, June, 11. https://fagbladet3f.dk/artikel/3f-og-arbejdsgivere-i-faelles-aktion-mod-piratbusser. Accessed on August 21, 2019.

Gans, Joshua. 2016. *The Disruption Dilemma.* Boston, MA: MIT Press.

Guides.dk. 2019. http://www.guides.dk/uk/About-Guides.dk.html. Accessed on July 1, 2019.

Holm, Jesper, and Berit Charlotte Kaae. 2017. *Definitionsramme for Bæredygtig Bynær Naturturisme.* Roskilde: Roskilde Universitet.

Johannessen, Jon-Arild. 2019. *The Works Place of the Future: The Fourth Industrial Revolution, the Precariat, and the Death of Hierarchies.* London and New York: Routledge.

Larsen, J., and J.W. Meged. 2013. Tourists co-producing guided tours. *Scandinavian Journal of Hospitality and Tourism* 13 (2): 88–102. https://doi.org/10.1080/15022250.2013.796227.

Meged, Jane W. 2017. Guides crafting meaning and identity in a flexible working life. *Scandinavian Journal of Hospitality and Tourism* 17 (4): 374–387. https://doi.org/10.1080/15022250.2017.1330845. Accessed on August 26, 2019.

Meged, Jane W., and Mathilde D. Christensen. 2017. Working Within the Collaborative Tourist Economy: The Complex Crafting of Work and Meaning. In *Collaborative Economy and Tourism: Perspectives, Politics, Policies and Prospects*, edited by Szilvia Gyimóthy and Diane Dredge, 203–220. Cham: Springer International Publishing.

Meged, Jane W., and Malin Zillinger. 2018. Disruptive network innovation in free guided tours. *Scandinavian Journal of Hospitality and Tourism* 18 (3): 303–318. https://doi.org/10.1080/15022250.2018.1497317. Accessed on August 26, 2019.

55 Minutter. 2019. København truet af overturisme. Radio 24/7, juni 19. https://www.24syv.dk/programmer/55-minutter/53175005/kobenhavn-truet-af-overturisme?start=0&fbclid=IwAR3vCKpl8JM42xfeXdklHRPpqhB-fbKm1ggrS6jF-8sRsbQ3vdFXLEDzAfuY. Accessed on July 5, 2019.

Nyheder.tv2.dk. 2017. Uber lukker idanmark. http://nyheder.tv2.dk/business/2017-03-28-uber-lukker-i-danmark-0. Accessed on May 15, 2019.

Standing, Guy. 2014. Why the precariat is not a "bogus". *Open Democracy*, March 4. https://eprints.soas.ac.uk/18275/1/Why%20the%20precariat%20is%20not%20a%20bogus%20concept.pdf. Accessed on August 21, 2019.

Veijola, Soile. 2010. Introduction: Tourism as work. *Tourist Studies* 9 (2): 83–87. https://doi.org/10.1177/1468797609360748. Accessed on October 31, 2019.

Visitdenmark Destinationsmonitor. 2018. Destinationsmonitor Januar–December 2018, Status. https://www.visitdenmark.dk/api/drupal/sites/visitdenmark.com/files/2019-04/Destinationsmonitor%20status%202018.pdf. Accessed on April 15, 2019.

World Tourism Organization and United Nations Development Programme. 2017. *Tourism and the Sustainable Development Goals – Journey to 2030.* UNWTO: Madrid.

Weaver, Adam. 2005. Interactive service work and performative metaphors: The case of the cruise industry. *Tourist Studies* 5: 15–27. https://doi-org.ep.fjernadgang.kb.dk/10.1177%2F146879760506271З. Accessed on June 5 2019.

3

Tourism Employment and Education in a Danish Context

Ida Marie Visbech Andersen

Introduction

In this chapter, employment conditions in the Danish tourism sector will be introduced. The chapter will give the reader insights into Denmark as a tourism destination and the economic contribution of tourism to the Danish economy.

The chapter will further elaborate on the Danish welfare system focusing on trade and labour unions and the 'flexicurity' model in the labour market and its meaning for employment conditions in general and in the tourism sector.

The chapter will highlight some of the main challenges and trends in the tourism sector such as low and insufficient education levels, digitalization and illustrate these using a case study of a Danish development project established to give small to medium-sized tourism enterprises (SMEs) better digital competencies.

I. M. V. Andersen (✉)
UCL University College Denmark, Odense, Denmark
e-mail: idav@ucl.dk

A. Walmsley et al. (eds.), *Tourism Employment in Nordic Countries*,
https://doi.org/10.1007/978-3-030-47813-1_3

Tourism in Denmark

Denmark is the smallest geographically of the Nordic countries but with around 5.8 million inhabitants the country has a high population density. Denmark is known for its pristine beaches, its monarchy and for being a safe and environmentally friendly tourism destination (VisitDenmark 2018). In 2015 Denmark had a 46% market share of the total amount of foreign tourists bed nights in the Nordic Region and the main markets are Germany, Norway, Sweden and The Netherlands (VisitDenmark 2016). The national destination marketing organization, VisitDenmark, operates with three main typologies: nature and coastal tourism, big city tourism and meeting tourism, the largest of these being nature and coastal tourism which contributed with 37.4 million overnight stays in 2017 (VisitDenmark 2018).

Tourism is becoming an increasingly important economic sector and growth factor in Denmark and has received increasing attention in the past 10–20 years. Tourism is not one single sector but stretches beyond the traditional tourism product to other sectors such as transportation, retail, education and other services (Riley et al. 2002). Many of the traditional tourism, but not exclusively tourism-related sectors in Denmark, still receive the majority of their income from tourism. In 2015, 64% of the total turnover in the hotel and accommodation sector was directly related to tourism, making it the sector most heavily dependent on tourism in Denmark (VisitDenmark 2017). A significant portion of employment within the restaurant sector, travel services, transportation, culture and experiences are also linked directly to tourism (VisitDenmark 2017). The total tourism expenditure in 2017 was 128 billion Danish Korner (dkkr.), where foreign tourist spending accounted for 43% of the total tourist spending (VisitDenmark 2019). The tourist expenditure is divided between the different sectors and traditional tourism sectors such as restaurants, transportation and accommodation accounted for 53% of tourist spendings in 2017 (VisitDenmark 2019).

Reports have shown that Denmark is experiencing a positive and stable rise in the number of jobs generated by tourism (The National Tourism Forum 2018). The total effect (direct, indirect and induced) of tourism expenditure in 2017 generated 161,000 fulltime jobs meaning that 1 million dkkr. spent on tourism created on average 1.26 jobs (VisitDenmark 2019). Tourism accounts for 5.5% of total employment in Denmark and direct tourism expenditure in 2017 generated 111,800 fulltime jobs in all sectors and 72% of these jobs were created directly in the tourism sector (VisitDenmark 2019). Tourism generates jobs in all parts of Denmark (see Table 3.1) with the largest concentration centred around the main capital region of Copenhagen.

Tourism has received more attention since the mid-2000s, where tourist arrivals in Denmark started to drop. From 2007 to 2012 Denmark experienced a decline in foreign tourist arrivals whereas Europe as a whole still experienced growth (The Danish Government 2014). A new law was passed about the development of tourism effectively restructuring the tourism sector and the government presented a plan to ensure growth and development in especially the coastal and nature tourism segment. The Growth Plan for Danish Tourism focuses on four main areas (The Danish Government 2014):

1. Better regulations of the public tourism development efforts
2. Raising Service and quality levels in Danish tourism

Table 3.1 Total employment derived from tourism in 2017 distributed into regions

	Total derived employment from tourism		Share of total Full-time equivalents (FTEs) in region
	Full-time equivalent (FTEs)	Percentage	
Capital Region	60,416	38	5.7
Region of Southern Denmark	32,302	20	5.5
Region of Middle Jutland	28,883	18	4.4
Region of Northern Jutland	21,007	13	7.4
Region of Sealand	18,346	11	5.7
Total	160,961	100	4.7

Source Table adapted from (VisitDenmark 2019)

3. Increase growth in city and meeting tourism
4. Coastal and Nature tourism must be developed

The growth plan also outlined a new structure focusing on VisitDenmark's three main tourism types: nature and coastal tourism, big city tourism and meeting tourism. In 2016 a new government approved a national strategy with additional initiatives to further drive the development in Danish tourism (The Danish Government 2016). The 10 key efforts are largely centred around facilitating and creating good quality experiences for the tourist, better marketing efforts and digitalization of the tourism sector. Only one of the key efforts are linked directly to employment. The strategy points to the importance of creating innovation and growth by ensuring that the tourism sector has access to competent employees both for delivering service and for business development (The Danish Government 2016). The Danish tourism sector has seen large changes in the past decade and has gone from being a sector considered less important to one of the most talked-about in recent years. To understand the foundation and prerequisites for employment in the tourism sector, there are some key conditions affecting the Danish labour market and welfare system in general that needs further elaboration.

The Danish Welfare System

Denmark, along with the Nordic countries, is generally characterized by a comprehensive welfare state financed through taxes and social security contributions (Bévort et al. 1992; Christiansen and Petersen 2010; Gustavsen 2007; Kolm and Tonin 2015; Torfing 1999). Denmark and Sweden spend more than any other OECD country on measures that are designed to reintegrate unemployed into the labour market or to upgrade their employment-specific skills (Jochem 2011). There are debates about whether the Nordic welfare model can survive the combined pressure of an ageing population, immigration, globalization and financial crisis (Kvist and Greve 2011).

In order to understand the employment conditions in the Danish tourism sector, we need to take a closer look at how the Danish welfare system and labour market function. The Danish system builds on historical agreements and is well known for the "Danish model" which includes both the model of collective bargaining and flexicurity (Knudsen and Lind 2018).

The Bargaining Model

The Danish labour market has its historical roots dating back to The Great Compromise (Septemberforliget) in 1899 (Kristiansen 2014). In 1898 the employer side formed a joint confederation as a response to an increasingly successful struggle where workers and their unions fought for higher wages, shorter working hours and against degrading conditions (Lind and Knudsen 2018). The trade unions decide to fight back this joint confederation and in 1899 a relatively small strike among workers was answered by a massive lockout of more than 40,000 workers by the joint employer confederation (Lind and Knudsen 2018). The conflict lasted over three months and ended in the famous September Agreement of 1899, where both sides had to recognize the rights of the opposing parties to organize themselves and to be acknowledged as a bargaining partner (Lind and Knudsen 2018). The strategic aim of the September Agreement was to create a labour market based on high salaries and good physical working conditions among others in exchange for more highly efficient production which was achieved through a closely linked institutional system of cooperation, negotiation and bargaining (Bévort et al. 1992).

Today all groups of workers from unskilled, craftsmen to academics are represented in different unions. This is also reflected in an increasing number of students attending middle and long-term education which is visible in the member development of the different trade unions associated with these educational pathways. Since the 1980s over 75% of the total labour force has been a member of a trade union (Knudsen and Lind 2018). One of the explanations for these high numbers is the close connection to unemployment benefit funds offered by the trade unions since the 1970s (Knudsen and Lind 2018). On the other side, employers are organized in different associations and together with the trade unions determine the rules regulating collective bargaining (Knudsen and Lind 2018). The Danish labour market is characterized by an equal balance of power (Bévort et al. 1992). This unique collaboration of Danish employer associations and trade unions and their collective bargaining agreements are one of the fundamentals for the way the Danish labour market functions and still plays an important role in both national and local bargaining (Lind and Knudsen 2018). The collective bargaining model works well in Denmark with support from the flexicurity model.

The Flexicurity Model

Another central characteristic to Denmark's welfare system is its flexibility (Wills 1989). Since the 1990s the Danish 'flexicurity' model has been highlighted as an ideal approach to economy and labour markets conditions (Knudsen and Lind 2018). The Danish 'flexicurity' model thus combines 'flexibility' and 'security' and suggests a situation of conditions where companies can employ, dismiss and deploy employees according to the needs of the organization (Knudsen and Lind 2018). At the same time, workers are guaranteed a decent living standard in case of unemployment along with access to education to help them get back into the labour market (Knudsen and Lind 2018). The flexibility in the Danish labour market is a clear competitive advantage (Wills 1989) and the flexicurity model thus offers both companies the flexibility to adapt to the market situations while simultaneously providing a certain level of security for workers. Conceptually, the 'flexicurity' model contains more elements than 'the bargaining model'. Flexicurity embraces a combination of elements from collective bargaining on the one hand and employment and welfare policies from the state on the other (Lind and Knudsen 2018). The key features of the model are in other words bargaining power, government policies and compensation from the welfare state for problems caused by the bargaining model

(Knudsen and Lind 2018). As a result, there are very few legal complications in dismissing an employee in Denmark compared to other European countries (Bévort et al. 1992), an issue often associated with tourism due to its strong seasonality in many countries.

While the bargaining model is primarily valued as a model that helps create a prosperous economy, the flexicurity model is valued for its ability to simultaneously strengthen the economy in terms of productivity and competitiveness of the enterprises and the total employment in society at the same time (Lind and Knudsen 2018). The conceptualizations of Danish flexicurity contain assumptions of mutual agreement, stating that the system functions in ways that serve everybody well: employers, employees and society as a whole (Lind and Knudsen 2018). Therefore, Danish companies have a high degree of flexibility primarily as an effect of the bargaining model but also based on social security and the right to education, which are both features of the flexicurity model (Gustavsen 2007; Knudsen and Lind 2018). This flexible system now covers around 85% of agreements on the employment area (Knudsen and Lind 2018). The two central models of bargaining and flexicurity shape the Danish labour market and thus set the rules and regulations of employment in the tourism and service sector as well. Due to these two models, it is therefore relatively cheap for a company in Denmark to recruit employees with the necessary skills and, conversely, to get rid of those employees whose skills are obsolete, or for any other reason (Bévort et al. 1992).

Employment Conditions in the Danish Tourism Sector

When talking about tourism and thus tourism employment, the first issue that arises is the very definition of tourism. Because tourism itself cannot be defined as a single sector (Riley et al. 2002) any attempt to define tourism employment therefore becomes difficult. Traditionally the tourism sector includes a large part of the service, transportation, accommodation and restaurant sector but can also include part of other sectors such as retail. Denmark's economy is dominated by services with more than 70% of the private sector employed in services (Danish Statistics 2019), a number that has been steadily growing since the mid-1960s when the service and the industrial sectors were roughly the same size. The entire service sector is also the largest export sector measured in the number of jobs created; nearly 400,000 in direct and indirect jobs (The Danish Chamber of Commerce 2018a). The tourism and service sector benefits from the Danish welfare system (Hjalager 2005) and is directly influenced by both the bargaining and flexicurity models.

Tourism and the service sector are known for being a hard and at times tedious sector to work in with less than favourable working conditions. There is much research conducted on various elements affecting employees working in tourism and service such as the service encounter, the physical and mental conditions of the service sector and emotional labour (Riley et al. 2002; Wirtz and Lovelock 2018). However, the conditions of the Danish labour market make it relatively easy for tourism and service companies to hire or dismiss employees as needed and at the same time giving employees the right to unemployment benefits and education if they are a member of a union. This is one of the explanations of Denmark's high employment rate. The employment rate for 15–64-year-olds was 75.5% in the 4th quarter of 2018, which is the fifth-highest in Europe only topped by the Netherlands, Sweden, Germany and Estonia (Danish Statistics 2019). The conditions for working in the tourism sector are thus relatively safe and secure for both employees and employers.

There are, however, still issues in tourism employment. The first issue is education levels. According to the national strategy for tourism from 2016, one of the key issues in Danish tourism is the need for better production and quality (The Danish Government 2016). In order to achieve such an ambitious goal, the necessity for a skilled and well-educated workforce is essential. A large part of the tourism sector faces issues surrounding education and skill levels among its employees (Bound and Lin 2013; Kyriakidou and Maroudas 2010). This is partly because jobs in the tourism and service sector are often the access point to the labour market for many young people, people with lower level education or immigrants with trouble finding jobs (Danish Statistics 2018; HORESTA 2018). The high number of immigrants and young people working in the sector can be seen as positive factor, as the employment of these groups is beneficial for society as a whole. The tourism sector's ability to hire employees with different ethnic backgrounds contributes to solving the challenge of integration faced by society at large (The Danish Government 2016). Research also shows that immigrants and young people are less likely to be part of a labour union (Danish Statistics 2018) and are therefore not entitled to the full benefits made possible by the bargaining and flexicurity models. This means that the security and benefits given by the Danish models of bargaining and flexicurity both work in favour but also against the sector as hiring seasonal workers and dismissing them are relatively easy, putting the sector at risk of creating a constant intake of unskilled workers each season. There seems to be an oxymoron in the sense that tourism is relying on seasonal workers, which are often young people and immigrants but at the same time these groups of people can be argued to have lower education levels or education from different sectors than tourism, thus creating a ripple effect of constant entry-level training and thereby maintaining the low education levels that challenges the sector to begin with. This creates long term issues with education levels and the continuous education of existing employees, as many are only guaranteed employment part-time. This leads to a growing need for education of the tourism workforce. Not only the current employees but also the newly educated, unskilled workers and immigrants. The Danish model can thus be seen as compromising both negative and positive factors, depending on the perspective taken, as hiring and dismissing employees is relatively easy but at the same time, there is no need for companies to invest in training and continuous education, because many employees are only seasonal or part-time. One could discuss whether the sector is partly to blame for its own challenges with low education levels. Furthermore, as mentioned above many young people and immigrants are underrepresented in trade unions making them ineligible to the full benefits provided by the bargaining and flexicurity models including financial support and the right to paid education.

The second issue with tourism employment is geography. Statistics show that the largest limitation to productivity in the tourism sector is lack of demand followed by lack of labour (StatBank Denmark 2019).

Job opportunities in the tourism sector are unevenly distributed. There are large differences between the number of tourism jobs among the municipalities and the size of the tourism sector in the different municipalities (The Danish Chamber of Commerce 2018b) and the tourism sector is still heavily affected by seasonality. During the high season, 6000 additional jobs are created (The Danish Chamber of Commerce 2018c), an indication of many seasonal jobs. Municipalities in the northern part and the western part of Denmark are more dependent on tourism than other parts of Denmark (The Danish Chamber of Commerce 2018b). Local economies in these municipalities rely more on tourism and the expenditure generated by tourists to create employment and income opportunities for the local residents. Often there are limited job opportunities in the smaller municipalities, which makes it difficult to attract skilled workers and keep them, as many SMEs can only offer part-time jobs during the high season. This is a challenge faced in many countries, not only Denmark. Research indicates that smaller towns in Denmark specialized in service jobs in the local labour market had negative socioeconomic development regarding employment growth from 2008 to 2013 (Hansen 2016). This issue was also noted in the strategy developed by the government in 2016. The strategy points to difficulties in attracting a qualified workforce especially outside the larger cities and issues securing the many part-time and seasonal jobs still dominating the tourism sector and the need to make the tourism sector attractive through job opportunities and career development supported by education options (The Danish Government 2016). This leads back to the issue regarding entry-level jobs in the sector often occupied by immigrants and young people.

Education and Employment

Many Danish companies continue to experience difficulties attracting employees since the financial crisis in 2008. In 2009 approximately 7000 positions were unoccupied and the number had risen to 20,000 unoccupied positions in 2017 (The National Tourism Forum 2018). The tourism and hospitality sector experiences similar tendencies. In the hotel and restaurant sector, the number has risen from 540 unoccupied positions in 2008 to approximately 1780 positions in 2017 (The National Tourism Forum 2018). Companies were in other words finding it problematic to locate and attract the right candidates and in many regions, the demand for skilled labour is higher than the supply. Inevitably, employment is linked with education and the lack of skilled labour supply is despite Denmark's reputation for having a well-renowned education system ranging from primary school to high schools and universities, still an issue.

Tourism education needs to balance three imperatives: the need to promote individual development, the need to advance knowledge and the need to be practical and relevant for the sector (Riley et al. 2002). Many tourism jobs are practical trainee jobs such as waiter, chefs and event coordinator with the majority of the education focused on practical experience and 'on the job' training. However, many aspects regarding tourism such as management and planning need higher levels of education (Riley et al. 2002) and cannot necessarily be learned only through practical experience. Another dilemma is the possibility to build a professional career in tourism. The tourism sector, does not have a distinct career system with clear internal career paths: staff come to tourism with quite varied backgrounds and sometimes irrelevant professional educations and leave it for a range of other activities (Hjalager and Andersen 2001). This is a major challenge for the sector. The last imperative mentioned by Riley et al. (2002) is the need to advance knowledge. This can be done through either practical training or education.

In the tourism sector the training and professional development of staff is the most important factor that directly or indirectly affects the competitive power of the tourism enterprise, as educated employees enable the development of new technologies, innovative products and services, aiming to increase competitiveness of the tourism product and company (Bahçelerli 2018). Furthermore, there is a growing need and demand for tourism employees with knowledge and skills of the digital world, as the large intensification in both development and application of digital technologies in all aspects of society is evident (Boes et al. 2016; Foroudi et al. 2017). Many employees in SMEs have limited Information Technology (IT) knowledge and skills as they lack a clear understanding of how advanced IT can improve performances and drive their business (Law et al. 2009). Today, employment in the tourism and service sector requires a different set of skills and knowledge than 20 years ago. Some skills are still essential, such as conflict management, customer service and communication but as the world is experiencing a growing digitalization and changing consumer behaviour (Law et al. 2009) the tourism and service sector must adapt and embrace this digital advancement. There is a need for digitalization in the Danish tourism sector (The National Tourism Forum 2016) and the following case study illustrates some of the challenges faced by SMEs in Denmark and the need for upgrading employees' digital competencies through an competency development project.

Case Study: Digital Growth Development Project

(All information in this case study is unpublished material from the author)

Background of the Project

This development project is an EU funded project. Digital Growth development project (2017–2020) is designed to deliver knowledge, skills and competencies to SMEs in southern Denmark and thereby strengthen their ability to compete in a growing digital world. The project is run in different regions of Denmark with each their special digital focus. This case study is only focused on the project in the region of Southern Denmark managed by UCL University College Denmark, as this project has an overall focus on delivering digital competencies to help SMEs grow their business by using and implementing digital tools and knowledge. The project is constructed around five independent rounds each consisting of between 15 and 20 companies with a total number of participants ranging between 25 and 40. Each round consists of six workshops during a one year period including various optional networking and individual counselling opportunities for the participants and their companies.

The project in Southern Denmark is aimed directly at SMEs in the tourism and hospitality sector or the service sector in general. The participating companies are from the following sectors: Hotels, restaurants, hostels, campsites, attractions, activities, retail stores and travel agencies. The participants have gained new knowledge in various topics such as Facebook, Instagram, video and picture editing, Google Analytics, content marketing, digital strategies, digital customer journeys and neuromarketing. Approximately 70 SMEs have been part of the project and roughly 53 companies have completed the full one year period of the project. 14 interviews were made with former participants from the first four rounds of the project after they had completed the project.

Discoveries

The interviews revealed several interesting findings. Firstly, all informants mention the level of knowledge they have gained from the project. Participants mention that they feel like they have a much larger overview of the different digital opportunities after participating in the project. When asked what value they have gained from the project, most informants spoke of the clarification and knowledge about the different digital technologies and tools as one of the main values experienced during their participation:

> *The value from the project lies in the overview we have gotten and knowing where to focus our digital efforts*
>
> *Digital growth has helped create a structure in our planning and a digital overview*
>
> *The project has created value in clarifying and given me new knowledge*

Technology and the level of information develops day by day and becomes increasingly complicated to handle (Bosco 2007) so the clarification gained by the participants can be seen as a positive effect of the project.

Another important aspect was to investigate the level of competencies gained by the participants. Hamel and Prahalad (1994) define core competence as a bundle of skills and technologies that enable a company to provide a particular benefit to the customers. Most of the Informants found it difficult to voice actual competencies but easier to talk about what knowledge and skills they had gained from the project:

The most important has been to gain new knowledge about the different opportunities and knowing how they can be used

The competencies has been focused on the use of individual tools

I am making a marketing and content plan and I have become much more structured in including everything such as facebook, print ads and everything

Some participants mentioned how they have started using the tools and knowledge gained from the project thus starting to build actual competencies according Hamel and Prahalad's (1994) definition. Developing competencies can take time and one informant summaries this:

I now feel a drive (becuase of this project) to develop more competencies within the digital field

This shows that many of the informants experienced some level of skill and knowledge enhancement and even though they could not mention specific competencies, the project seemed to have help turn existing and new knowledge into more concrete skills.

Digitalization is an important factor for growth for many SMEs but often seen as a major challenge (Buhalis and Egger 2008). Participants felt somewhat strongly that participating in the project had moved the process of digitalization in their company and that combined with the clarification of the digital possibilities, the project has contributed to an increased digitalization for many of the participants and their companies:

Being part of the project has definitely contributed to an increased digitalization of the company

The project has contributed with digitalization in a marketing perspective

We would not have gotten to where we are now if not for this project

Many of the informants also talk about how they can see the project as a way to create new collaborations with different companies but that it takes time. Some of the companies involved have created new collaborations in the form of different projects with the education provider (UCL) both as concrete case for students in various subjects and by offering new internship positions for the students at UCL.

This case study supports existing research stating that education and digital competencies are two critical issues but SMEs are lacking behind (Bound and Lin 2013; Kyriakidou and Maroudas 2010; Law et al. 2009). An important factor for the competitiveness of the tourism sector is having the right competencies to ensure product and business development (The Danish Government 2016). The national strategy further talks about adapting the continuous education opportunities for companies to support the development in the labour market (The Danish Government 2016). The tourism sector consists of many SMEs which require new competencies and counselling and there is a need to create better and more specific offers aimed at SMEs in capitalizing the digital opportunities (The Danish Government 2016). Dealing with how learning and training are structured to deliver skills and knowledge to SMEs is a matter of major interest (Kyriakidou and Maroudas 2010). This case study demonstrates the need for continuous education and knowledge about digitalization within SMEs in the tourism sector. The project illustrates one way of delivering education to employees in the tourism and service sector on terms fitting the structure of SMEs and thus helping them in gaining valuable new knowledge and skills to cope with the challenges of tomorrow.

Read more about the project here: www.digitalvækstkultur.dk.

Discussion and Conclusion

Denmark has in recent years experienced a growing interest and focus on the tourism sector. Tourism has become a political issue and also an approach to achieve economic growth and employment opportunities for many rural areas. Denmark is a well-known tourist destination with white clean beaches and a friendly atmosphere. Tourism creates approximately 5% of total employment in Denmark or around 161,000 jobs.

These jobs are mainly in the traditional tourism sectors such as accommodation, restaurants and transportation. Employment opportunities are distributed across the entire country with some geographical differences. The labour market conditions in Denmark creates a unique setting for the tourism and service sector due to the influence of the bargaining and flexicurity models. Welfare policies and these models shape and structure the labour market and are the foundation for the operation of tourism and service enterprises and their employees. The tourism sector has access to continuous learning and education but there are still gaps in ensuring employees in the sector that have the right skills and knowledge about digitalization, which is a key growth factor in today's society.

There are still many issues surrounding tourism employment such as seasonality and general working conditions and there is a dilemma in terms of battling low education levels, seasonality and the way the Danish models of bargaining and flexicurity influence the labour market in Denmark. As long as tourism is so heavily dependent on seasonality it seems like an impossible task to lift education levels much while still relying on the current groups of workers, namely young people and immigrants. Despite the political focus on tourism development, there is still limited focus on how to deliver these strategies and further improve the sector in regard to education, continuous education and seasonality. So regardless of the national focus on tourism in Denmark, few plans are specifically addressing how to deliver the education that SMEs need. The Digital Growth Project case study, described above, is one example of how SMEs can upgrade their knowledge and skill set to overcome the knowledge gap within one area, namely digitalization.

The employment conditions in the Danish tourism and service sector are generally good but there are still issues concerning how to battle seasonality, education and career improvements for employees and how to leverage the continuous reliance on both the youth and immigrants while still ensuring these groups can benefit from the Danish models. The future of tourism employment is dependent on the development in both continuous education and training of these two groups of the workforce but also on employees in SMEs. A greater focus should be placed on SMEs' ability to transform their employees' skills and knowledge within digitalization and application of digital tools and solutions in driving their businesses forward. The case study of the Danish development project showed potential in terms of creating a platform for learning and delivering new knowledge and skills to SMEs and thereby help them develop new digital competencies that can be utilized in driving and growing their businesses and thus altogether creating better conditions for tourism employment in Denmark.

References

Bahçelerli, Nesrin Menemenci. 2018. Strategy for lifelong learning in vocational schools of tourism education. *Quality & Quantity* 52 (1): 43–58.

Bévort, F., Storm Pedersen, J., and Sundbo, J. 1992. Human resource management in Denmark. *Employee Relations* 14 (4): 6–20.

Boes, Kim, D. Buhalis, and A. Inversini, 2016, smart tourism destinations: Ecosystems for tourism destination competitiveness. *International Journal of Tourism Cities* 2 (2): 108–124.

Bosco, James. 2007. *Life-Long Learning: What? Why? How?* 1–8. Kalamazoo: Western Michigan University.

Bound, Helen, and Magdalene Lin. 2013. Developing competence at work. *Vocations and Learning* 6 (October): 403–420.

Buhalis, Dimitrios, and Roman Egger (eds.). 2008. *ETourism Case Studies: Management and Marketing Issues in ETourism*. Oxford: Butterworth Heinemann.

Christiansen, N.F., and K. Petersen. 2010. The Dynamics of social solidarity: The Danish welfare state, 1900–2000. *Scandinavian Journal of History* 26 (3): 177–196.

Danish Statistics. 2018. Immigration in Denmark 2018. https://www.dst.dk/da/Statistik/Publikationer/VisPub?cid=32561. Accessed on June 24, 2019.

Danish Statistics. 2019. Facts about the labourmarket. https://www.dst.dk/da/Statistik/bagtal/2019/2019-02-15-fakta-om-arbejdsmarked. Accessed on June 24, 2019.

Foroudi, P., S. Gupta, A. Nazarian, and M. Duda. 2017. Digital technology and marketing management capability: Achieving growth in SMEs. *Qualitative Market Research: An International Journal* 20 (2): 230–246.

Gustavsen, Bjørn. 2007. Work organization and 'the Scandinavian model'. *Economic and Industrial Democracy* 28 (4): 650–671.

Hamel, Gary, and C.K. Prahalad. 1994. Competing for the future. *Harvard Business Review*, July 1. https://hbr.org/1994/07/competing-for-the-future. Accessed on June 24, 2019.

Hansen, Kalle Emil Holst. 2016. Local labour markets and socio-economic change: Evidence from Danish towns, 2008–2013. *European Planning Studies* 24 (5): 904–925.

Hjalager, Anne-Mette. 2005. Innovation in tourism from a welfare state perspective. *Scandinavian Journal of Hospitality and Tourism* 5 (1): 46–62.

Hjalager, Anne-Mette, and Steen Andersen. 2001. Tourism employment: Contingent work or professional career? *Employee Relations* 23 (2): 115–129.

HORESTA. 2018. Danish tourism, April 13. https://www.horesta.dk/vores-branche/dansk-turisme/. Accessed on June 15, 2019.

Jochem, Sven. 2011. Nordic employment policies—Change and continuity before and during the financial crisis. *Social Policy & Administration* 45 (2): 131–145.

Knudsen, Herman, and Jens Lind. 2018. The "Danish Models" of labour market regulations and their status after recent changes. *New Zealand Journal of Employment Relations* 43 (1): 83–98.

Kolm, Ann-Sofie, and Mirco Tonin. 2015. Benefits conditional on work and the Nordic model. *Journal of Public Economics, the Nordic Model* 127 (July): 115–126.

Kristiansen, Jens. 2014. *Den kollektive arbejdsret*, 3rd ed. Copenhagen: Jurist-Og Økonomforbundets Forlag.

Kvist, Jon, and Bent Greve. 2011. Has the Nordic welfare model been transformed? *Social Policy & Administration* 45 (April): 146–160.

Kyriakidou, Olivia, and Leonidas Maroudas. 2010. Training and development in British hospitality, tourism and leisure SMEs. *Managing Leisure* 15 (April): 32–47.

Law, Rob, Rosanna Leung, and Dimitrios Buhalis. 2009. Information technology applications in hospitality and tourism: A review of publications from 2005 to 2007. *Journal of Travel & Tourism Marketing* 26 (5–6): 599–623.

Lind, Jens, and Herman Knudsen. 2018. Denmark: The long-lasting class compromise. *Employee Relations* 40 (4): 580–599.

Riley, Michael, Adele Ladkin, and Edith Szivas (eds.). 2002. *Tourism Employment, Analysis and Planning*. Clevedon: Channel View Publications.

StatBank Denmark. 2019. Limitations to production. https://www.statistikbanken.dk/kbs2. Accessed on May 25, 2019.
The Danish Chamber of Commerce. 2018a. Denmark is a service and knowledge society. https://www.danskerhverv.dk/siteassets/mediafolder/downloads/analysenotater-2018/analysenotat—danmark-er-et-service–og-videnssamfund.pdf. Accessed on May 25, 2019.
The Danish Chamber of Commerce. 2018b. How large are the tourism and experience economy sector in the municipalities. https://www.danskerhverv.dk/siteassets/mediafolder/downloads/analysenotater-2018/nr.-57-hvor-meget-fylder-turismeoplevelsesokonomi-i-landets-kommuner.pdf. Accessed on May 25, 2019.
The Danish Chamber of Commerce. 2018c. 6,000 more jobs in hotels and restaurants during the summer. https://www.danskerhverv.dk/siteassets/mediafolder/downloads/analysenotater-2018/40-6.000-flere-jobs-i-hoteller-og-restauranter-om-sommeren.pdf. Accessed on May 25, 2019.
The Danish Government. 2014. Growth plan for Danish tourism. https://em.dk/media/9333/aftale-vaekstplan-for-turisme.pdf. Accessed on March 10, 2019.
The Danish Government. 2016. National strategy for Danish tourism. https://em.dk/publikationer/2016/national-strategi-for-dansk-turisme/. Accessed on March 10, 2019.
The National Tourism Forum. 2016. Digitalization in the tourism sector. http://em.dk/publikationer/2016/digitalisering-i-turismeerhvervet/. Accessed on March 10, 2019.
The National Tourism Forum. 2018. Statusanalysis of the development and competiveness of tourism. https://em.dk/ministeriet/arbejdsomraader/erhvervspolitik-og-planlaegning/turisme/. Accessed on July 10, 2019.
Torfing, J. 1999. Workfare with welfare: Recent reforms of the Danish welfare state. *Journal of European Social Policy* 9 (1): 5–28.
VisitDenmark. 2016. Tourism in Denmark. https://www.visitdenmark.dk/. Accessed on January 14, 2019.
VisitDenmark. 2017. Tourism employment. https://www.visitdenmark.dk/. Accessed on January 14, 2019.
VisitDenmark. 2018. Tourism in Denmark. https://www.visitdenmark.dk/. Accessed on January 14, 2019.
VisitDenmark. 2019. The economic effects of tourism in 2017. https://www.visitdenmark.dk/corporate/videncenter/turismens-okonomiske-betydning. Accessed on July 10, 2019.

Wirtz, Jochen, and Christopher Lovelock. 2018. *Essentials of Services Marketing*. Essex: Pearson Education.
Wills, Mathew. 1989. Flexibility: Denmark's competitive advantage. *European Business Review* 89 (2).

4

Tourism Work: Public Management of the Tourism Workforce in Finland

Anu Harju-Myllyaho, Maria Hakkarainen and Mari Vähäkuopus

Introduction

In this chapter, we discuss how tourism and especially tourism labour and tourism work are understood and governed by public institutions in Finland, and how this is realized in everyday human resource management practices such as recruitment of tourism companies. We introduce viewpoints to public management in general and in the Finnish context through development projects. We also discuss the main reasons for the practical and structural challenges in managing skilled labour.

A. Harju-Myllyaho (✉) · M. Vähäkuopus
Lapland University of Applied Sciences, Rovaniemi, Finland
e-mail: anu.harju-myllyaho@lapinamk.fi

M. Vähäkuopus
e-mail: mari.vahakuopus@lapinamk.fi

M. Hakkarainen
University of Lapland, Rovaniemi, Finland
e-mail: Maria.Hakkarainen@ulapland.fi

A. Walmsley et al. (eds.), *Tourism Employment in Nordic Countries*,
https://doi.org/10.1007/978-3-030-47813-1_4

In the context of this chapter, we discuss *public management* of tourism labour, but we could also use the term *governing*, which is conducted together with public and private actors (Hall et al. 2009). Tourism researchers have earlier noted the role of public actors in tourism. In Nordic countries, the role of the state in tourism has been both a direct and an indirect influencer. In Finland specifically, government steers the development of tourism with different strategies and programmes by directing funds and resources as well as by defining the relationships between public and private sectors. The public sector also regards tourism as a source of economic and social development; this is especially the case in rural or remote areas where tourism often takes centre stage in economic development plans. Tourism is thus an essential tool for economic governance, although the means vary over time (Telfer 2015; Hall et al. 2009, p. 53).

In many destinations, tourism is a major source of income and employment for local communities (Sharpley 2015, p. 7). Even if the statistics show that tourism is a significant employer at the regional and destination levels, the economic impacts of tourism are still difficult to specify. This is caused by the diversity and cross-sectoral nature of tourism (Ladkin 2011, pp. 1138–1139; Veijola et al. 2013.)

In this chapter we use the term *tourism workforce* to refer to employees and potential employees in the tourism industry, in companies providing services to tourists, and in public or third-sector organizations (Baum et al. 2016a, p. 3).

The industry is labour-intensive; the success of companies and the industry is highly dependent on gaining a skilled workforce. This puts pressure on the role of public administration in securing the prosperity of the industry. The challenge in gaining skilled labour requires attention and agile actions. However, it is important to recognize the complex nature of tourism work and all the presumptions that affect the circumstances in the industry.

Tourism as an Industry in Finland

According to a report by the Ministry of Economic Affairs and Employment of Finland (MEAE), in recent years, tourism has grown and internationalized faster than other industries. The tourism industry's share of GDP has stayed at 2.5%, but its importance is more visible compared to other industries. In terms of GDP, the tourism industry is larger than the forestry and food industries. The impact of tourism on GDP varies by region however; for example, in Åland its share of GDP is 15.6%, while the corresponding figure is 5.7% in Lapland and 2.8% in Uusimaa (Jänkälä 2019, p. 7).

The MEAE (2019a) report presents some of the important indicators of the industry in terms of workforce. The turnover of accommodation and catering services, programme services, passenger traffic and rental/leasing was 19.7 billion Euros. Together, these sectors covered 112,000 person-years in tourism companies directly. In addition, the tourism industry uses large amounts of rental labour, which is not included in statistics. It was estimated that the number of agency workers would have been about 6000 full-time jobs in 2015–2016 (Jänkälä 2019, p. 8). While the World Tourism Organisation UNWTO has estimated every tenth job in the world was, either directly or indirectly, allied to the tourism industry in 2017 (UNWTO 2018), in Finland, the year 2017 the corresponding figure was 5.5% (MEAE 2019a).

In Finland, while tourism has grown considerably in total demand (with one billion from 2016 to 2017), the development has not had an impact on employment, since it has grown by only 1% (Visit Finland 2019, p. 10). This could mean that the income is made more efficiency gains, there is not enough workforce to fill the open positions or the use of staff agencies has increased. It could also be that the growth has been met with the existing workforce. In addition to direct tourism work, in Finland, tourism generates a wide range of indirect work such as the retail and service sector, the social and welfare sector, and construction. Increased seasonal tourism work also has an impact on areas outside the tourist centres as new job opportunities arise. Increased and diversified job opportunities have brought new labour to the regions as well as new temporary and permanent residents (Hakkarainen 2017; Satokangas 2013).

Many of the changes and prevailing practices of working life—precarious nature of work, emphasis on personality and emotional work, for instance—become more visible in tourism work (Veijola et al. 2008, 2013). As pointed out by Hakkarainen (2017), although tourism brings work and services to rural and peripheral regions, the prevailing perceptions towards tourism work among other jobs affect the position and appreciation of tourism work. Thus, the industry directly and indirectly offers a wide range of jobs for both genders in Lappish ski resorts. However, the industry is not valued highly, and tourism work is not always accepted as "a proper job". Such views reflect the values and professional identities of different industries (Hakkarainen 2017). In recent years, the industry's reputation has suffered, and the number of study places in tourism has decreased both in secondary and tertiary level educational institutions (Vesterinen and Hakkarainen 2013, p. 22). In addition, people with tourism degrees tend to leave the industry if there are no career structures for them (PAM 2019b). Especially skilled chefs and people with university degrees in applied sciences are considered potential career changers.

In this chapter we discuss how tourism work and its challenges have been addressed in recent years within a Finnish context. Specifically, we explore how tourism work is governed by public institutions in Finland using a Finnish government's tourism employment initiative (MatkailuDiili) and other development projects as case examples. MatkailuDiili is a core development project funded by the MEAE to find solutions for the challenges faced concerning a shortage of tourism labour. We describe how the initiatives organized through MatkailuDiili answer the call to find solutions for problems facing the tourism companies.

Public Tourism Management in Finland

Tourism is an abstract phenomenon practised and studied in various fields (Edelheim and Ilola 2017, p. 23; Gunn and Var 2002, p. 4). Although tourism has established itself as an industry (Edelheim and Ilola 2017, pp. 12–13), it does not have the administrative structures or strong labour representative bodies compared to traditional Finnish industries such as forestry or agriculture. In this sense, since tourism lacks its own ministry, government-owned businesses and representative bodies, it does not have similar power to influence public decision-making nor the representation to influence the public economy. Instead, public management of tourism has been scattered between numerous ministries such as the MEAE, the Ministry of Agriculture and Forestry (MAF), the Ministry of the Environment (ME), the Ministry of Transport and Communication (MTC) and the public organizations run by them such as Business Finland (BF) and the Finnish Transport agency (FTIA; see Fig. 4.1). Tourism policy and stakeholder groups exist at different levels from the international to the regional. The international level is represented by organizations such as the OECD, the EU and the Arctic Council. The second level is national with the governmental bodies and the ministries. Different cross-sectoral organizations take part in tourism development, but they do not focus solely on the tourism industry.

Tourism development is guided by the national tourism strategy (2015–2025), which has generated various development initiatives and actions, financed by instruments under the MEAE (2019b). Due to the scattered and cross-sectoral nature of tourism, the MEAE cooperates with other ministries. An example of this is the cross-administrational tourism working group (MiniMatka), which aims for more efficient information exchange and preparation of dossiers in cooperation. The years 2016–2018 also witnessed a cooperation group connecting 35 members from tourism entrepreneurs to central actors in the industry to support the development work conducted in the ministry (MEAE 2019c). The ministry also takes part in tourism-related issues in the EU and other international instances.

Among other reasons, the scattered administration has led to developing the industry in a project-based manner in which the development activities have been limited in terms of time and somewhat disconnected and dependent on the government. Representation in tourism is covered by 2.5 full-time equivalent posts within the MEAE. The ministry employs one senior advisor in tourism and two junior advisors, one of whom is part-time. Narrow public structures in terms of the industry are partially due to the strongly cross-sectoral nature of the industry. In the narrow sectors, development is highly efficient,

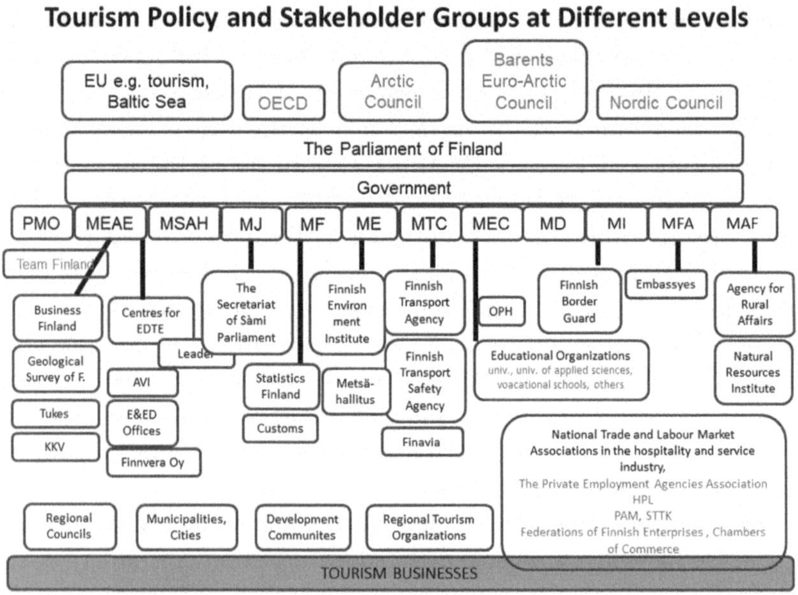

Fig. 4.1 Tourism stakeholders (Kyyrä 2019). Abbreviations of the ministries: PMO = Prime Minister's Office; MFA = Ministry for Foreign Affairs; MJ = Ministry of Justice; MI = Ministry of the Interior; MD = Ministry of Defence; MF = Ministry of Finance; MEC = Ministry of Education and Culture; MAF = Ministry of Agriculture and Forestry; MTC Ministry of Transport and Communications; MEAE Ministry of Economic Affairs and Employment; MSAH = Ministry of Social Affairs and Health; ME = Ministry of the Environment

but this is problematic from a broader perspective because it hinders the development of wider cross-sectoral entities, such as housing and infrastructure.

In addition to the MEAE, one significant national actor in the field of tourism is a unit financed by the government, Visit Finland. Its purpose is to promote Finland abroad (Business Finland 2019). Project-based work and development has considerably increased through the EU's regional development policies. Public support for tourism granted 500 million Euros during the years 2007–2013. Tourism is supported through European Regional Development Fund (ERDF), European Social Fund (ESF), agricultural funds, Business

Finland programmes and direct funds from the ministries. The public subsidies granted to tourism are roughly equal to the proportion of the tourism cluster of the GDP (MEAE 2019d).

The values of the decision-makers are always behind the political decisions in the field of tourism. Changes in ideological and sociopolitical climate have an impact on tourism planning (Dredge and Jenkins 2011, p. 2; Hall 1994, p. 51). In other words, the political decision-making is not purely rational and straightforward but value-based and complex. The parliamentary system in Finland is based on four-year election cycles, which challenges systematic, long-term development as the changes in government bring new values to the table as a consequence. Although the political decision-making in Finland is connected to the current political atmosphere, tourism has long been on the political agenda of Finnish society. The first official tourism strategy was published in 2006, but before that, there were instances taking part in tourism development. Various strategies and programmes have noted the tourism workforce, mainly from the viewpoint of developing skills and knowledge (see e.g. Finnish Tourism Strategy 2010).

The tourism areas in Finland show a great deal of variety from cities to very sparsely populated, rural and wilderness destinations. Visit Finland, the public tourism promotion organization, has divided the country into four different tourism areas, each of which has its own special features and strengths. The different areas share a common challenge, *seasonality*, which also calls out the public sector. Seasonality and short peak seasons are typical for tourism in Finland and especially in Lapland. Tourism entrepreneurs should make full-year earnings in just a few months. Thus, the development of year-round tourism could lead to a significant increase in tourism. In the future, profitable tourism business could consist of 9–10 months (instead of the current 6 months), leaving time to recover from peak seasons as well (Rantala et al. 2019, pp. 12, 23, 60). In the regions, where the seasonal changes are strong, entrepreneurs and the local community might not even have an interest in extending the tourist season, although the goal is to balance the seasonality in the long run (Rantala et al. 2019, p. 12).

The Multidimensional Challenges in Managing the Tourism Workforce in Finland

Regarding the tourism workforce, the political guidance in Finland is related to maintaining the vitality of regions and creating services and jobs for sparsely populated and remote regions. Due to the intense growth of tourism, there are plenty of jobs available, yet there is a shortage of skilled labour. To illustrate the growth angle, in 2018 international tourism grew by almost 6% compared to the year before which itself had seen a large increase in tourists with the share of international overnight stays growing by 22% (Statistics Finland 2019a). This makes the challenge structural, a matter of workforce mismatch and structural unemployment. There are several reasons behind the challenge of labour mismatch in the Finnish tourism industry—profitability, seasonality, lack of housing and segregation.

As an industry, tourism is heavily labour-dependent, and the profitability of the industry is weak, leading to wage levels remaining low. According to the Finnish Hospitality Association MaRa Ry, the profitability of Finnish hospitality companies is the second-lowest in the EU. MaRa Ry, has taken a stand on profitability by demanding the reduction of alcohol taxes (see Mara Ry 2019). The majority of the tourism workforce is under collective agreements between the federation of hotel, restaurant, and leisure service business owners, the Finnish Hospitality Association MaRa, and the Service Union United PAM (PAM 2019a). Salary as well as terms and conditions of work are based on the collective labour agreement between the employers' organization and the labour union.

Seasonality, as mentioned above, is one of the defining features of much tourism in Finland, especially in Lapland; it presents a two-way problem for recruitment. On one hand, it is difficult to hire people with high-level skills and capacities, who are willing to commit to a place. On the other hand, it has led to high costs associated with the persistent need to train new employees (Rantala et al. 2019). For example, hundreds of new employees are recruited in Lapland annually (e.g. Hakkarainen 2017). Gaining seasonal employees/workforce is also hindered by a lack of housing and irregular contracting as well as zero-hour contracts (the number of working hours might depend on snowfall, etc.). A significant number of the workers are young; in Finland, 30% of tourism workers are under 26 years old (Stat.fi 2019a), which is a strong indication that a job in the tourism industry is seen as a "stepping stone" or "walk through industry", not a long-lasting career choice. This is one of the side effects of seasonality and its by-products: long days, irregular working hours and low education level.

Seasonality is indeed a challenge, but it also has some upsides such as interesting working environments, versatility and the social nature of the work (Duncan et al. 2013; Hakkarainen 2017; Vaugeois and Rollins 2007, p. 637). For some, working in tourism allows one to pursue the desired lifestyle and take part in leisure activities in the destination (Tuulentie and Hakkarainen 2014).

In addition to strong segregation, there are certain curiosities in tourism work in Finland. For instance, a significant amount of people in Finland work on the so called *zero-hour contracts*. In 2018, approximately 106,000 people reported that they had this kind of contract (Stat.fi 2019a). Tourism and hospitality are renowned for using zero-hours contracts whereby according to Statistics Finland (2019b) hospitality uses the most zero-hour contracts in Finland. While some of the workers wish to have flexible contracts, some are put in difficult situations, where they would like to work more hours to increase their income, but there are not enough available hours for them.

Tourism work in Finland is often seen as service work occupied by women (Veijola et al. 2008, 2013). This is an accurate assessment at the national level, but it varies between regions. The industry in Lapland offers employment regardless of gender. Generally, tourism in Finland is a significantly segregated industry; four out of five employees are women in tourism and recreation, and two out of three are women in hospitality (Keski-Petäjä and Witting 2018). Some reasons for the mentioned recruitment issues can be found in this gender segregation. Most employees being women and today's family politics combined might encourage leaving the industry due to family matters. Tourism activities are often based on attractions in sparsely populated areas, where people are not ready to move for a job, as there might also be a lack of services such as day-care for children. Even tourism offers a dynamic working environment; a career change is often seen as a favourable option for a worker with a family, because work usually is very physical and entails working in shifts.

According to Keski-Petäjä and Witting (2018), segregation is considered a challenge in Finland since the gap between the number of men and women in the labour market has not decreased, and in some industries, it has increased. The reasons behind this kind of development are interesting from many perspectives, but within the topic of this article, it is noteworthy that segregation stiffens the labour market and decreases employee versatility within industries (Keski-Petäjä and Witting 2018). Consequently, this decreases the possibility that the open positions are filled. Stiffness of the market might also add to the workforce challenges from another perspective. As suggested by Baum (2018, p. 875), the employee view of tourism work is dependent on how the employee sees the future career prospects of the industry. Thus, an industry with high horizontal segregation along with low wages and vertical segregation is a potential factor in labour supply not meeting demand.

Managing Tourism Workforce Through Development Projects

Project MatkailuDiili has taken an important step towards tourism workforce issues by acknowledging the importance of the tourism workforce in Finland. According to Baum, policymakers do not actively address tourism workforce issues sufficiently (Baum 2018, p. 875), and the discussion and research concerning workforce has been dominated by the hospitality sector (Baum et al. 2016b). Thus, it is very beneficial that tourism work has been acknowledged in the public management and that there is funding available for finding solutions to the challenges. However, it should be noted that, according to standard industrial classification, hospitality industries are included in tourism.

Having said that, it seems that the approach to tourism workforce in Finland has been mainly based on economic and industrial development viewpoints. As pointed out by tourism researchers Hall (1994), Telfer (2015) and Hakkarainen (2017) developing tourism has long been a synonym for a developing economy locally and regionally. Especially in regional development, the guiding thought has been that by developing tourism, it is possible to increase employment and support the economies of peripheral areas. For instance, new tourism centres and infrastructures have been built based on the so-called "trickle down" model, the idea that economic development would spread well-being to the whole region (Hakkarainen 2017, pp. 24–25; Hall 1994, p. 112; Telfer 2015, p. 167). This has been the common line of thought in Finland as well. Concerning tourism workforce issues, the former Minister of economic affairs and employment articulated the issue by stating: "*Development of the tourism industry is one of the government's strategic spearhead projects and gaining skilled workforce cannot become the bottleneck of the tourism companies' growth. We support the tourism companies and tourism job seekers with a new tourism service guarantee [MatkailuDiili]*" (MatkailuDiili 2018). As part of a government's programme, a separate programme of actions/initiatives MatkailuDiili was implemented. Thus, the profound basis behind tourism workforce development is supporting the economy and putting the emphasis on attracting workers rather than considering the aforementioned challenges of the work itself.

In addition to MatkailuDiili, the workforce challenges have been addressed in the tourism industry in Finland with regional and national development projects that are partly funded by the EU. In the context of Lapland, for instance, the following projects exist:

- Asiakkaan Aika [Time of the customer], (European Social Fund, ESF) tourism companies' productivity is developed by paying attention to employees' and entrepreneurs' well-being at work. (RR tietopalvelu 2019)
- Polut Pohjoisen Kasvuun [Paths to growth in the North] (ESF) has also focused on improving well-being in the industry. The approaches have included both developing skills and knowledge (new customer groups for example) and well-being at work. (see Redu 2019)
- Talent Boost (European Regional Development Fund, ERDF) has had a different approach, focusing on gaining workforce (also other industries besides tourism) to Lapland. (House of Lapland 2019)

- On the national level, project NewWow (ESF) has tackled the challenge of modern and mobile work, which takes place in irregular hours and multiple locations, a situation that puts pressure on gaining new tools for ensuring well-being at work. In Lapland, the focus has been on tourism work. (see NewWow 2018)

The goal in most of these projects has been in gaining skilled workforce to join the industry and to prepare the industry in order to keep the staff. The approaches vary, which is beneficial; the reasons for the workforce challenges are not simple, and therefore there is no single solution for them either. Multiple approaches are needed to address the tourism workforce challenges. MatkailuDiili adds to these projects at the national level. The aim has been to answer the call of industry for support in finding and gaining skilled workforce by polishing the image of the industry with positive marketing, strengthening the cooperation between tourism actors, and providing a service guarantee and regional pilots (see MatkailuDiili 2018).

A large proportion of the project was based on piloting new services for companies and unemployed individuals in 2018 and 2019. The pilots were put out to tender as a so-called "black box model" meaning that the companies could suggest the services without special limitations. The initiatives in 2018 included four agile employment pilots and eight service pilots, for example, an introduction programme for international workforce by Mood of Finland, an online course for employers, and a Sesongista sesonkiin service [From season to season]. Both services have continued with the option for the year 2019. Concerning the agile employment, Etärekry [Distant recruitment] and Huippufirma [Top firm] pilots were continued in 2019. Other pilots were not continued. However, there was a new tendering in 2019 (MatkailuDiili 2019).

The pilot services have been presented in different MatkailuDiili events, and there is a report on the pilots conducted in 2018. There were nine companies nationally involved in the *Top Firm* pilot. No recruitment occurred during the pilot, but according to the report, the results appeared positive. Eight companies participated in the *Distant recruitment pilot*, most of them from Northern Finland. Concerning this service, the companies felt that their voices were heard. In the pilot conducted by Mood of Finland, introductory material in English was produced for companies for leading today's multicultural workforce. There were 17 participants in the pilot kick-off workshop. During the same pilot, an introductory programme was developed for (possible) work-related immigrants to familiarize them with the Finnish work culture before coming to Finland.

There were seven Spanish chefs and three waitresses/waiters involved in a work-related immigration pilot. The *From Season to Season* pilot aim was to even out the seasonality by offering lumbermen tourism work. There were 39 participants in the pilot in 2018, and 13 of them were employed. The three pilots mentioned above were continued in 2019.

According to MatkailuDiili's website, concerning other pilots, there were also marketing campaigns (for instance, a campaign targeted to immigrants including radio and outdoor marketing). There were also some pilots concerning virtual platforms where jobseekers and employees could meet. A platform called *ElämysDuuni* [ExeperienceJob] is no longer in use and cannot be found online. Altogether, 27 pilots were conducted around Finland in 2018 (MatkailuDiili 2019).

The emphasis of MatkailuDiili has been in finding new service models enabling the tourism companies and workforce to meet and connect and that would take into consideration the special features of the regions. The services were organized by private actors chosen with a public tendering process. This kind of procedure is in line with Dredge and Jenkins' (2011, p. 4) description of private sector involvement in policymaking, which is rationalized with the efficiency of the private sector. In Finland, services have been privatized in different sectors for similar reasons. For instance, in the tourism-related industry of transportation, this is done to increase the performance of the sector.

Overall, the pilots of MatkailuDiili may be regarded as having been successful experiments. The concept of lean development or agile experimenting means learning quickly from mistakes and proceeding with fast development cycles (see Ries 2011). Thus, even if a pilot has not shown significant results and has not been successful, it does not mean that the pilot has not served a purpose for further development if the results are put to use. It could be suspected, for instance, that new virtual platforms might not be the best possible solutions since the platform is no longer in use. It might be useful to look into how the industry operates and what are the common existing platforms where information is already currently shared. Getting users on new platforms is difficult and in addition, the maintenance of platforms after the end of external financing is always uncertain.

MatkailuDiili is a much-needed intervention to address the lack of skilled employees in tourism. However, there is a danger that the project paints a picture of tourism work as having low skill requirements that can be performed by anyone. In reality, the situation is quite the contrary; for instance, Koikkalainen et al. (2016, p. 44) describe a safari guide's work as a combination of various aspects that require skill, persona and stamina. According to Baum et al. (2016a), actions implemented in development projects might also unintentionally replicate the image of the work as meaningless positions and irregular work relationships. This is misleading because the prospects in the industry develop and are in line with wider social, economic and technological changes, and the industry provides expert work as well (see Baum et al. 2016a, p. 15). Although much of the desired tourism workforce in Finland is related to frontline service work and guiding, the career prospects and valuation still play an important part in motivating people to join the industry.

However, human resource management has clear shortcomings in the tourism industry, which means that the workforce challenges cannot be overcome solely by "attracting" workforce to join the industry. The situation could be described as a severe need of education and training, and not only educating new workforce to join the industry but also the challenge of improving skills in human resource management as well as developing leadership and management practices.

In our discussion we have again come across the perennial issue of the poor image of tourism work. The work itself should be considered as meaningful, and the workers should feel respected and valued. Promoting the industry to prospective workers with leisure time possibilities, which the workers might not even be able to enjoy, does not serve the purpose of sustainable human resource management. On the contrary, attracting employees with false promises might do the opposite. Thus, support is needed in addition to image marketing for developing working conditions in the tourism industry as a whole. Otherwise, the existing (unsustainable) image of tourism work and workforce is easily replicated, not confronted. This does not apply to all projects and initiatives, and different people interpret the messages through their own lenses.

On the other hand, for instance, the *Paths to growth in the North* project (a regional development project focused on well-being in the industry) organized a seminar "*Alive through the winter season*", which implies that the industry is so hard for the workers and the entrepreneurs that the workers might burn out, replicates the familiar picture of tourism work. The name of the seminar also tells an inside story known by everyone in the industry. The toughness is one of the elements that create a sense of pride and community within the industry. As Baum et al. (2016a, p. 15) also noted, the people who work in the industry do not feel the same way about it as people outside the industry. In this sense, the tourism workers live in a "bubble" that is difficult to describe from outside.

Concluding Thoughts

We have outlined how Finland's expanding tourism industry faces a number of challenges with regard to the recruitment and retention of tourism workers. MatkailuDiili and other development projects have successfully raised the attention of the issue, whereby it was also recognized that some structural challenges such as the lack of decent housing, insufficient hours and low salaries play a role in exacerbating labour shortages. Companies that took part in the MatkailuDiili pilots felt that their troubles were heard, and the projects have brought different actors around the same table. Companies eagerly took part in the project and other workforce-related initiatives and surely gained new tools and viewpoints on recruitment. For instance, it is likely that the lumbermen were not a part of the workforce politics of the tourism companies before they took part in that specific MatkailuDiili pilot. More research should be conducted in terms of the effectiveness of the different public initiatives on workforce management practices in companies.

To the existing and possible employees, the project certainly opened doors to the industry. Through development projects, it is possible to find new opportunities in the tourism industry or gain new prospects and widen the employment horizon through combining short-term employment and seasonal work in different industries. The regional development projects funded by structural EU funds commonly aim to support people in getting employed as well as the well-being of the existing workers and entrepreneurs. Often, the success of these projects is measured by the number of participants or jobs. The projects and pilots are important because they bring many different viewpoints and actors together.

In this chapter, we have described how MatkailuDiili and other development projects reflect tourism work. The overall impression is that they seem to reproduce the picture of tourism work that tourism as an industry is trying to detach itself from. In addition to attracting fresh workforce to the industry, it would be useful to consider how to ensure that the existing workforce would stay. What are the possible practices and benefits that would convince and ensure the workforce to stay from season to season? Is the tourism worker necessarily young, fun and dynamic? What are the elements that potential workers would appreciate? Are expressions such as "Spectacular possibilities in the restaurant industry" or "Experience Job" credible when, at the same time, the newspapers are writing about salaries and working conditions being poor—or when there is a seminar called Hengissä kevääseen [Alive through the winter season]?

Overall, MatkailuDiili has played an important part in responding to the challenge of tourism workforce by promoting tourism as an industry with a lot of employment possibilities. In addition to communication, supporting the employees in building their career paths on their strengths could be one motivational factor in gaining and keeping skilled workforce. Moving forward, when developing tourism and trying to assist companies to find skilled workforce and skilled individuals in finding jobs suitable for their talents, attention should be paid on how the words we use and our practices and conventions create reality and norms. At a minimum, the replicated customs and conventions should be recognized, when discussing tourism work and workforce.

Tourism work should be recognized as an experience that is defined from inside the industry, not from outside, as well as the fact that the people working in the industry frequently see tourism work differently from those on the outside.

References

Baum, Tom. 2018. Sustainable human resource management as a driver in tourism policy and planning: a serious sin of omission? *Journal of Sustainable Tourism* 26(6): 873–889. https://doi.org/10.1080/09669582.2017.1423318.

Baum, Tom, Anna Kralj, Richard N.S. Robinson, and David J. Solnet. 2016a. Tourism workforce research: A review, taxonomy and agenda. *Annals of Tourism Research* 60: 1–22. https://doi.org/10.1016/j.annals.2016.04.003. Accessed on August 11, 2019.

Baum, Tom, Catherine Cheung, Haiyan Kong, Anna Kralj, Shelagh Mooney, Hải Nguyễn Thị Thanh, Sridar Ramachandran, Marinela Dropulić Ružić, and May Ling Sio. 2016b. Sustainability and the tourism and hospitality workforce: A thematic analysis. *Sustainability* 8: 809–809. https://doi.org/10.3390/su8080809; www.mdpi.com/journal/sustainability. Accessed on August 11, 2019.

Business Finland. 2019. Visit Finland. Last modified n.d. https://www.businessfinland.fi/suomalaisille-asiakkaille/palvelut/matkailun-edistaminen/visit-finland/. Accessed on July 16, 2019.

Dredge, Diane, and John Jenkins (eds.). 2011. New spaces in of tourism planning and policy. In *Stories of Practice: Tourism Policy and Planning*, edited by Diane Dredge and John Jenkins. Surrey: Ashgate.

Duncan, Tara, David G. Scott, and Tom Baum. 2013. The mobilities of hospitality work: An exploration of issues and debates. *Annals of Tourism Research* 41: 1–19. https://doi.org/10.1016/j.annals.2012.10.004. Accessed on August 11, 2019.

Edelheim, Johan, and Heli Ilola. 2017. Matkailututkimus, matkailija ja matkailu. In *Matkailututkimuksen avainkäsitteet*, ed. Johan Edelheim and Heli Ilola. Rovaniemi: Lapland University Press.

Finnish Tourism Strategy [Suomen matkailustrategia]. 2010. Suomen matkailustrategia 2020. 4 hyvää syytä edistää matkailutoimialojen kehitystä. http://www.visitfinland.fi/wp-content/uploads/2013/04/Matkailustrategia_020610.pdf. Accessed on August 11, 2019.

Gunn, Claire, and Turgut Var. 2002. *Tourism Planning: Basics, Concepts, Cases*. New York: Routledge.

Hakkarainen, Maria. 2017. Matkailutyön ehdot syrjäisessä kylässä. PhD thesis (Acta Universitatis Lapponiensis 357). Rovaniemi: University of Lapland.

Hall, C. Michael. 1994. *Tourism and Politics: Policy Power and Place*. New York: Wiley.

Hall, C. Michael, Dieter K. Müller, and Jarkko Saarinen. 2009. *Nordic Tourism: Issues and Cases*. Bristol: Channel View Publications.

House of Lapland. 2019. Talent boost -ohjelma. Last modified May 14, 2019. https://www.lapland.fi/fi/house-of-lapland/kumppanit-ja-ohjelmat/elinvoimaohjelma/. Accessed on August 11, 2019.

Jänkälä, Susanna. 2019. Työ- ja elinkeinoministeriön julkaisuja 2019:3 TEM Toimialapalvelu. Kevät 2019 Toimialaraportit Matkailu. http://julkaisut.valtioneuvosto.fi/bitstream/handle/10024/161292/TEM_3_2019_Matkailun_toimialaraportti.pdf?sequence=1&isAllowed=y. Accessed on August 11, 2019.

Keski-Petäjä, Miina, and Mika Witting. 2018. Alle viidennes opiskelijoista opinnoissa joissa tasaisesti naisia ja miehiä – koulutus-alojen eriytyminen jatkuu. Last modified November 29, 2019. http://www.stat.fi/tietotrendit/artikkelit/2018/alle-viidennes-opiskelijoista-opinnoissa-joissa-tasaisesti-naisia-ja-miehia-koulutusalojen-eriytyminen-jatkuu/?fbclid=IwAR3_3FL-Bz-n83grL3M0pGR7N3Af8AMwH1Gv9xxvyzNWjKbooh2RJtROBMFg. Accessed on July 14, 2019.

Koikkalainen, Saara, Jarno Valkonen, and Heikki Huilaja. 2016. Drunken sled dogs: Celebration, alcohol use and teamwork in nature tourism guiding. *Journal of Comparative Research in Anthropology & Sociology* 7(1). https://doaj.org/article/6269ef93c299483bbf5491c2d26642dc.

Kyyrä, Sanna. 2019. Tourism policy and stakeholder groups at different levels. Ministry of Economic Affairs and Employment.

Ladkin, Adele. 2011. Exploring tourism labour. *Annals of Tourism Research* 38: 1135–1155. https://doi.org/10.1016/j.annals.2011.03.010. Accessed on August 11, 2019.

Mara Ry. 2019. Alkoholiveron korotus siivittää alkoholimatkailua Baltiaan. Last modified June 3, 2019. https://www.mara.fi/ajankohtaista/tiedotteet/alkoholiveron-korotus-siivittaa-alkoholimatkailua-baltiaan.html. Accessed on October 13, 2019.

MatkailuDiili. 2018. Matkailun kohtaannon toimenpideohjelma - Matkailudiili. Last modified n.d. https://www.matkailutyo.fi/matkailudiili/. Accessed on August 11, 2019.

MatkailuDiili. 2019. Pilottiyhteenveto vuodesta 2018 ja ajatukset vuoteen 2019. Last modified March 6, 2019. http://www.ely-keskus.fi/documents/10191/30003218/ESITYS_pilottikokeilut_2018_yhteenveto.pdf/e0aebd5f-e94f-4c8f-9754-66ad99afe34b. Accessed on August 11, 2019.

MEAE. 2019a. Finnish tourism in numbers. Ministry of Economic Affairs and Employment. https://tem.fi/en/finnish-tourism-in-numbers. Accessed on July 16, 2019.

MEAE. 2019b. Tourism is growing and undergoing renewal. Ministry of Economic Affairs and Employment. Last modified n.d. https://tem.fi/en/tourism. Accessed on July 16, 2019.

MEAE. 2019c. Tourism actors. Ministry of Economic Affairs and Employment. Last modified n.d. https://tem.fi/en/tourism-actors. Accessed on July 16, 2019.

MEAE. 2019d. Public subsidies for developing tourism. Ministry of Economic Affairs and Employment. Last modified n.d. https://tem.fi/en/public-subsidies-for-developing-tourism. Accessed on August 11, 2019.

NewWow. 2018. NewWoW Crafting - mobiilin ja monipaikkaisen työn muotoilusta tuottavuutta ja hyvinvointia. Last modified August 8, 2018. https://www.turkuamk.fi/fi/tutkimus-kehitys-ja-innovaatiot/hae-projekteja/newwow-crafting-mobiilin-ja-monipaikkaisen-tyon-mu/. Accessed on August 11, 2019.

PAM. 2019a. Collective agreements. Last modified n.d. https://www.pam.fi/en/work/collective-agreement.html. Accessed on August 12, 2019.

PAM. 2019b. Matkailualan houkuttelevuus työllistäjänä riippuu alan arvostuksesta, palkkauksesta ja työhyvinvoinnista. https://www.pam.fi/uutiset/matkailualan-houkuttelevuus-tyollistajana-riippuualan-arvostuksesta-palkkauksesta-ja-tyohyvinvoinnista.html. Accessed on June 2, 2020.

Rantala, Outi, Suzanne de la Barre, Brynhild Granås, Gunnar Thór Jóhannesson, Dieter K. Müller, Jarkko Saarinen, Kaarina Tervo-Kankare, Patrick T. Maher, and Maaria Niskala. 2019. Arctic tourism in times of change: Seasonality. In *TemaNord* 528. Copenhagen: Nordic Council of Ministers.

Redu. 2019. Polut pohjoisen kasvuun, Lapin yrityksiä ja työhyvinvointia kehittämässä -hanke. n.d. https://www.redu.fi/hankesivut/Polut-pohjoisen-kasvuun-/Etusivu. Accessed on August 11, 2019.

Ries, Eric 2011. The lean startup: How constant innovation creates radically successful businesses. http://free.epubebooks.net/ebooks/books/the-lean-startup.pdf. Accessed on August 13, 2019.

RR tietopalvelu. 2019. Asiakkaan aika. Last modified n.d. https://www.eura2014.fi/rrtiepa/projekti.php?projektikoodi=S20995. Accessed on August 11, 2019.

Satokangas, Pasi. 2013. *Matkailulla maakunta menestyy – Matkailun tulo- ja työllisyysvaikutukset 12 lappilaisessa kunnassa vuonna 2011*. Rovaniemi: Lapin korkeakoulukonserni.

Sharpley, Richard. 2015. Tourism: Vehicle for development. In *Tourism and Development: Concepts and Issues*, 2nd ed., edited by Richard Sharpley and David J. Telfer. Bristol: Channel View Publications.

Statistics Finland. 2019a. Statistics Finland's PxWeb databases. https://pxnet2.stat.fi/PXWeb/pxweb/fi/StatFin/StatFin__lii__matk/statfin_matk_pxt_11j1.px/.

Statistics Finland. 2019b. Suomen virallinen tilasto (SVT): Työvoimatutkimus. Palkansaajien Nollatuntisopimukset 2018. Helsinki: Tilastokeskus. http://www.stat.fi/til/tyti/2018/15/tyti_2018_15_2019-06-03_tie_001_fi.html. Accessed June 2, 2020.

Telfer, David J. 2015. Tourism and regional development issues. In *Tourism development: Concepts and Issues*, 2nd ed., edited by Richard Sharpley and David J. Telfer. Bristol: Channel View Publications.

Tuulentie, Seija, and Maria Hakkarainen. 2014. Matkailu pohjoisen maaseudun työnä. In *Hyvinvointia metsästä*, ed. Liisa Tyrväinen, Mikko Kurttila, Tuija Sievänen, and Seija Tuulentie, 141–152. Helsinki: Suomalaisen Kirjallisuuden Seura.

UNWTO. 2018. Why tourism? Last modified n.d. https://www2.unwto.org/content/why-tourism. Accessed on August 11, 2019.

Vaugeois, Nicole, and Rick Rollins. 2007. Mobility into tourism refuge employer? *Annals of Tourism Research* 34: 630–648. https://doi.org/10.1016/j.annals.2007.02.001. Accessed on September 11, 2019.

Veijola, Soile, Maria Hakkarainen, and Jatta Nousiainen. 2013. Matkailu työnä. In *Matkailututkimuksen lukukirja*, ed. Soile Veijola, 173–185. Rovaniemi: Lapin yliopistokustannus.

Veijola, Soile, Anu Valtonen, Jarno Valkonen, Seija Tuulentie, Outi Rantala, Maria Hakkarainen, and Minni Haanpää. 2008. Töissä tunturissa: Tutkijatulkintoja elämäkerta-aineistosta. In *Töissä tunturissa: Ajatuksia ja kirjoituksia matkailutyöstä*, ed. Jarno Valkonen and Soile Veijola, 20–69. Rovaniemi: Lapland University Press.

Vesterinen, Nina and Hakkarainen, Maria. 2013. Matkailualan koulutuksen ja tutkimuksen ohjaus Suomessa. In *Oppiva matkailu. Puheenvuoroja matkailualan koulutuksen kehittämisestä*, edited by Minni Haanpää, Maria Hakkarainen, and Heidi Kaihua, 19–31. Rovaniemi: Lapland University Press.

Visit Finland. 2019. Matkailutilinpito. Matkailun talous- ja työllisyysvaikutukset. https://www.businessfinland.fi/contentassets/4b07e15186484a69b62e991ed85a6c45/matkailutilinpito_2016-2017.pdf. Accessed on July 15, 2019.

5

A Potential Treasure for Tourism: Crafts as Employment and a Cultural Experience Service in the Nordic North

Outti Kugapi, Maria Huhmarniemi and Laura Laivamaa

Introduction

Imagine yourself travelling abroad and listening to interesting tales of talented local crafters while visiting their homes. Imagine a situation where you could observe the local cultural heritage shown through craft-based tourism activities. Imagine that you could make your own souvenir with a local crafter and bring it home with the knowledge of the local culture gained from the workshop. This can all happen in the near future if the expectations of business development and the

O. Kugapi (✉)
Multidimensional Tourism Institute, University of Lapland, Rovaniemi, Finland
e-mail: outi.kugapi@ulapland.fi

M. Huhmarniemi · L. Laivamaa
Faculty of Art and Design, University of Lapland, Rovaniemi, Finland
e-mail: maria.huhmarniemi@ulapland.fi

L. Laivamaa
e-mail: laura.laivamaa@ulapland.fi

A. Walmsley et al. (eds.), *Tourism Employment in Nordic Countries*,
https://doi.org/10.1007/978-3-030-47813-1_5

needs of tourists come together, as discussed by the Finnish Ministry of Education and Culture (2018) and Sandell and Skarveli (2016). The growing number of tourists arriving in Nordic countries (Regional Council of Lapland 2018) has affected the figures of new enterprises: according to the Finnish Trade Register, during 2018, more than 1000 new enterprises were established, and most of them were based on tourism (Lapin Kansa 2019).

Tourism in Lapland is often criticised of being too seasonal since the winter months represent high season in the area (House of Lapland 2018; Rantala et al. 2011). Seasonality can cause issues for both employers and employees as it can lead to an unevenly divided workforce and a lack of skilled employees (e.g., Chen and Wang 2015; Rantala et al. 2011, 2019; Tuulentie and Heimtun 2014). There is also an ongoing political and local discussion about whether employment in Northern Finland and Sweden should be based on industrial use of natural resources or the tourism industry (e.g., Similä and Jokinen 2018) and also on the negative impact that outdoor tourism activities have on other livelihoods, such as reindeer herding (e.g., Mällinen and Simo 2014). Although seasonality has been researched from many perspectives during recent years, there is still a lack of concrete actions to face the issues that seasonality causes (Rantala et al. 2019). Nevertheless, seasonality can also be seen as a positive thing as many tourism entrepreneurs argue that they need the low season for themselves, to relax after the high season and to repair their equipment. This means that there is a chance for other companies to create businesses—the same companies do not need to operate their businesses all year round if they do not want to (see also Baum and Hagen 1999).

In addition, local authorities have suggested that tourism development in the near future should be focused on creating new products and services for the summer months when hotels, tourism companies and restaurants lack customers. This would make tourism more sustainable, also economically, and create year-round employment. Another way of doing things in a more sustainable way is by adding cultural, creative and craft-based services to Nordic tourism (see OECD 2014) as most of the tourism services in Finnish Lapland are combined with nature and motor-based activities, which are highly dependent on the weather conditions (Rantala et al. 2011). This would not only improve the employment rate for artists but also the quality of tourism services. For example, crafted souvenirs and craft-based services can present and maintain cultural heritage and increase the cultural sustainability and authenticity of Nordic tourism as also discussed by Banks (2010).

However, it needs to be acknowledged here that the shift from being a crafter to a service provider is not easy. Crafters might need new skills to be able to design and facilitate craft-based services for tourism industry (see Huhmarniemi and Jokela 2019), get used to work with people and get education about legislation of tourism industry for example. In an ideal situation, this would lead to year-round jobs and more sustainable ways of living in, staying in and visiting Nordic countries. This is in line with many visions, for example the vision of Lapland's tourism development (Regional Council of Lapland 2015) and can be accomplished through business-oriented training, addressed, for example, by the Handmade in Lapland (2018) project funded by European Social Fund, on which the authors of this chapter are working.

While tourism, as one of the biggest sectors providing livelihoods in Lapland, has been researched from many points of view, the combination of tourism and art is still mainly unresearched. Some previous research has looked at winter art and environmental art for tourism sites (see e.g., Hiltunen 2009; Huhmarniemi and Jokela 2019; Jokela et al. 2013, 2014), art-based practices applied to the tourism context (e.g., Erkkilä-Hill 2017) and knitted mittens as travel souvenirs (Kugapi and Höckert 2018). Also, for a wider perspective, Duxbury and Richards (2019a, p. 5) suggest more research should be carried out into the companies who offer creative tourism experiences. The *Handmade in Lapland* project and this chapter are creating an important opportunity to fill the research gap in this field of tourism studies, which is also addressed in Duxbury and Richards (2019b, pp. 184–185).

The data used in this chapter were collected in two ways. First, the authors interviewed fifteen handcrafters and tourism Destination Management Organisations (DMO) in Lapland in late 2018 and early 2019 to get an overall picture of the craft industry, companies' development needs, expectations from the tourism industry and companies' knowledge about tourism. After the interviews, the authors organised four service design workshops during spring 2019, where eighteen participants from the fields of craft and tourism participated. The main aim of the service design workshops was to gather participants' needs and expectations related to tourism issues and the prospective content of the continuing education programme.

In next section, we introduce craft labour and discuss how it is combined with cultural sustainability, inspecting these through the lenses of creativity and employment. Then, in section "Scenarios as a Tool for Developing Creative Tourism in Lapland", we discuss service design as a methodological tool for creating scenarios for craft-based tourism. In section "Developing Creative Tourism with Crafters in Finnish Lapland", we discuss the findings from the workshops and then conclude with some remarks in section "Well-Maintained Networks for Crafters".

Culturally Sustainable Craft Labour in Tourism

> Craft *noun:* skill and experience, especially in relation to making objects; a job or activity that needs skill and experience, or something produced using skill and experience. (Cambridge Dictionary, n.d.)

In addition to this definition, Adamson (2007) defines craft labour as also including internal and external motivational factors, whereas Banks (2010, p. 309) adds the need for keeping 'creative freedom and meaning in work' in workshop models. According to Banks (2010), craft labour is often overlooked and not carefully researched in academic literature: the focus has been more on artists, not craft workers. Craft is often associated with creativity, which is a broader term, including the film industry and theatre among others (Banks 2010; OECD 2014; Richards 2011). In this chapter, we especially focus on craft production and understand crafters to be artists, designers and handcrafters, to adjust the definition to the situation in Finnish Lapland, where practices in art, craft and design have traditionally been understood as the same (Härkönen et al. 2018). This concept emphasises the creative skills of the crafters. The diversity of livelihoods can include commission work, income from sales and unemployed and part-time employees in a variety of jobs while producing art, crafts or designs (Huhmarniemi and Jokela 2019; see also Banks 2010). Crafters can also be defined as lifestyle entrepreneurs who are not seeking wealth and financial independence, but instead are focused on improving their quality of life (Marcketti et al. 2006).

When discussing crafts, craft labour and employment, it is a good idea to look at the definitions through the lens of sustainable development, which became a recognised concept in 1987 when the World Commission on Environment and Development used it in their report, *Our Common Future* (WCED 1987). Since then, the concept has evolved and has been studied in many fields, including tourism, where the aim was to involve local communities in tourism development (Swarbrooke 2002). Sustainable tourism is defined by the World Tourism Organisation (WTO) in the following way: 'Tourism that takes full account of its current and future economic, social and environmental impacts, addressing the needs of visitors, the industry, the environment and host communities' (UNEP and WTO 2005, p. 12). Sustainability is divided into four dimensions: economic, ecological, social and cultural. Occasionally, the latter ones are discussed together. Sustainable tourism 'conserves ... built and living cultural heritage and traditional values, and contributes to inter-cultural understanding and tolerance' (UNEP and WTO 2005, p. 11), which is the idea at the core of most crafts.

In Soini and Birkeland's (2014) thorough research on cultural sustainability, they argue that the concept can be examined through heritage, vitality, economic viability, diversity, locality, ecocultural resilience and ecocultural civilisation (see also Siivonen 2003). Often these aspects are also linked to strengthening cultural identity (Friedman 1994), cultural revitalisation (Matahaere-Atariki 2017) and knowledge creation in the changing world (e.g., Siivonen 2003). Local community members should approve the changes in the culture, and they should be included in planning tourism services and sites when these interfere with their culture (e.g., de Bernardi et al. 2018; Lee and Jan 2019), although Simpson (2008) points out that inclusion does not assure economic benefits for local communities (see also Müller and Viken 2017; Swarbrooke 2002). Unfortunately, cultural sustainability in tourism lacks attention even though tourists and their interests seem to be changing. It is visible that tourists wish to acquaint themselves with the local culture, support sustainability and experience new things, such as local handicrafts and design products, more than before (see e.g., Cornelisse 2018; Duxbury and Richards 2019a; Remoaldo et al. 2019; Sthapit and Björk 2017; Veijola and Strauss-Mazzullo 2019). Anyhow, most tourists who currently come to Lapland for a holiday, buy ready-made touristic packages of outdoor activities, and, therefore, potential job opportunities for the creative sector have not yet been created (Huhmarniemi and Jokela 2019).

In this chapter, we consider locally created craft-based tourism as part of cultural tourism and culturally sustainable development in Lapland. Moreover, we seek guidance from creative tourism, which has been said to be developed from cultural tourism and as a reaction to the development of mass tourism (Duxbury and Richards 2019a, p. 1). Creativity can reproduce cultural tourism by adding creativity and tourist involvement to it (Miettinen 2007; Richards and Raymond 2000; Richards and Wilson 2006). In other words, tourists actively participate in events, local activities and discussions, and the cultural elements are not interpreted by some other person (i.e. a tourist guide). Direct involvement in creative tourism decreases the consumption of *tourist art*, which refers to souvenirs that have primarily been made to satisfy the customer and whose relation to the original culture has thinned (Graburn 1976). Behind the concept, there are debates about the artistic styles of souvenirs and whether tourism destroys the original culture and environment of a travel destination (Hume 2013). Since 2000, the discussion around creative tourism has expanded and moved from production-focused experiences to more relational, co-created ones (Duxbury and Richards 2019a; Miettinen et al. 2019). Thus, creative tourism offers something for all the senses: something to see, touch, visit and do in creative settings (Richards 2011). Nevertheless, it must be acknowledged here that craft as a method does not necessarily guarantee cultural sustainability. While many elements of Indigenous cultures, such as techniques and symbols, have been exploited for souvenirs (see Schilar and Keskitalo 2018), the same could be done in workshops too. Rather than repeating visual language or methods of traditional (Indigenous or nonindigenous) crafts, the craft-based workshops could focus on connecting and mediating.

Scenarios as a Tool for Developing Creative Tourism in Lapland

The *Handmade in Lapland* project conducted fifteen interviews with handicraft and design entrepreneurs, tourism DMOs and design retailers in late 2018 and early 2019 in Finnish Lapland to research the state of tourism in the creative industry and the potential of adding craft-based services to the market. During spring 2019, the project organised three service design workshops in different areas of Finnish Lapland where this topic was discussed with a variety of actors, both from tourism and creative areas. Altogether, eighteen crafters and entrepreneurs took part in the workshops. Moreover, nine crafters were interviewed about their needs and expectations related to the continuing education programme. All participants in the workshops were living and working in Lapland, and thus were locals.

Service design, as an approach, was a very natural choice when building the planning phase of the continuing education programme. Service design is a research field where cultural, social and human interaction are connected (Miettinen and Valtonen 2012). It provides methods and tools that help to concretise and understand the wants and needs of a customer, in this case, those of potential participants of the continuing education programme (Miettinen et al. 2014). The use of co-creation methods in early phases of the project ensured the commitment of the participants but also helped to promote the upcoming continuing education programme. Srivastava and Verma (2012) have described the concept of co-creating value as a systematic and structured process where the collaboration of all stakeholders related to the development process establishes a value for both the company and customers and, in this case, both project employees and potential participants of the continuing education programme.

The aim of the service design workshops in the *Handmade in Lapland* project was to form the basis of the continuing education programme and discover the needs and expectations related to the tourism issues of handicrafts and design entrepreneurs. The structure of all the workshops was similar: Every event started with a warm-up, then participants discussed the present situation of craft labour and craft-based tourism with a SWOT analysis. The events continued with the generation of ideas for the future of craft-based tourism in Lapland, which offered a basis for creating the path of the continuing education programme. The path followed the customer journey model (Mager 2009; Miettinen and Koivisto 2009) and included the expectations of participants related to, for example, potential specialists, places and themes.

Scenarios are useful tools for visualising the early phases of new concepts, and they complemented other methods used during the *Handmade in Lapland* workshops. Scenarios are essentially stories about people and how they operate in different situations and contexts (Heinilä et al. 2005). Scenarios have been seen as storytelling related to the dimensions and interactivities of products under development in industrial design (Keinonen et al. 2004) and in service design, context scenarios are usually used to help the customer visualise the service concept (Miettinen 2011). Moreover, scenarios help to open up the demands of all possible stakeholders and are very useful when planning new functions: the iterative process needs tools that assist the evaluation and development work (Heinilä et al. 2005). In the workshops, they enabled interesting conversations as participants co-created scenarios about the situation of craft-based tourism in Lapland after five years, first in small groups, and then by discussing together with all the workshop participants. Even though the main aim of the workshops was to map out the near future of craft-based tourism via innovative ideation methods, such as scenarios and tailored customer journeys (Mager 2009), the outcomes of the tasks and conversation highlighted themes, needs and expectations that are already solvable through the continuing education programme.

The outcomes of the scenarios and interviews follow the framework for creative tourism development discussed in Duxbury and Richards (2019a). The framework opens up which elements should be taken into consideration when designing creative tourism (Duxbury and Richards 2019a, p. 5). There are three elements represented as circles emanating from the creative core (learning). These elements are: co-creation, consuming (creative spectacle) and the enabling environment (to include items such as policies, platforms, networks and infrastructure). Moreover, networks, communities, crafters and infrastructure also affect creative tourism development, and these aspects should be taken into consideration when adding craft labour to tourism (Duxbury and Richards 2019a, p. 5). This framework has been the basis for our analysis in section "Developing Creative Tourism with Crafters in Finnish Lapland".

Developing Creative Tourism with Crafters in Finnish Lapland

The framework by Duxbury and Richards (2019a, p. 6) inspired us to analyse craft labour in tourism in Finnish Lapland from certain perspectives as shown in Fig. 5.1. First, we concentrate on the issues that are important for the crafters in finding employment in the field of tourism in Finnish Lapland (the outer dimension). From there, we move towards the core and discuss the possible options for increasing the use of handicrafts and craft-based services in Lapland's tourism.

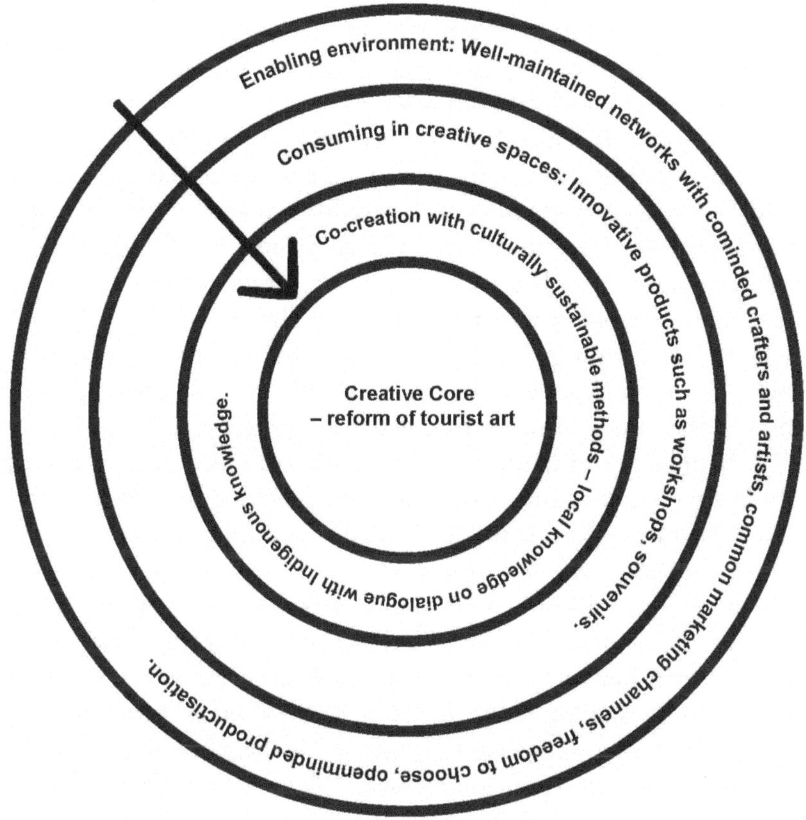

Fig. 5.1 Aspects of craft-based tourism development in Finnish Lapland

Well-Maintained Networks for Crafters

> In Rovaniemi, there is no point in doing crafts in the case that you are not willing to benefit from tourism.

By reading that quote from one interview, it becomes understandable that tourism is already impacting the craft industry in Lapland—some crafters even point out that their businesses would not be successful without tourism. Nevertheless, the data show that the crafters do not necessarily have the capability, know-how and networks to get the best out of the potential cooperation and business-to-business strategies, which is also emphasised in Banks (2010) and Remoaldo et al. (2019). In Finnish Lapland, currently, there are no existing networks or associations bringing together designers and crafters or to promote work opportunities for them. Furthermore, national unions have not yet expanded their activities into the North.

It is necessary to remember that networks are not so easy to manage and maintain. In one workshop, it became clear that the network should be small-scale in order to be well-maintained as a too large a network becomes hard to handle, and companies might not find the network relevant enough to their needs. One interviewee also brought up the issue of economic responsibilities within the network; the rules should be clear enough for everyone, as also discussed in Simpson (2008). However, many crafters emphasise the fact that creating something together and belonging to a network brings good things with it, such as happiness and creativity (see also Richards 2011). Even competition is seen more as an opportunity rather than as a challenge—it is seen as a chance to upscale the services and learn from others: '*We create experiences together with these people.*'

The network could offer a common marketing channel for crafters. Workshop participants visualised a co-created workbook that would offer an insight into the craft-based activities available in the area. This would make it easier for guests to choose the best service for their needs. Also, by appointing one person to manage the network and be in contact with tourism entrepreneurs, it would help the crafters to find the correct markets for their products. In fact, finding the correct person from the tourism sector is seen as a challenge by many crafters. Crafters point out that, sometimes, they might have the perfect tourism service ideas, but they cannot find the correct person to contact, or they are too scared to move forward without network support. Therefore, in summary, a well-maintained creative tourism network is needed in Lapland in order to support the crafters in becoming acquainted with the tourism industry.

Artists and Makers—Challenges and Opportunities

Although crafters seem to be aware of the potential to increase their earnings in the tourism sector, there are various challenges to overcome to make it happen. First, the shift from being a crafter to a service provider is not easy. Crafters might need to change their way of working and get used to working with people as many crafters are used to working alone or with teams of experts, not with clients such as travellers and tourist groups. Crafters need new skills to be able to design and facilitate craft-based services and extend their expertise to the design of holistic service concepts, such as craft-based interior design for hotels or other tourism environments (Huhmarniemi and Jokela 2019). According to Rantala et al. (2019, p. 36), involvement in tourism needs time and experience, and sometimes there is even a need to work full-time to get the necessary capability. This brings challenges for crafters as some are afraid that if they get involved in tourism, they will lose valuable time for their productive and creative work (see also Banks 2010). The data show that one solution to this could be the network we discussed earlier. The network could offer much-needed craft-based services for tourists every day, but the services would be offered based on the timetables of the crafters—not those of the tourists. This way, the individual craft companies would not have to be available for tourists every day, but instead, for example, only once a week, and the rest of the week would be reserved for their creative work.

Second, there is the challenge of growing the business. According to the data, crafters are usually microentrepreneurs who do not want to grow and create employment for many people. Instead, the crafters would like to create employment just for themselves: they want to earn extra income with tourism but keep the creative work as their main activity. This is seen as a challenge from a retailer's point of view. According to one retailer:

> Many crafters just do a few crafts for their own pleasure and do not even understand how valuable the products are that they have in their hands … We cannot force these microentrepreneurs and say to them: "do more, sell us more, expand. If it sells well, bring us more immediately", because they all have their backgrounds behind them. Sometimes, we need to wait a long time before we get the products – we kind of live in the moment.

However, among crafters, microentrepreneurship is seen as a solution for improving quality of life for themselves and their communities (see Marcketti et al. 2006). Crafters do not even necessarily want their products and services to be successful for the masses; instead, they prefer to sell only to selected customers (see Jóhannesson and Lund 2018).

Third, crafters hope that the employment will be all year round and that they will have a steady income, to be economically sustainable, but, unfortunately, the work is usually so-called project work without any hope of continuity and steadiness (see Banks 2010). In tourism, this means seasonal changes in work: during the high and low seasons, there is a big difference in the demand for tourism services. Nevertheless, if the cultural sector gains more importance in summertime, which is considered low season in Finnish Lapland, perhaps the seasonal changes would not be so challenging for crafters.

Fourth, productisation is seen as maybe the biggest challenge as the crafters do not really have the knowledge of what can be sold to tourists: '*It is important to think about the productisation, it is a keyword. What can you productise?*' Moreover, crafters are unsure of how to market their craft-based activities to tourists: '*How can we promote our things here as being much more valuable and sophisticated than elsewhere?*' It is important to acknowledge here that tourism services that are based on crafts and culture need to be adjusted based on tourist groups and their interests. In one workshop, crafters visualised how services could be promoted to family tourists, for example, during Christmas, which is a busy tourist season for families. This originates from the assumptions that children are interested in crafts and that other family members are happy if their children are satisfied. Also, one interviewed DMO pointed out that, for example, Chinese guests are not satisfied by any knitted wool socks, and that instead, they want the knitted material to have ornaments, Fair Isle patterns or to be small in scale. This is a perfect way to promote the cultural heritage of the area, to include specific information about the history through crafts and make tourism more culturally sustainable. The potential earning methods in the tourism field are thus diverse.

Consuming in Creative Spaces

In the future, crafters see the interaction with tourism companies as being more intensive: the tour operators and safari companies could also bring customers to crafters. This develops a need for creative spaces at tourist destinations, as noted in the data. In Lapland, crafters' studios are often small and located outside of city centres and other tourism destinations, and crafters ponder whether tourism entrepreneurs could offer the space for crafters to run their craft-based workshops, for example, during summer when the safari companies are less occupied. A common creative space located where tourists wander would make it easier for both tourists and DMOs to find the creative tourist services. In fact, one interviewed DMO pointed out the problem of finding the crafters now that tourists demand new services based on culture (see also Duxbury and Richards 2019a, p. 1; Veijola and Strauss-Mazzullo 2019). In one workshop, the participants even started visualising a craft village in the Arctic Circle in Rovaniemi, where artists, crafters, tourists and locals could meet, create and do things together. In an ideal situation, this collaboration could lead to sustainable development, products and services based on local culture and wise usage of spaces and locations. Moreover, if the craft-based services were offered in some other place rather than in crafters' own studios, it would be easier for crafters to separate the time for their creative work and tourism.

Co-creation with Culturally Sustainable Methods

Among crafters, there is also a call for a common label for crafts and craft-based services produced in the area. This can be seen as one kind of infrastructure that could help the local industry. According to the data, it is controversial that many souvenir shops in Lapland are selling imported souvenirs while tourists are looking for more locally made products. There are many tourists who ask what products are originally made in Lapland and might not know what to look for. Also, one retailer stated that the most important thing for them is to sell locally produced and not imported products. Luckily, there is one certificate, the Sámi Duodji, which guarantees the origin of the product (see also de Bernardi et al. 2018), but a common label could also be developed for other locally produced crafts. This, together with the aforementioned workbook, that would offer an insight into the craft-based activities available in the area, could increase sales of local goods and decrease those of imported souvenirs.

In fact, souvenirs cannot be disregarded as they are the most common craft products in Finnish Lapland (see e.g., Sthapit and Björk 2017). For souvenirs, the production and marketing differ from other designs for a variety of reasons: the costs should be low, and the production volume must be great. Thus, the expectation of growth is laid on services rather than on souvenirs: artisans would, for example, sell their own designs, organise workshops or create combinations of these. Nevertheless, craft as a method does not necessarily guarantee cultural sustainability as the exploitation of (Indigenous) cultures can happen through workshops too. For example, it can be questioned whether Santa-themed craft workshops are promoting the local culture or not. Ethnicity is often expressed through handicrafts, which brings the discussion of cultural ownership of techniques and symbols to the table (Schilar and Keskitalo 2018). For example, the production of *Duodji*—the Sámi handicraft—is widely discussed from this perspective. In addition, the topics of how much tradition can be changed and what the core of the culture is discussed (see e.g., Guttorm 2015; Kramvig and Flemmen 2019). Minnakhmetova et al. (2019) argue that transforming traditions calls for sensitivity. Therefore, it can be said that Indigenous knowledge should be studied and applied when developing the concept of craft-based services in Lapland.

Creative Core—Impact on Tourist Art and Culturally Sustainable Development

Among crafters, a sceptical attitude towards souvenir production and other collaboration with tourism can be noted. The aforementioned concept of *tourist art* by Nelson Graburn (1976) refers to souvenirs which have primarily been made to satisfy the customer and whose relation to the original culture has thinned. In Lapland, this can mean, for example, that souvenirs only make a weak reference to authentic crafts and cultural heritage, and crafters are worried about this. However, the change of handicrafts and art to souvenir art can be seen as a process where artefacts begin to serve new purposes and, at the same time, reform the traditional culture (Hume 2013).

An essential question is whether souvenir and tourism service production can be sustainable and revitalise and reconstruct local cultures and traditions. Luckily, the crafters in Lapland are aware of this discussion and willing to create something more sustainable for the souvenir market or, in fact, create more experiences than products. As Heldt Cassel and Maureira (2017) conclude, authentic cultural practices are constantly negotiated and traditions are reinvented in relation to the expectations of guests as well as for creating economic benefits for the hosts. This is seen in the vision for creative tourism development in Finnish Lapland:

> In the future, Lapland will be a well-known destination for craft-based tourism. Travellers will come here to buy craft kits and materials as well as to participate in workshops and other services.

The data show that creative tourism as well as handmade souvenirs in Lapland is currently negotiated by many stakeholders. The *Handmade in Lapland* project is designed to create a space for common discussion among crafters, and, therefore, it can be said that the future looks positive.

Towards a Craftier Future

Data from the service design workshops and interviews show that crafters experience tourism as being necessary to their livelihoods, but the current benefit from tourism has become more about product sales—the souvenirs. This brings challenges as souvenirs need to be low priced and mass production lacks cultural sustainability: these destroy the aura of tourist art through cheapness and lack of quality and cultural sensitivity. In addition, crafters who are lifestyle entrepreneurs and microentrepreneurs and give priority to creative work are often unable, or unwilling, to produce large numbers of their products and hesitate to commit to tourism collaboration. Thus, sometimes unreasonable expectations are placed on craft-based services and creative tourism.

The data show that networking, together with common marketing material, joint spaces for producing creative tourism services and a common label would support crafters in their start-up businesses or in finding employment. It can be said that networks are needed among the crafters as well as between tourism stakeholders and crafters. In addition, continuing education is needed for improving skills and capacities for creating new services and stepping into the tourism sector. Nevertheless, the overall expectation for the potential of crafters and tourism collaboration is positive—means of crafters intervening in tourism are diverse.

Craft-based tourism is seen as one solution towards sustainable development in the tourism field in Lapland, due to the potential of year-round services. Existing infrastructure for winter tourism could also be used in the summer, by the same or partner entrepreneurs. When craft-based products and services are authentic, rooted in local culture and designed in interaction with local communities, the services can be seen as culturally sustainable. A culturally oriented tourism industry that includes craft-based services has the potential to increase appreciation of crafting traditions and crafted goods and textiles as well as knowledge of the cultural heritage of Lapland. The aura of tourism art could be shifted from exploitation of the visual language of local (indigenous and nonindigenous) crafts to souvenirs and craft-based services as mediators of traditions, cultural values and discussions. The potential of Indigenous knowledge in relation to the significance of crafts could be researched more in the future, along with an evaluation of already existing craft-based services and the impact of the continuing education that will be carried out through the *Handmade in Lapland* project.

References

Adamson, Glenn. 2007. *Thinking Through Craft*. Oxford: Berg.

Banks, Mark. 2010. Craft labour and creative industries. *International Journal of Cultural Policy* 16 (3): 305–321. https://doi.org/10.1080/10286630903055885.

Baum, Tom, and Laura Hagen. 1999. Responses to seasonality: The experiences of peripheral destinations. *International Journal of Tourism Research* 1 (5): 299–312.

Cambridge Dictionary. n.d. "Craft" meaning. https://dictionary.cambridge.org/dictionary/english/craft. Accessed on May 2, 2019.

Chen, Joseph S., and Wei Wang. 2015. Foreign labours in arctic destinations: Seasonal workers' motivations and job skills. *Current Issues in Tourism* 18 (4): 350–360. https://doi.org/10.1080/13683500.2014.894499.

Cornelisse, Miranda. 2018. Understanding memorable tourist experiences: A case study. *Research in Hospitality Management* 8 (2): 93–99.

de Bernardi, Cecilia, Outi Kugapi, and Monika Lüthje. 2018. Sámi Indigenous tourism empowerment in the Nordic countries through labelling systems: Strengthening ethnic enterprises and activities. In *Tourism and Ethnodevelopment: Inclusion, Empowerment and Self-Determination*, edited by Ismar Borges de Lima and Victor T. King, 200–212. Oxon: Routledge.

Duxbury, Nancy, and Greg Richards. 2019a. Towards a research agenda for creative tourism: A synthesis of suggested future research trajectories. In *A*

Research Agenda for Creative Tourism, edited by Nancy Duxbury and Greg Richards, 1–14. Northampton, MA: Edward Elgar Publishing.
Duxbury, Nancy, and Greg Richards. 2019b. Towards a research agenda for creative tourism: Developments, diversity and dynamics. In *A Research Agenda for Creative Tourism*, edited by Nancy Duxbury and Greg Richards, 182–192. Northampton, MA: Edward Elgar Publishing.
Erkkilä-Hill, Jaana. 2017. Walking on water, living adventurously: Travelling laboratories for artistic thinking. In *Relate North: Practising Place, Heritage, Art & Design for Creative Communities*, ed. Timo Jokela and Glen Coutts, 132–148. Rovaniemi: Lapland University Press.
Friedman, Jonathan. 1994. *Cultural Identity & Global Process*. London: SAGE.
Graburn, Nelson H.H. 1976. Introduction: Arts of the fourth world. In *Ethnic and Tourist Art*, edited by Nelson H.H. Graburn, 1–32. Berkeley: University of California Press.
Guttorm, Gunvor. 2015. Contemporary Duodji—A personal experience in understanding traditions. In *Relate North: Art, Heritage and Identity*, ed. Timo Jokela and Glen Coutts, 60–76. Rovaniemi: Lapland University Press.
Handmade in Lapland. (2018). Project application. https://www.eura2014.fi/rrtiepa/projekti.php?projektikoodi=S21344. Accessed on May 2, 2019.
Härkönen, Elina, Maria Huhmarniemi, and Timo Jokela. 2018. Crafting sustainability: Handcraft in contemporary art and cultural sustainability in Lapland. *Sustainability* 10 (6): 1907. https://doi.org/10.3390/su10061907.
Heinilä, Juhani, Hanna Strömberg, Jaana Leikas, Veikko Ikonen, Netta Iivari, Timo Jokela, Kari-Pekka Aikio, Ilari Jounila, Jettie Hoonhout, and Natasja Leurs (eds.). 2005. *User Centred Design Guidelines for Methods and Tools*. Tampere: VTT Information Technology; University of Oulu, Department of Information Processing Science; Philips Research, Philips Applied Technologies. The Nomadic Media Consortium. https://www.vtt.fi/inf/julkaisut/muut/2005/UCD_Guidelines.pdf.
Heldt Cassel, Susanna, and Teresa Miranda Maureira. 2017. Performing identity and culture in Indigenous tourism—A study of Indigenous communities in Québec, Canada. *Journal of Tourism and Cultural Change* 15 (1): 1–14. https://doi.org/10.1080/14766825.2015.1125910.
Hiltunen, Mirja. 2009. *Yhteisöllinen taidekasvatus. Performatiivisesti pohjoisen sosiokulttuurisissa ympäristöissä* [Community-Based Art Education: Performative Activities in Northern Environment and Society]. Acta Universitatis Lapponiensis 160. PhD dissertation, University of Lapland. Rovaniemi: Lapland University Press.

House of Lapland. 2018. 10 facts about tourism in Lapland. https://www.lapland.fi/uploads/2018/04/428a8f68-pdf_10-facts-about-tourism-in-lapland-2018.pdf. Accessed on June 26, 2019.

Huhmarniemi, Maria, and Timo Jokela. 2019. Environmental art for tourism in the arctic: From handicraft to integrated art and reform on artists' skills. *Synnyt/origins, Special issue on Arctic Arts Summit* 1: 63–80. https://wiki.aalto.fi/pages/viewpage.action?pageId=151504259. Accessed on June 26, 2019.

Hume, David L. 2013. *Tourism Art and Souvenirs: The Material Culture of Tourism*. Routledge Advances in Tourism. Oxon: Routledge.

Jóhannesson, Gunnar Thór, and Katrin Anna Lund. 2018. Creative connections? Tourists, entrepreneurs and destination dynamics. *Scandinavian Journal of Hospitality and Tourism* 18 (sup. 1): S60–S74. https://doi.org/10.1080/15022250.2017.1340549.

Jokela, Timo, Elina Härkönen, and Antti-Jussi Yliharju (eds.). 2014. *Snow Design in Lapland—Initiating Cooperation*. Rovaniemi: Lapland University Press.

Jokela, Timo, Glen Coutts, Maria Huhmarniemi, and Elina Härkönen (eds.). 2013. *Cool—Applied Visual Arts in the North*. Rovaniemi: Lapland University Press.

Keinonen, Turkka, Janne Andersson, Jukka-Pekka Bergman, Sampsa Piira, and Mikko Saaskilahti. 2004. Mitä tuotekonseptointi on? [What is concept design for products?]. In *Tuotekonseptointi* [Concept Design for Products], edited by Turkka Keinonen, Janne Andersson, Jukka-Pekka Bergman, Sampsa Piira, and Mikko Sääskilahti. Helsinki: Teknologiateollisuus ry.

Kramvig, Britt, and Anne Britt Flemmen. 2019. Turbulent Indigenous objects: Controversies around cultural appropriation and recognition of difference. *Journal of Material Culture* 24 (1): 64–82. https://doi.org/10.1177/1359183518782719.

Kugapi, Outi, and Emily Höckert. 2018. Knitting and purling care: Entangled stories of green-and-white mittens. Paper presented at the Nordic Symposium of Tourism and Hospitality Research, Alta, Norway, September 24–26.

Lapin Kansa. (2019). *Lappiin perustettiin yli tuhat uutta yritystä* [Almost 1000 New Enterprises Established in Lapland], January 28, 2019, A4.

Lee, Tsung Hung, and Fen-Hauh Jan. 2019. Can community-based tourism contribute to sustainable development? Evidence from residents' perceptions of the sustainability. *Tourism Management* 70: 368–380. https://doi.org/10.1016/j.tourman.2018.09.003.

Mager, Birgit. 2009. Introduction to service design: Digital communications tool. Culminatum Innovation 2009. http://www.share2solve.org/introsd/start/Main.html. Accessed on May 1, 2009.

Mällinen, Maria-Riitta, and Sarkki Simo. 2014. Tekemällä matkailun kulttuuriseen kestävyyteen: Ongelmia ja niiden hallintaa Utsjoen kirkonkylässä [Doing culturally sustainable tourism: Problems and solutions for their governance in Utsjoki, Finland]. *Matkailututkimus* 11 (2): 25–41.
Marcketti, Sara B., Linda S. Niehm, and Ruchita Fuloria. 2006. An exploratory study of lifestyle entrepreneurship and its relationship to life quality. *Family and Consumer Sciences Research Journal* 34 (3): 241–259.
Matahaere-Atariki, Donna. 2017. *Cultural Revitalisation and the Making of Identity Within Aotearoa New Zealand.* Te Puni Kōkiri. https://www.tpk.govt.nz/en/a-matou-mohiotanga/culture/cultural-revitalisation-and-the-making-of-identity. Accessed on May 12, 2018.
Miettinen, Satu. 2007. Designing the creative tourism experience: A service design process with Namibian craftspeople. PhD dissertation, University of Art and Design Helsinki.
Miettinen, Satu (ed.). 2011. *Palvelumuotoilu. Uusia menetelmiä käyttäjätiedon hankintaan ja hyödyntämiseen* [Service Design: New Methods for Collecting and Applying Users Knowledge]. Helsinki: Teknologiainfo Teknova.
Miettinen, Satu, and Anu Valtonen (eds.). 2012. *Service Design with Theory: Discussions on Change, Value and Methods.* Rovaniemi: Lapland University Press.
Miettinen, Satu, and Mikko Koivisto. 2009. *Designing Services with Innovative Methods.* Kuopio and Helsinki: Kuopio Academy of Design and University of Art and Design Helsinki.
Miettinen, Satu, Jaana Erkkilä-Hill, Salla-Mari Koistinen, Timo Jokela, and Mirja Hiltunen. 2019. Stories of design, snow and silence: Creative tourism landscape in Lapland. In *A Research Agenda for Creative Tourism*, ed. Nancy Duxbury and Greg Richards, 69–83. Northampton, MA: Edward Elgar Publishing.
Miettinen, Satu, Simo Rontti, and Jaana Jeminen. 2014. Co-prototyping emotional value. In 19th DMI: Academic Management Conference. Design Management in an Era of Disruption, London, UK, 2–4 September.
Ministry of Education and Culture. 2018. Taide ja kulttuuri osana alueiden kehitystä; Näkymä vuoteen 2025 [Art and Culture Part of Destination Development: Vision to Year 2025]. Helsinki: Publications of Ministry of Education and Culture 2018:20. Retrieved from http://urn.fi/URN:ISBN:978-952-263-566-2.
Minnakhmetova, Ramilya, Svetlana Usenyuk-Kravchuk, and Yulia Konkova. 2019. A context-sensitive approach to the use of traditional ornament in contemporary design practice. *Synnyt/origins, Special issue on Arctic Arts*

Summit 1: 49–62. https://wiki.aalto.fi/pages/viewpage.action?pageId=151504259. Accessed on June 26, 2019.

Müller, Dieter K., and Arvid Viken. 2017. Indigenous tourism in the arctic. In T*ourism and Indigeneity in the Arctic*, edited by Arvid Viken and Dieter K. Müller, 3–15. Bristol: Channel View Publications.

OECD. 2014. *Tourism and the Creative Economy*. OECD Studies on Tourism. Franc: OECD Publishing.

Rantala, Outi, Anu Valtonen, and Vesa Markuksela. 2011. Materializing tourist weather: Ethnography on weather-wise wilderness guiding practices. *Journal of Material Culture* 16 (3): 285–300.

Rantala, Outi, Suzanne de la Barre, Brynhild Granås, Gunnar Thór Jóhannesson, Dieter K. Müller, Jarkko Saarinen, Kaarina Tervo-Kankare, Patrick T. Maher, and Maaria Niskala. 2019. *Arctic Tourism in Times of Change: Seasonality*. Tema Nord 2019, 528. Copenhagen: Nordic Council of Ministers.

Regional Council of Lapland. 2015. Lapin matkailustrategia 2015–2018 [Tourism Strategy of Lapland 2015–2018]. http://www.lappi.fi/c/document_library/get_file?folderId=2265071&name=DLFE-25498.pdf. Accessed on June 15, 2019.

Regional Council of Lapland. 2018. Lapin suhdannekatsaus 2018 [The Economic Outlook of Lapland 2018]. http://www.meri-lappi.fi/assets/Uploads/Lapin-suhdannekatsaus-2018.pdf. Accessed on May 20, 2019.

Remoaldo, Paula, Olga Matos, Isabel Freitas, Hélder da Silva, Vítor Ribeiro Lopes, Ricardo Gôja, and Miguel Pereira. 2019. Good and not-so-good practices in creative tourism networks and platforms: An international review. In *A Research Agenda for Creative Tourism*, ed. Nancy Duxbury and Greg Richards, 167–181. Northampton, MA: Edward Elgar Publishing.

Richards, Greg. 2011. Creativity and tourism: The state of art. *Annals of Tourism Research* 38 (4): 1225–1253. https://doi.org/10.1016/j.annals.2011.07.008.

Richards, Greg, and Crispin Raymond. 2000. Creative tourism. *ATLAS News* 23: 16–20.

Richards, Greg, and Julie Wilson. 2006. Developing creativity in tourist experiences: A solution to the serial reproduction of culture? *Tourism Management* 27: 1209–1223. https://doi.org/10.1016/j.tourman.2005.06.002.

Sandell, Terry, and Lila Skarveli. 2016. Mapping exercise: How could creative industries foster innovation in tourism in the Northern dimension area? Country Report ± Finland. The Northern Dimension Partnership on Culture. https://www.ndpculture.org/studies/mapping-exercise-how-could-creative-industries-foster-innovation-in-tourismin-the-northern-dimension-area. Accessed on December 15, 2018.

Schilar, Hannelene, and E.Carina Keskitalo. 2018. Ethnic boundaries and boundary-making in handicrafts: Examples from Northern Norway, Sweden and Finland. *Acta Borealia* 35 (1): 29–48. https://doi.org/10.1080/08003831.2018.1456073.

Siivonen, Katriina. 2003. Mikä on kulttuurisesti kestävää? Kulttuuri arkisena vuorovaikutuksena ja matkailutuotteena [What is culturally sustainable? Culture as everyday interaction and as tourism product]. *Kulttuurintutkimus* 20 (2): 9–15.

Similä, Jukka, and Mikko Jokinen. 2018. Governing conflicts between mining and tourism in the arctic. *Arctic Review* 9: 148–173. https://doi.org/10.23865/arctic.v9.1068.

Simpson, Murray C. 2008. Community benefit tourism initiatives—A conceptual oxymoron? *Tourism Management* 29 (1): 1–18. https://doi.org/10.1016/j.tourman.2007.06.005.

Soini, Katariina, and Inger Birkeland. 2014. Exploring the scientific discourse on cultural sustainability. *Geoforum* 11: 213–223.

Srivastava, R.M., and Shubhra Verma. 2012. *Strategic Management: Concepts, Skills and Practices.* New Delhi: PHI Learning.

Sthapit, Erose, and Peter Björk. 2017. Relative contributions of souvenirs on memorability of a trip experience and revisit intention: A study of visitors to Rovaniemi, Finland. *Scandinavian Journal of Hospitality and Tourism.* https://doi.org/10.1080/15022250.2017.1354717.

Swarbrooke, John. 2002. *Sustainable Tourism Management.* Oxon: CABI Publishing.

Tuulentie, Seija, and Bente Heimtun. 2014. New rural residents or working tourists? Place attachment of mobile tourism workers in Finnish Lapland and Northern Norway. *Scandinavian Journal of Hospitality and Tourism* 14 (4): 367–384. https://doi.org/10.1080/15022250.2014.967998.

UNEP and WTO. 2005. Making tourism more sustainable: A guide for policy makers. http://www.unep.fr/shared/publications/pdf/DTIx0592xPA-TourismPolicyEN.pdf. Accessed on June 28, 2019.

Veijola, Soile, and Hannah Strauss-Mazzullo. 2019. Tourism at the crossroads of contesting paradigms of arctic development. In *The Global Arctic Handbook*, ed. Matthias Finger and Lassi Heininen, 63–81. Cham: Springer.

World Commission on Environment and Development (WCED). 1987. *Our Common Future*. Oxford: Oxford University Press. https://sustainabledevelopment.un.org/content/documents/5987our-common-future.pdf. Accessed on June 19, 2019.

6

Hardworking, Adaptive, and Friendly: The Marketing of Volunteers in Iceland

Jónína Einarsdóttir and Guðbjörg Linda Rafnsdóttir

Introduction

In the contemporary globalised world, the labour market can be a challenging place for young people. In 2017, the unemployment rate was, on average, 16.7% among 15–24-year olds in the EU and up to 35–40% in Italy, Spain, and Greece (Powell 2018). At the same time, the lowest youth unemployment rates in Europe were found in Germany (6.6%) and Iceland (6.5%) (Statistics Iceland 2017). Besides, over 40% of the youth participating in the labour market in Europe are caught in a cycle of low-paid or unpaid temporary jobs (Kalleberg 2018). The lives of young people are not characterised by unemployment and insecure work alone but by mobilities as well, and these may

J. Einarsdóttir (✉) · G. L. Rafnsdóttir
University of Iceland, Reykjavík, Iceland
e-mail: je@hi.is

G. L. Rafnsdóttir
e-mail: glr@hi.is

A. Walmsley et al. (eds.), *Tourism Employment in Nordic Countries*,
https://doi.org/10.1007/978-3-030-47813-1_6

intersect (Skaptadóttir and Rancew-Sikora 2016). Some young people travel around the world as volunteers or voluntourists during gap years and holidays (Lyons et al. 2012; Wearing and McGehee 2013). Some do so due to unemployment in their own home country. That allows them to gain skills and experience for the labour market, which might enhance their employability, even though the jobs performed by volunteers are rarely related to their education (Dean 2014; Kamerāde and Paine 2014; Paine et al. 2013).

Hustinx et al. (2010) argue that there is no integrated theory on volunteerism due to the complexity of the phenomenon, which spans across a wide variety of activities, organisations, and sectors. Furthermore, they show that academic disciplines vary in their focus, with sociology focusing on the motivations for volunteering and economics classifying volunteering as unpaid work (Hustinx et al. 2010). Voluntourists are defined as those who 'volunteer in an organised way to undertake holidays that might involve aiding or alleviating the material poverty of some groups in society, the restoration of the environments, or research into aspects of society or [the] environment' (Wearing 2001, p. 1). Definitions of voluntourism, commonly known as a form of alternative tourism, vary in their focus on the volunteering experience as compared to the touristic one (Wearing and McGehee 2013). Similarly, in general volunteers comprise a diverse group, active in a wide range of contexts, and thus, the term 'volunteer' is difficult to define (Bussell and Forbes 2002; Hustinx et al. 2010).

The volunteers' motivations vary and range from humanitarian and ethical to more egoistic ones or a mixture of both. Altruistic motivations involve helping people in need, cultural and historical restoration, and environmental conservation (Wearing and McGehee 2013). Scholars use the term 'voluntourism' when the primary motivation for volunteering is to travel to other countries with minimal expenditure, thus, combining volunteering and travelling. Voluntourism has been promoted as a profound experience that improves 'global citizenship' and allows the volunteers to 'become more involved in changing the world' (McGehee and Santos 2005, p. 775). Selfish motivations include enjoyable travelling without being a conventional tourist, gaining relevant work experience, and enhancing their CVs (Wearing and McGehee 2013; Dlaske 2016; Lyons et al. 2012; McGloin and Georgeou 2016).

Some scholars argue that voluntourists active in low-income communities often exploit their hosts for their self-actualisation through colonial discourses and inherent neocolonial power relations (Bandyopadhyay and Patil 2017; McGloin and Georgeou 2016; Mostafanezhad 2013). In contrast, there is some concern regarding the exploitation of voluntourists in high-income countries. Dlaske (2016) studied voluntourism in Lapland, Finland, mediated through Workaway.info. She argues that '[a]lthough the Workaway initiative positions itself as non-capitalist if not anti-capitalist, the study shows how the workawayer gradually gets languaged into a self-responsible, freely choosing, enterprising self, geared towards the requirements of the contemporary neoliberalised world of work' (p. 415). Market-based rationalisation characterises the daily life of the volunteers, who are mostly Western and young, accept to work long days with no days off at an art centre or a guesthouse. They stay during few weeks or some months 'in a chain of stays at different Workaway hosts–a chain which provides them with a way to travel across the world' (2016, p. 420). The stay allows limited cultural exchange and English is the language of communication. Despite the anti-capitalist rhetoric, Dlaske concludes, Workaway.info is supporting neoliberal values and functions as a provider of cheap labour.

The labour market participants (the employees and employers unions) in Iceland argue that volunteers illegally replace employees in regular jobs and, therefore, their recruitment is a breach of labour market regulations (ASÍ and SA, n.d.). According to the terms agreed upon by the Icelandic Federation of Trade (SA) and the Icelandic Confederation of Labour (ASÍ), those who run economic activities for profit are not permitted to recruit individuals for unpaid labour. The breaking of labour market agreements by hosts, particularly within the areas of tourism and farming, has gotten media attention in recent times. News reports have likened the exploitation of volunteers to be akin to modern slavery (Sæmundsdóttir 2018).

Research on volunteering and voluntourism in Iceland is scarce. Studies on the eco-village Sólheimar indicate that the volunteers' cheap labour is required 'to produce profitable and environment-friendly goods and produce' (Prince and Ioannides 2017, p. 354). At the same time, volunteers are eager to engage in activities beyond the economic profit of the home, which is the cause for their discontentment (Prince and Ioannides 2017; Prince 2019). After analysing the volunteering tasks listed in announcements at two webpages that facilitate recruitment of volunteers, HelpX and Workaway, the authors concluded that almost all are included in collective labour agreements (Rafnsdóttir et al. 2019). Thus, the concern of labour market partners regarding breaches of contracts is well-founded. Based on interviews with volunteers, the authors (forthcoming) conclude that the volunteers have the characteristics of voluntourists with mainly self-centred motives. They appreciate the flexibility of the recruitment procedure, abandon difficult hosts, and move with ease between locations and countries.

In this chapter, we aim to shed light on the marketing of voluntourism in Iceland. Grimm and Needham (2012) point out that such marketing, including studies on internet webpages and their specific content, is under-researched. Volunteering in high-income countries has also received limited attention (Wright 2013). Our study fills these gaps in the literature. Iceland is a fascinating country for a study like this, as it has one of the most rapidly expanding tourist industries in Europe (Romei and Murphy 2017) and one of the lowest unemployment rates in the world—3% in 2016 compared to 8.1% in the EU (Statistics Iceland 2017). Besides, unions for employees and employers have been working actively against the increasing number of volunteers in the country (Rafnsdóttir et al. 2019). More specifically, we focus on the marketing of volunteers by analysing the tasks they are required to perform as voluntary acts and the hosts' requirements regarding the volunteers' personalities, skills, and the attractions offered to them. The following questions will be addressed: Are the tasks requested likely to benefit the volunteers in any way, for instance, enhance their CVs? Do the hosts expect the volunteers to be altruistic, eager to benefit others or the environment? Which attractions are highlighted?

Methods

To shed light on the marketing process, we examined the advertisements for volunteers in Iceland, the requirements demanded by the hosts regarding the tasks to be performed, skills and personality of volunteers, and the attractions they offer to prospective volunteers. The data for this research was derived from the webpages www.workaway.info and www.helpx.net. These sites were pointed out as useful on the Icelandic labour market by an employee in one of the largest labour unions in Iceland, which is involved in breaches of labour agreements by volunteers. Both webpages appeal to people interested in travelling; Workaway is 'the site for travel, volunteering, and cultural exchange' (Workaway n.d.), while HelpX appeals to those who 'are taking a gap year, backpacking around the world, or simply looking for an alternative holiday' (HelpX n.d.). Thus, the webpages attribute statuses of a volunteer and a tourist, or so-called voluntourists, to the applicants.

The hiring process begins when a host publishes a profile on a webpage that is open to everybody. It is possible to search for the advertisements based on countries or spheres of engagement. The volunteers make a descriptive profile and can be contacted by hosts who have paid for their account and have the premium membership. Volunteers can also contact hosts and gain access to their full details and reviews after paying for the premium membership.

On both websites, the minimum age for volunteering is 18 years. The only exceptions are made in case a parent accompanies a minor and the host approves such an arrangement. There is no maximum age for volunteers; however, the hosts can state their preferences regarding the volunteer's age. According to both websites, volunteers need to be healthy and fit enough to participate. Both websites mention the average number of hours that the volunteer should be ready to work for—5 hours for Workaway and 4 hours for HelpX, five days a week. However, both admit that these may vary according to the hosts.

All the online advertisements from hosts on Workaway and HelpX on 27 February 2017 and 27 February 2018 were copied and saved. In 2017, there were a total of 244 ads—173 from Workaway and 73 from HelpX. Of the 246 ads, two were excluded from the analysis, since these were published twice on HelpX. In 2018, the number of ads had increased to 295—223 from Workaway and 72 from HelpX. Some of the ads asked for more than one volunteer; Rafnsdóttir et al. (2019) conclude that in 2017 and 2018, these two websites advertised for over 1000 volunteers in Iceland. In January 2018, 195,500 people were employed in the Icelandic labour market. Thus, would the volunteers be included, they would make up to 0.5% of all the employed individuals in the country.

We downloaded the advertisements into Atlas.ti, a software for qualitative analysis. In line with the research aims, we analysed all the downloaded ads. We read the ads several times after which the content was sorted, and recurring themes were identified. First, we explored the tasks the volunteers were requested to perform. Then, we focused on the hosts' demands regarding the volunteers' personality and skills. Finally, we examined the compensation offered, including the accommodation, food, and the attractions highlighted. When directly citing the ads, we did not correct the language or the spelling.

When opening the ads for Iceland on these websites, the volunteer immediately sees one photo per ad at Workaway and 1–6 photos for each ad on HelpX. The front photographs for 78 ads were randomly selected from the 223 ads on Workaway, and all the photographs on 20 ads were randomly chosen from the 72 ads on HelpX for 2018. The content was analysed, and the result was summarised (Table 6.1).

Table 6.1 Workaway and HelpX photo analysis

Photos/Website	HelpX (20 ads)	%	Workaway (78 ads)	%
Family/hosts	12	15	20	26
Nature	23	28	14	18
Farm or house	16	20	14	18
Children	7	9	3	4
Children + animal(s)	6	7	8	10
Adults + animal(s)	6	7	6	8
Animals	7	9	5	6
Northern lights	1	1	2	3
Swimming	2	2	2	3
Urban	2	2	4	5
Total photos	82	100	78	100

Results

Voluntary Work

Each volunteer position advertised included diverse tasks, which were not described uniformly.

The task listed by most ads, about 76.6%, was housework. Generally, this entailed cleaning, laundry, shopping, and cooking. Some ads were vague in their descriptions; for instance, they wanted someone to work 'just normal household things' (HelpX 2017). Nearly 41.7% of the ads requested help with childcare, which included picking children from day-care or school, helping them with homework, and teaching foreign languages. The age range of the children ranged from toddlers to early teens. 'Elderly care' was listed in eight ads. Childcare and housework often overlapped; in 2017, there was a 93% overlap. Typically, a host was in need of 'a hardworking and open-minded person to help out with 2 children 2 year old boy and 1 year old girl and helping out with some light housework' (HelpX 2018).

In 13.6% of the ads, language practice was listed as a task. The hosts often wanted to practise their language skills. In some cases, language practice for a language other than English was requested. In such cases, one of the hosts in the family used to be a foreigner and the language requested was her/his native language. Often, they wanted the volunteer to practise the language with their children. Some of these ads even mentioned the possibility for the volunteer to practise their Icelandic.

Tasks relating to farming and animal care were included in 67% of the ads. A majority of these ads stemmed from sheep farms or mixed farms with sheep, horses, and cattle. As is typical with farming, the types of tasks differed, depending on the season. In spring, farmers wanted help with the lambing season, while in summer, haying was the most essential task. In fall, collecting the sheep from the mountains back to the farm took priority, and in winter, tasks such as taking care of the animals at the farm, feeding, and mucking the stables were required. One host confirmed that the volunteer would 'learn working with animals. How you can become farmer and so mutch more' (Workaway 2018). In 30% of the ads, the hosts asked for help with gardening, which included helping with the vegetable garden and cutting down trees, for instance.

In 43% of the ads, the hosts sought help with construction, including building, renovation, and general maintenance such as painting and fencing. Commonly, farmers sought help with construction, especially fencing and general maintenance of their farm. Some hosts asked for skilled staff, such as a qualified carpenter, for a construction project: 'We would need help for rebuilding this summer house. It is a work for a carpenter or a person who knows how to work with wood' (HelpX 2018). Another host sought a volunteer with experience or skills in 'plumbing, carpentry, painting and decorating, electrical wiring' (Workaway 2017).

The volunteers were expected to perform touristic services in 32.3% of the ads. Roughly 23% of the ads listed tasks categorised as restaurants and accommodation. For instance, one host was in need of 'an efficient and intelligent person to assist in the guesthouse to clean and replace the beds' (Workaway 2017). Work with horse rentals was often listed with restaurants and accommodation services. However, in 5.4% of the ads, horse rentals had no other touristic activities listed. Hosts running horse farms typically requested for volunteers with some experience with horses and detailed knowledge at times. In one ad, the volunteer was required to have specific skills such as the ability to 'lead a horse from the partner position, do a head down exercises, do a nose to flank exercises (have to horse kiss the stirrups), do yielding to driving aids exercises, exercise a horse by lunging' (Workaway 2017). Additionally, 3.7% of the ads wanted volunteers to work with tourist services other than restaurants and accommodation or horse rentals. These included dog-sledging, canoes, or bike rentals.

Roughly 6.3% of the ads listed help with computer work and marketing in addition to other tasks. In one ad, the volunteer was expected to help with cleaning the premises, meet with the tourists and to take part 'in the art and culture program we run in the house. Those projects could be from overseeing workshops, running the cinema, working on theatre shows, helping out with festivals, marketing etc.' (Workaway

2017). This ad overlaps with the ads requesting help with 'arts and crafts', which comprise about 4.6% of the ads. In all these ads, 'arts and crafts' was an additional task, almost entirely in the tourist services. In one of the ads, the host ran a hostel that 'is a multi-functional artist run company that hosts travellers, produces performances of all kind, offers artists working spaces and residencies and gets involved in local community outreach projects' (Workaway 2017).

The term 'other' was included in 23.4% of the ads. Occasionally, the hosts marked the alternatives 'Help with Eco project' and 'Charity work' in a list of predefined tasks.

Accommodation and Food

In exchange for their voluntary work, the hosts provide accommodation. The facilities offered to the volunteers varied greatly, ranging from a tent or caravan to a separate house shared by other workers or volunteers. In almost 80% of the ads from 2017, the volunteers were promised access to a private room within the family home. In 7% of the ads, the hosts offered staff houses. For instance, a host with a guesthouse with 27 rooms would 'offer food and accommodation. You would live in your own cozy summerhouse together with your co-workers from all over the world' (HelpX 2017). In 14% of the ads from 2017, there is no mention of accommodation.

Only about 25% of the ads from 2017 mentioned the provision of food for the volunteer. The reason for not mentioning food might stem from the fact that both of the websites assume that the volunteer will receive accommodation and food in exchange for their labour, although there are some exceptions, as outlined above. With regard to food, some hosts accurately described their eating habits and preferred food. One host family mentioned that 'we eat tipical icelandic food like fish, meat, pasta, bread, cereals, milk products' (Workaway 2018). Another host asked for somebody 'who is over 35 years old, very handy, energetic and positive and it would be great if that someone is interested in cooking wholesome food rich in vegetable (not Vegan though, since I like meat a lot!)' (Workaway 2018).

Where meals were offered, some hosts put forward demands regarding the diet of the volunteers—7% of the ads asked vegetarians or vegans not to apply. One ad mentioned that '[i]f you are Vegan there is no bother for you to send us a mail!!!' (Workaway 2017). Another one stated that 'it is not a house to come to if you don't eat all' (Workaway 2017). Three of the hosts mentioned that while vegetarians could join them, they would have to prepare their meals. Another host specified that the volunteer must at least eat fish. One family declared that they were vegans but did not list any food demands. Some hosts did not request specifically for meat eaters but mentioned that they offer mostly meat. One of the hosts described their family eating habits: 'We always have lunch together, and the food is diverse but meat will pretty much always be involved' (Workaway 2017).

The Volunteer's Characteristics

The hosts put forth diverse demands with regards to the duration of staying, gender, age, personality traits, and lifestyle of the prospective volunteer. Some wanted a minimum duration of stay to be half a year, while others offered 1–2 weeks.

Hosts who needed help with babysitting and cleaning of their home tended to ask for female volunteers. For instance, one host asked for female volunteer 'of any age with experience with children and house-keeping, German-speaking'. One host who wanted a volunteer to take care of children and housework requested for a clean criminal record. A Scout centre wanted any volunteer over the age of 18 regardless of their gender, religion, sexual orientation, and ethnicity, although '[a]ll volunteers must provide a recent criminal background check, from their home country, due to Icelandic child protection laws' (Workaway 2017).

The hosts valued volunteers who were arduous, well-mannered, and friendly. A family needed an 'easy going' person who 'is willing to get her hands dirty if needed' (HelpX 2018). Another host was 'looking for fun' and asked for a volunteer with the following characteristics and skills: 'Honest, hard worker, fast learner, flexible, mature, clean and tidy, non-smoker, good cleaner, good on the computer, good communication skills, easy with food and like to be in the heart of the city and travel' (HelpX 2017). Another host listed the following qualities: 'The helper has to be responsible and sensitive to other human beings. Self-initiative, cleanliness and precision are required' (HelpX 2018). Some demanded volunteers who did not indulge in smoking, alcohol, or drugs.

In some cases, the hosts outlined in detail their demands and described the challenging living conditions of the volunteer: 'We are looking for cheerful persons, that can handle hard work in though conditions. Living in a close space with no privacy'. The host also posted a list of questions: 'Your age, height and weight...When would you like to join us? For how long would you like to stay with us? What is your first motivation to come to us? What are your expectations?' (Workaway 2017). Others did not bother about the preferences of the prospective volunteers. One host family outlined the importance of the volunteer's willingness to adapt to the 'Icelandic way'. They have hosted volunteers for over 15 years:

> Pleace notice our demand, are after having had problems with volunteer workers in our house – ... with time and scedule – that people read and know the times – and the rutine work we set up - and that they try to understand that we are working with nature, – so when the nature calls = like calf birth or rain/snow when we plan something that needs dry hours ... and please know sometimes you need to put yourselfs in other persons shue's ... to all candidat to our farm at volunteer work – ... bring understanding, patience and try to learn the Icelandic way – it might not be your way in normal day but it is our way and you come to our house! – Having said this... – it is said for a reason, – for things were understood and – ... we try to do our best to welcome you – that is a promiss –... but to feel welcome you also have to chip in! :) OK? (Workaway 2018)

The Host Family

The hosts were keen to mention the importance of family life and wanted the volunteers to become a part of the family. What that entailed differed from host to host. One host explained that '[i]n the farm there is a lot of activity, friends and family visiting, tourists and guests coming for dinner/lunch/stay over and riding tours. In the house everyone takes part and helps out with everything and the atmosphere is warm and friendly' (Workaway 2018). Another host underlined that the volunteer would 'learn icelandic way and enjoy family life' (Workaway 2018).

Most of the hosts described themselves, and their family and the location of their home and suggested recreational activities for the volunteers. However sometimes hosts do not involve family members, both in terms of photos and descriptions, at least in the ads, which are open to everybody. Others provide some details of the family members. The age and gender are outlined, and occasionally, hosts mention when some family members have a disability. At times, animals are included in the list of family members. Some hosts describe their personalities, mostly in favourable terms: 'We are an active and open-minded family' (Workaway 2018). Parents talk about the kids that the volunteers are expected to take care of: 'Our girl is very energetic, a lot of fun, artistic and a girl that is very curious about life. Our son is very down to earth … loves lego's, watching tv, relaxing, cuddling and meeting friends' (Workaway 2018). Few hosts post family photographs along with the ads.

At Workaway, there is one photo that the interested volunteer can see before opening up the ad, wherein additional photos are posted, most often 1–8 photos or more. The analysis of the content of randomly chosen front photos for 78 ads from 2018 shows that the images of family, with or without children, are the most common, followed by the surrounding nature, the farm, the house of the family, children alone or in company with animal(s), adults with animals, or only animals. At HelpX, 1–6 photos are posted per ad, all visible at once. Occasionally, no photos are posted.

In short, the websites mostly contain images of the family, home, nature, and animals. While a photo of family members (some or all) is the most commonly posted at Workaway, photos of surrounding nature are most frequent at HelpX. These two images, in addition to photos of the farm or the home, are the three most common ones. Circa 90–95% of all photos show family members, surrounding nature, and animals.

The photos have different quality—some are professional, while others are quite amateur. Most are taken during summer months, while a few show the surrounding nature and the home during winter.

Nature, Sagas, and Travel

The most commonly listed attraction was the Icelandic nature. In 73% of the ads from 2017, the hosts described the environment in the vicinity—its beauty and purity. Nearly 58% mentioned natural landmarks in the proximity as something the volunteers could enjoy, such as waterfalls, national parks, natural pools, and hot springs. In 5% of the ads, the hosts mentioned the Sagas of the Icelanders and pointed out that their home was situated where it all happened: 'We are located in the north of Iceland, in the valley of Icelandic saga' (HelpX 2017).

As with nature and natural landmarks, the hosts highlighted the midnight sun and northern lights, since they set Iceland apart, as something unique. Almost 4% of the ads mentioned either the midnight sun or the northern lights concerning the season: 'If you come during the winter months, it is the perfect place for catching a view of the northern lights. Don't forget to download your northern lights app, so that you can be continuously updated on potential sightings!' (Workaway 2017).

In 27% of the ads, the hosts mentioned that the volunteers would be able to enjoy arts and culture in the neighbourhood. Hosts who are living within the capital area or in bigger towns mentioned museums, art shows, and nightlife. A pair of hosts boasted that they used to urge 'people to get involved with the local culture, join the yoga group, train volleyball or go to the swimming pool' (Workaway 2017). Hosts who run cultural programmes within their establishment encouraged volunteers to participate in the same.

In 12% of the ads, the hosts mentioned the possibility for the volunteer to enjoy tranquillity, especially when the host lived in a remote area. The possibility of getting away from the 'busy city life', getting in touch with themselves, and reading and relaxing was expressed in one ad: 'Therefore it is a good opportunity for you, whom love to read and write in the quietness to come and enjoy for exchange of all helping hand in the house' (HelpX 2017).

The hosts often highlighted that the volunteer could enjoy what Iceland has to offer. In 40% of the ads in 2017, the hosts mentioned travel as spare time activity; in some cases, the host suggested that the volunteer could travel with the family. While some hosts offered a car that the volunteer could occasionally borrow for their travels, others mentioned the public transport system and the ease of mobility.

In 2017, the hosts promoted outdoor activities that were in proximity in 36% of the ads. These activities differed depending on the hosts' location. While hiking and walking paths were mentioned most commonly, some ads included other activities such as river rafting and kayaking. Some hosts emphasised, in particular, on horse riding and birdwatching as outdoor activities. It is worth noting that these two activities did not always coexist with other outdoor activities. Taking care of the horses was used as an attraction while also being a task. The ads that mentioned birdwatching typically pointed out that the hosts lived close to a bird colony.

Discussion

This chapter explores the marketing of volunteers to Iceland through an examination of the tasks they are required to perform, the accommodation and food provided, the hosts' requirements regarding the volunteers' personalities, experience, skills, and the attractions offered to the volunteers. The study is based on analyses of ads posted by Icelandic hosts that aimed to recruit volunteers through the websites Workaway and HelpX on 27 February 2017 and 2018. Are the tasks requested likely to benefit the volunteers in any way by enhancing their CVs? Do the hosts expect the volunteers to be altruistic, eager to benefit others or the environment? What attractions are highlighted the most?

The work requested by the hosts included diverse tasks, and the ones mentioned most often involved housework, childcare, farming/animal care, and construction. Further, a little over 25% of the ads involved work in restaurants and guesthouses. While housework, childcare, and unqualified work in restaurants and guesthouses are frequent and not particularly likely to enhance the volunteers' CVs, work in farming/animal care in the Icelandic context is likely to be a unique experience among young people. The hosts at times demanded former working experience and, occasionally, qualified skills in construction work and the training of horses. Thus, the volunteers recruited can add a line of qualified work in their CVs. However, the voluntary experience in itself, being a sign of individual initiative and personal qualifications, is probably more valuable for the CV than the tasks performed (Kamerāde and Paine 2014; Paine et al. 2013).

As with the tasks requested, the accommodations provided were diverse and varied from tents to a separate staff house. About 25% of the ads mentioned food, and some volunteers would have a separate kitchen from the host family. Many of the ads mentioning that the volunteer would eat with the family underlined that they commonly served meat; in some of the ads, vegetarians and vegans were asked not to apply. When the hosts formulate their offerings of accommodation and food and the demands made, the frankness is at times noteworthy. Many of the ads were amateur, written in poor English, and characterised by a straightforward, even rude, tone. This resonates with Dlaske's (2016, p. 435) description of the hosts' feedback on Workaway from guesthouses in Lapland, which 'is entirely personal, direct and painstakingly detailed evaluation (and criticism) not only of the (lack of) skills and competencies of the volunteers, but also of their attitudes, personal characteristics and occasional mood swings'.

The Icelandic hosts were concerned about competencies and the personal character of volunteers. Some hosts seemed to prefer no volunteer than having one who did not suit their preference and was unable to adapt to the 'Icelandic way' in terms of behaving and working; at times mentioned in the ads, without an explanation. However, some of the most common personal characteristics asked for were hardworking, amiable, and flexible volunteers with healthy lifestyles.

In trying to appeal to the volunteers, most of the hosts advocated the same attractions as tourist bureaus when announcing Iceland as a tourist destination, such as beautiful nature and fresh air. Most of them (86%) assumed that volunteers wanted to come to Iceland for the opportunity to visit exotic places, learn about the Sagas, and enjoy nature, fresh air, and tranquillity. They advocated the same attractions as the tourist industry (Lund 2013). Therefore, most of the hosts attributed the status of a tourist to the volunteer. Often, the hosts were willing to support them in exploring what Iceland has to offer. In only 14% of the ads, the hosts did not mention any appeal to attract prospective volunteers.

So far, we have referred to the travelling individuals as volunteers, as do the ads that were analysed. Indeed, HelpX calls them helpers. However, according to most definitions of volunteers mentioned above and the conditions outlined in the ads, the terms tourist, migrant, worker, or a combination of these better describe their status. Thus, although the examined websites underline their role in facilitating travel through voluntary work (HelpX n.d.; Workaway n.d.), which might legitimise the term voluntourist, their status as volunteers may be challenged. This conclusion accords with Skaptadóttir and Rancew-Sikora (2016, p. 8), who argue that 'the boundaries between tourism and migration are not always clear cut. Tourists may take on temporary work either for wages or as volunteers, and it is not uncommon for migrants to arrive to new places initially on tourist visas'.

Most definitions of volunteering involve little to no pay, the free choice of the volunteer, little or no prior contact between beneficiaries and the volunteer, and that the work be performed for a charity or non-profit organisation while informal volunteering is, at times, accepted (Hustinx et al. 2010; Cnaan et al. 2011). All these requirements, except the last one, apply to these websites; neither can the hosts be described as charities or non-profit organisations, nor do the tasks performed fall within the framework of so-called informal volunteering. According to the Icelandic Confederation of Labour (ASÍ), volunteering is accepted only in the case of charities and cultural and humanitarian activities. None of the ads listed charity, artwork, or cultural activities as the only tasks required, and these tasks were rarely seen in the ads. The literature also highlights environmental conservation as an appropriate task for volunteerism, a task that was rarely requested and, even then, never as the only task required. Almost all the tasks listed in the ads fell under the category of profit activities, being economic activities covered by collective labour agreements in Iceland (Rafnsdóttir et al. 2019). This fact calls for further research. The volunteers do not sign an employment contract and do not get paid according to collective agreements. Being unregistered, they are permitted to stay in the country for up to three months. Their position raises the question whether it allows employers to discriminate them. Further, the volunteers are hidden for the labour market authorities and fall outside all statistic on labour market participants.

The strength of this chapter lies in the analyses of all the online ads posted by Icelandic hosts that appeared on Workaway and HelpX on 27 February 2017 and 27 February 2018. We have presented a good overview of the expectations and needs of Icelandic hosts who sought volunteers through these two websites at the chosen time. Thus, we responded to a gap in the literature about volunteering in high-income countries (Wright 2013) and the marketing of volunteers (Grimm and Needham 2012). However, there are more websites for recruiting volunteers, thus we did not include all the marketing of volunteers to Iceland listed on 27 February 2017 and 27 February 2018. One limitation is that the results rest entirely on the deduction about the volunteers' status from requirements stated by the hosts as expressed in the ads. Examining the actual experiences of this group of the mobile workforce would be a valuable addition.

Conclusion

In their marketing, hosts expect the volunteers to be looking for spectacular natural surroundings, animals, and family life and be willing to learn the 'Icelandic way' of doing work. Thus, the hosts, in particular farmers, are upfront about the seasonality of workload. Assuming that volunteers are, above all, tourists, the hosts attempt to attract them with typical tourist bureaus.

The volunteers mostly perform unskilled tasks such as domestic chores, childcare, and unqualified work within the tourism industry. However, some engage in farming/animal care, which is likely to be an innovative experience and rare addition to the CV. Further, some volunteers are recruited to perform skilled work in the fields of construction and training of horses. Thus, the volunteering work in itself is likely to be the most important experience for most.

The marketing of volunteers indicates that they are expected to be flexible and hardworking individuals who are willing to perform a broad range of tasks. They are not seen as altruistic individuals but rather as voluntourists looking for an adventure, and the ads highlight the same attractions as the tourist bureaus when announcing Iceland as a tourist destination.

Acknowledgements We are grateful to MARK, the Center for Diversity and Gender Studies and the School of Social Sciences, for the grants we received for this research. We also thank Ragnhildur Guðmundsdóttir and Ástrós Anna Klemensdóttir for their assistance.

References

ASÍ, and SA. n.d. Are you working illegally as a volunteer? It is a violation of Icelandic Law and workers right to let volunteers replace employees in a regular job. http://www.sgs.is/wp-content/uploads/2017/06/A5_einbl_Sjalfboda_ens_HQ.pdf. Accessed on September 30, 2019.

Bandyopadhyay, Ranjan, and Vrushali Patil. 2017. "The White woman's burden"—The racialized, gendered politics of volunteer tourism. *Tourism Geographies* 19 (4): 644–657. https://doi.org/10.1080/14616688.2017.1298150. Accessed on September 30, 2019.

Bussell, Helen, and Deborah Forbes. 2002. Understanding the volunteer market: The what, where, who and why of volunteering. *International Journal of Nonprofit and Voluntary Sector Marketing* 7 (3): 244–257. https://doi.org/10.1002/nvsm.183. Accessed on September 30, 2019.

Cnaan, Ram A., Kathleen H. Jones, Allison Dickin, and Michele Salomon. 2011. Estimating giving and volunteering: New ways to measure the phenomena. *Nonprofit and Voluntary Sector Quarterly* 40 (3): 497–525. https://doi.org/10.1177/0899764010365741. Accessed on September 30, 2019.

Dean, Jon. 2014. How structural factors promote instrumental motivations within youth volunteering: A qualitative analysis of volunteer brokerage.

Voluntary Sector Review 5 (2): 231–247. http://dx.doi.org/10.1332/204080514X14013591527611. Accessed on September 30, 2019.

Dlaske, Kati. 2016. Shaping subjects of globalisation: At the intersection of voluntourism and the new economy. *Multilingua* 35 (4): 415–440. http://dx.doi.org/10.1515/multi-2015-0002. Accessed on September 30, 2019.

Grimm, Kerry E., and Mark D. Needham. 2012. Internet promotional material and conservation volunteer tourist motivations: A case study of selecting organizations and projects. *Tourism Management Perspectives* 1: 17–27. https://doi.org/10.1016/j.tmp.2011.12.007. Accessed on September 30, 2019.

Hustinx, Lesley, Ram A. Cnaan, and Femida Handy. 2010. Navigating theories of volunteering: A hybrid map for a complex phenomenon. *Journal for the Theory of Social Behaviour* 40 (4): 410–434. https://doi.org/10.1111/j.1468-5914.2010.00439.x. Accessed on September 30, 2019.

Kalleberg, Arne L. 2018. *Precarious Lives: Job Insecurity and Well-Being in Rich Democracies*. Cambridge: Cambridge University Press.

Kamerāde, Daiga, and Angela Ellis Paine. 2014. Volunteering and employability: Implications for policy and practice. *Voluntary Sector Review* 5 (2): 259–273. https://doi.org/10.1332/204080514X14013593888736. Accessed on September 30, 2019.

Lund, Katrín A. 2013. Experiencing nature in nature-based tourism. *Tourist Studies* 13 (2): 156–171. https://doi.org/10.1177/1468797613490373.

Lyons, Kevin, Joanne Hanley, Stephen Wearing, and John Neil. 2012. Gap year volunteer tourism: Myths of global citizenship? *Annals of Tourism Research* 39 (1): 361–378. https://doi.org/10.1016/j.annals.2011.04.016. Accessed on September 30, 2019.

McGehee, Nancy Gard, and Carla Almeida Santos. 2005. Social change, discourse and volunteer tourism. *Annals of Tourism Research* 32 (3): 760–779. https://doi.org/10.1016/j.annals.2004.12.002. Accessed on September 30, 2019.

McGloin, Colleen, and Nichole Georgeou. 2016. "Looks good on your CV": The sociology of voluntourism recruitment in higher education. *Journal of Sociology* 52 (2): 403–417. https://doi.org/10.1177/1440783314562416. Accessed on September 30, 2019.

Mostafanezhad, Mary. 2013. The politics of aesthetics in volunteer tourism. *Annals of Tourism Research* 43: 150–169. http://dx.doi.org/10.1016/j.annals.2013.05.002. Accessed on September 30, 2019.

Paine, Angela Ellis, Stephen McKay, and Domenico Moro. 2013. Does volunteering improve employability? Insights from the British Household Panel Survey and beyond. *Voluntary Sector Review* 4 (3): 355–376. http://dx.doi.org/10.1332/204080513X13807974909. Accessed on September 30, 2019.

Powell, Andy. 2018. Youth unemployment statistics. Briefing Paper Number 5871. House of Commons Library. https://dera.ioe.ac.uk/31765/1/SN05871__Redacted.pdf. Accessed on September 30, 2019.

Prince, Solène. 2019. Volunteer tourism and the eco-village: Finding the host in the pedagogic experience. *Hospitality & Society* 9 (1): 71–89. https://doi.org/10.1386/hosp.9.1.71_1. Accessed on September 30, 2019.

Prince, Solène, and Dimitri Ioannides. 2017. Contextualizing the complexities of managing alternative tourism at the community-level: A case study of a nordic eco-village. *Tourism Management* 60 (June): 348–356. https://doi.org/10.1386/hosp.9.1.71_1. Accessed on September 30, 2019.

Rafnsdóttir, Guðbjörg Linda, Jónína Einarsdóttir, and Ástrós Anna Klemensdóttir. (forthcoming). "Not like every other tourist with selfie stick": Voluntourism in Iceland. In Kristín Loftsdóttir and Unnur Dís Skaptadóttir (eds), *Mobility and Transnational Iceland: Current Transformations and Global Entanglements*. Reykjavík: Háskólaútgáfan.

Rafnsdóttir, Guðbjörg Linda, Jónína Einarsdóttir, and Ragnhildur Guðmundsdóttir. 2019. Heiðarlegir Vinnuþjarkar, Hreinlátir og Fljótir að Læra: Um Sjálfboðaliðastörf á Íslandi. *Íslenska Þjóðfélagið* 10 (2): 29–47. http://www.thjodfelagid.is/index.php/Th/article/view/159. Accessed on September 30, 2019.

Romei, Valentina, and Hannah Murphy. 2017. Boom, bust and boom again: Iceland's economy in charts. *Financial Times*, March 17. https://www.ft.com/content/26270ef4-0a35-11e7-97d1-5e720a26771b. Accessed on September 30, 2019.

Sæmundsdóttir, Sunna. 2018. Líkist Stundum Nútíma Þrælahaldi: Reglulega Þarf að Bjarga Erlendum Sjálfboðaliðum. *visir.is*, February 5. http://www.visir.is/g/2018180209358. Accessed on September 30, 2019.

Skaptadóttir, Unnur Dís, and Dorota Rancew-Sikora. 2016. Introduction: Blurring boundaries in mobility studies. In *Mobility to the Edges of Europe: The Case of Iceland and Poland*, 7–16. Warsaw: Scholar Publishing House.

Statistics Iceland. 2017. Iceland in figures 2017. Reykjavík: Statistics Iceland. http://www.statice.is/media/50481/icelandinfigures2017.pdf. Accessed on September 30, 2019.

Wearing, Stephen. 2001. *Volunteer Tourism: Experiences That Make a Difference*. Wallingford: CABI.

Wearing, Stephen, and Nancy Gard McGehee. 2013. Volunteer tourism: A review. *Tourism Management* 38: 120–130. http://dx.doi.org/10.1016/j.tourman.2013.03.002. Accessed on September 30, 2019.

Wright, Hayley. 2013. Volunteer tourism and its (mis) perceptions: A comparative analysis of tourist/host perceptions. *Tourism and Hospitality Research* 13 (4): 239–250. https://doi.org/10.1177/1467358414527984. Accessed on September 30, 2019.

7

On the Move: Migrant Workers in Icelandic Hotels

Margrét Wendt, Gunnar Thór Jóhannesson and Unnur Dís Skaptadóttir

Introduction

Issues of mobility shape much of our contemporary world, including the labour market. More and more people are on the move in search for employment abroad, which has resulted in an increased number of international labour migrants. Labour migrants are a highly diverse group, which makes it difficult to generalise about their experiences, opportunities and challenges they face as they engage in various types of

M. Wendt (✉) · G. T. Jóhannesson
Faculty of Life and Environmental Sciences,
University of Iceland, Reykjavík, Iceland
e-mail: maw6@hi.is

G. T. Jóhannesson
e-mail: gtj@hi.is

U. D. Skaptadóttir
Faculty of Sociology, Anthropology and Folkloristics,
University of Iceland, Reykjavík, Iceland
e-mail: unnurd@hi.is

A. Walmsley et al. (eds.), *Tourism Employment in Nordic Countries*,
https://doi.org/10.1007/978-3-030-47813-1_7

work. While most migrant workers travel for work, some seek employment for touristic pursuits, as they would not be able to travel without working at the same time. Others have the freedom to move and work where they like without economic or other types of restraints (Uriely 2001). While some migrants are employed in skilled and rewarding jobs, other experience precarious working conditions and suffer from exploitation (Janta et al. 2011b).

Tourism is labour-intensive, requiring a multitude of workers in various subsectors. It has become one of the world's largest generators of jobs, creating by estimate one out of every ten jobs (UNWTO 2018). Iceland has experienced a rapid increase in tourism in recent years. The number of international tourist arrivals increased from 488,000 in 2010 to 2.3 million in 2018 (Ferðamálastofa 2019). The growth of Iceland's tourism has resulted in a labour shortage and thus there has been a high demand for migrant workers in the sector. In 2018 there were 200,000 people working in Iceland, of which around 33,000 were employed in tourism (Statistics Iceland, n.d.a, b). Today, immigrants[1] account for around 30% of the employees in the Icelandic tourism industry, most of whom work within the accommodation and restaurant sector (Statistics Iceland, n.d.a). This means that approximately 10,000 immigrants are employed in tourism in Iceland.

Based on a case study of three hotels in rural Iceland, this chapter sheds light on experiences of migrant workers. Tourism studies are traditionally organised around binary concepts such as hosts and guests, ordinary and extraordinary, home and away/destination. There has been a tendency in the literature to gloss over the intrinsic diversity of people occupying these categories, their motivations and performances (Jóhannesson et al. 2015; Ren et al. 2018). Our objective in this chapter is to open up and delve into the experiences of migrant workers, describing the challenges and opportunities associated with working at rural hotels in Iceland. The focus is set on two occupations, namely receptionists and housekeepers. Of special importance for us it to tease

[1]According to Statistics Iceland, immigrants are defined as individuals who are born abroad and whose parents are both born abroad. Others are considered to have an Icelandic background.

out how socio-spatial organisation of different jobs at hotels shape opportunities and limitations for occupational and geographical mobility, which sustain and reinforce particular tourism labour dynamics more generally. Mobility can be regarded as an organisational principle of international labour, creating distinctions between groups of people (Hannam et al. 2006; Sheller and Urry 2006), shaping the opportunities they have to secure and sustain their wellbeing and job/career. In this regard we highlight the relational character of the job in question as decisive. Tourism work is not performed in a vacuum but in multilayered relations. Working as a receptionist or a housekeeper enacts a particular network that gives access to some options while preventing others. This case study also demonstrates that tourism employment requires various soft skills as well as knowledge opposing the view that tourism employment is a low-skilled sector. We highlight the important role of social networks for acquiring the necessary skills and knowledge to perform one's job.

The chapter starts with a discussion of the theoretical background of the case study, connecting tourism labour with the concept of mobility. Thereafter, the findings of the case study are presented, focusing on two different occupations within the hotels: the reception and housekeeping. We will present how these occupations impact the ability of migrant workers to engage in cross-cultural communication and learning in different ways. Finally, we outline how the two employment occupations affect migrant workers' professional as well as personal lives from a mobility perspective.

Mobility, Tourism Work and Migrant Workers

Tourism can be seen as iconic for the present world order, characterised by increased mobility of people, goods, knowledge and money. At the turn of the century, Zygmunt Bauman (2000) used the terms "liquid modernity" and "the tourism syndrome" to describe the human condition under these circumstances, underlining how many of the traditional institutions of modern society such as the family, the state and work had become more fluid and uncertain. He also discussed how people are becoming more conscious that their social relations may be temporary, constantly being subject to potential mobility and that they only hold until further notice (Franklin 2003). It follows that the responsibility to sustain these relations is increasingly placed on the shoulders of individuals rather than being secured by social structures or social institutions. It is, for instance, unlikely today that someone will work in the same job for all of his or her working life. In order to build up a decent life under these circumstances it is necessary, for better or worse, to be willing to acquire new skills, be willing to change and move for work.

In the wake of the broad societal changes which Bauman describes, the concept of mobility has received increasing attention in the social sciences, including tourism studies (Urry 2000; Sheller and Urry 2006). Ontological mobility underlines that movement is a basic condition of life and what we recognise as social structures is accomplished and maintained through various ordering activities (Law 2004; Latour 2005). Societal ordering needs to be done and enacted, which draws attention towards the ways people perform the relations necessary to lead their lives, what options they have to do so, what challenges they face while doing so and what implications their performance has (Law and Urry 2004; Hannam et al. 2006; van der Duim et al. 2012). This approach urges us as researchers to trace what happens on the ground where people are busy composing their lives, that is empirical mobility and performances. In the present context it entails opening up the idea of tourism work and ask questions such as: What does it mean to be working in tourism? How do different jobs create different configurations of mobility and immobility and how are particular labour dynamics created and sustained that affect the possibilities (migrant) workers have?

Working in tourism means different things to different people. Tourism is a labour-intensive industry and can create various jobs for the local population. However, working in tourism is often regarded as undesirable as many of the jobs are seasonal, perceived as low-skilled and arguably offer low status and low wages. Tourism employment is described as encompassing difficult and tiring working conditions, precarious employment and unclear career structures, resulting in a high turn-over rate for employment within the industry (Baum 2007; Devine et al. 2007; Janta et al. 2011b; Ladkin 2011; Joppe 2012).

Overall, the employment options and working conditions available in the tourism industry are therefore in many ways considered unattractive vis-a-vis other sectors. This in turn facilitates relatively easy access to jobs within the sector, which can be an advantage for migrant workers seeking employment. As a result, tourism businesses in general rely on a culturally diverse workforce (Baum 2007; Devine et al. 2007; Ladkin 2011; Joppe 2012). The reasons for why migrant workers may accept employment that the local workers find unattractive are diverse and based on various factors. Among reasons are the facts that migrant workers may regard their stay as temporary and that their employment options may be limited due to a lack of the host country's language or other required skills (Wojtynska 2012). Their skills or education are often not recognised in the host country, which means they have to accept other jobs that do not demand particular skill sets (Skaptadóttir and Loftsdóttir 2019). In addition, migrant workers often have a different point of reference when evaluating the work environment and conditions. They compare their employment in the host country to employment in their home country. In many cases, the labour conditions in the home country are comparably worse, resulting in the acceptance of the conditions in the host country. Often, wages are for instance higher in the host country compared to the country of origin. Thus, many migrant workers pursue work in the tourism industry, both temporarily as well as long-term, in order to improve their economic status or send remittances to their families in the home country (Baum et al. 2007; Wojtynska 2012; Piso 2014).

In addition to improving their economic status, tourism employment also creates possibilities for migrant workers to gain new experiences. Since the tourism industry has a low entry barrier, migrant workers may work in jobs which they have little or no professional experience in, thus gaining new work experience (Baum 2012). Working in the tourism industry also offers the opportunity to experience a new country and culture by combining working with travelling. Uriely (2001) presents a typology of working tourists and travelling workers, who differ with regard to the emphasis that they put on either work-related or touristic pursuits. The combination of tourism and labour results in blurred boundaries of terms such as "tourists" and "workers", "home" and "away" as well as "host" and "guest", especially with regard to short-term migrant workers (Baum 2007; Duncan et al. 2013). This underlines the importance of following the performances of people and how they shape the profile and dynamics of tourism rather than rest on predefined roles of particular groups of people.

Learning and acquiring new skills is also an important source of new experiences for migrant workers. By working in the tourism industry, they can develop and gain new skills, in particular cultural, social as well as linguistic skills (Janta et al. 2011b, 2012; Piso 2014). In their research on Polish workers in the United Kingdom (UK) tourism industry, Janta et al. (2011a) found that the tourism industry arguably has greater potential compared to other industries to positively influence the migrant workers' development of cultural competence as well as their integration due to the social relationships and contacts created within tourism. The Polish workers interacted with British guests and co-workers, as well as co-workers from various countries. These interactions create learning opportunities for migrant workers and their cultural competence can improve. In another paper Janta et al. (2012) shed light on how tourism employment can also positively influence migrant workers' English language skills. Their findings showed that frequent communication with guests was a key aspect that helped migrant workers develop proficiency in English, but those migrant workers who worked without direct customer contact reported that they were also able to improve their English skills by observing and listening to co-workers. The hotels' English-speaking environment proved to be advantageous for migrant workers wanting to improve their English skills (Janta et al. 2012). Migrant workers are often able to enter jobs in the tourism industry without any knowledge of the host language. However, a lack of linguistic skills contributes to social exclusion and limits information that the workers receive. Learning the host country language is therefore considered a key to adjusting to the migrant workers' new environment and lives

(Janta et al. 2012; Piso 2014), see also Linge et al.'s (2020) contribution in this volume.

In addition to learning the host language, tourism employment also opens up the possibility to learn about the host country. Employees engage in conversations with tourists and the topic of conversation is most often the destination which the guests are visiting. Therefore, knowledge about the destination is needed. In their study on migrant workers in hotels in Northern Ireland, Devine et al. (2007) found that certain information about the host country, including information on the local culture and attractions in the nearby area, helped migrant workers settle in Ireland. The knowledge about the host country is thus not only useful at work when communicating with guests, but it is also beneficial for migrant workers in their free time and assists them in settling in.

The recognition that tourism employment offers learning opportunities and assists in developing skills, in particular cultural, social and linguistic skills, challenges the notion that tourism is a low-skill industry. Baum (2007) for example points out that emotional and interpersonal skills are important in the hospitality industry. Soft skills are often neglected in the hierarchies of skills, where technical skills demand is in the foreground. However, soft skills are vital for the provision of quality tourism service (Burns 1997; Baum 2007; Devine et al. 2007; Duncan et al. 2013). Hotel employment requires that employees are able to engage with the guests from various cultural backgrounds, which demands communication skills. It is also beneficial if the employees are able to speak multiple languages. Another skill which employees in the tourism industry should possess is adaptive thinking, as they must often respond to guests' inquiries quickly and efficiently. In addition, employees must be able to manage their emotions when interacting with guests, so that positive emotions are highlighted for the benefit of the guests' experience (Baum 2007; Heimtun 2016). It is therefore important to move beyond the notion that tourism labour requires no skills. Instead, it must be recognised that employment in the tourism industry involves various soft skills (Baum 2007; Devine et al. 2007; Duncan et al. 2013). Although migrant workers may not possess all of these skills before starting work within the tourism industry, the social relations created at work can assist them in developing these skills, as outlined above.

It is important to note that social relations are also often the reason that migrant workers receive work in the tourism industry in the first place (Piso 2014). In their study on migrant workers in hotels in the UK, Baum et al. (2007) found that just over half of the international staff members participating in their study had found employment in a hotel due to recommendations from friends or family. Friends and family can be regarded as a valuable network, assisting the migrants with finding employment and "emotional support, companionship and socialization opportunities in and out of work" (Piso 2014, p. 12).

Overall, tourism work thus offers various possibilities to migrant workers, including the opportunity to improve their economic status, combining working and travelling and developing valuable skill sets. The relational character of the job shapes the experiences that migrant workers engage in, including the opportunities but also the limitations that their work environment generates. In the following, we present what opportunities and limitations are created through employment in three hotels in rural Iceland. Our case study set out to shed light on the experience of migrant workers in Icelandic hotels by inquiring about their motivations for pursuing employment in Icelandic hotels, their interactions with co-workers and guests, their daily tasks and their plans for the future. In addition, the hotel managers of the three hotels were interviewed which provided a managerial perspective. The findings are thus based on a total of thirteen qualitative interviews, of which ten were with migrant workers. The ten workers were aged between 25 and 50 years old and nine of them were female. Five were Polish, two came from Switzerland and the remaining participants came from Croatia, the Czech Republic and the United States. The workers, all of whom had lived in Iceland for at least three months, gave accounts of different experiences, providing an insight into how performing different jobs is shaped by and sustains a particular relational ordering of tourism work.

The Experience of Working in an Icelandic Hotel

Before moving to Iceland, only one of the participants had worked in the tourism industry, so tourism employment was a new professional experience for almost all of them. They expressed various motivations for moving to Iceland and working in tourism. Similar to the findings of other studies, many of the participants wanted to improve their economic status or were in search of new experiences. They all shared an interest in experiencing new countries, in particular Iceland. In addition, most of them reported that they found work at the hotel, because a friend or family member had worked or was still working there. The study therefore confirmed the importance of friends and family networks for securing employment.

The participants worked in different functional areas within the hotels. The majority of them worked in housekeeping (5), whereas the remaining participants worked in the restaurant (1) or the reception (3). One participant would help out where needed. Given the diverse tasks and duties available within a hotel, it did not come as a surprise that the experience of working in an Icelandic hotel varied between occupations. This difference was most evident between housekeepers and receptionists. Receptionists worked as front-line staff and described their work environment as hectic as well as mentally challenging. They engaged in frequent interaction with guests. Housekeepers, however, described that they only had limited guest contact. They worked in the background and they found their tasks to be low-skilled, repetitive and physically challenging.

All of the hotels had a mix of migrant workers and Icelandic workers. Most of the hotel guests were international and Icelandic guests were the exception. It is therefore safe to say that the hotels were characterised by an international environment. The findings clearly show that this international environment was perceived as a key benefit of hotel employment, in particular because it offers the employees a chance to meet and interact with people from various countries. The hotel guests as well as the co-workers had different cultural backgrounds and the hotel employment provided a basis for interaction. Through this interaction, the migrant workers were able to learn about different countries and cultures.

However, there was a clear difference in the ability to interact with people based on the participants' occupation. The relational affordances of the job of a receptionist or a housekeeper are crucial in that regard. As pointed out, receptionists were in direct contact with hotel guests, whereas housekeepers described that they rarely had the chance to speak to hotel guests. In addition, there was also a difference with regards to the communication between co-workers. The majority of Icelandic employees worked in the reception due to their ability to speak Icelandic. The hotel managers valued when receptionists were able to speak to Icelandic guests in their native language, despite the fact that the percentage of Icelandic guests was not high in any of the hotels. In contrast, there were only migrant workers in housekeeping. Therefore, Icelandic was commonly spoken between employees in the reception whereas those working in housekeeping spoke English among themselves, outlining a segregation of workers based on nationality in departments. In the following, we further explore the opportunities and challenges migrant workers experience as they engage in different jobs of a receptionist or a housekeeper.

Receptionists

Migrant workers who were employed as receptionists interacted both with guests as well as co-workers from various countries. When speaking with hotel guests, they would hear about the guests' home countries and often received travel tips, should they decide to travel to the guests' home country. The participants described that the same occurred when they spent time with their co-workers. They would get together and for example prepare meals, introducing each other to culinary traditions of their home countries. The receptionists learned from both co-workers and hotel guests what they saw as first-hand and trustworthy information about other countries. This information and knowledge was valuable to them, considering that many of them had sought out to work in the tourism industry in order to gain cultural experience.

Not only did the migrant workers learn about other countries but working in the hotel provided them with the opportunity to establish international contacts and opened up the possibility of visiting another country in the future. Some of the participants said that they would keep in touch with co-workers who had resigned from their hotel employment in Iceland, which made it possible for them to visit their former co-workers in his or her home country. As one of the participants said:

> It's just good to have the connections around the world. [...] We love travelling around so it's always good to know somebody who actually lives in a place where we want to go.

In addition, interaction with guests would often highlight ties to the migrant workers' home countries. The participants described how hotel guests often showed an interest in their background, asking them where they came from. This was generally perceived as positive by the workers. In addition, some hotel guests were from the same country as a migrant worker, giving both parties the opportunity to communicate in their mother tongue away from home. One of the participants described that when they talked to guests or co-workers from their own home country, they would get "that taste of home". In a largely international environment away from home, it would allow them to connect to their home.

Another advantage of communicating with guests as well as co-workers was that the migrant workers were able to improve their linguistic skills. Due to frequent communication with hotel guests as well as co-workers, they were able to improve their English skills. Those working in the reception were also able to develop Icelandic skills. As mentioned, hotel managers preferred to employ Icelanders as receptionists due to their Icelandic skills, but due to a lack of local labour migrant workers with sufficient English skills but without proficiency in Icelandic would also get employed as receptionist. However, working in a reception proved to be a way for practicing Icelandic. Certain tasks of the receptionists, such as answering Icelandic e-mails or phone calls, entailed that migrant workers encountered Icelandic in their work environment. As many of their co-workers were Icelandic, they would also learn words and phrases in Icelandic from them. The participants described that learning Icelandic was difficult and they found it necessary to take a course in Icelandic in order to learn the language fluently, but their work environment provided them with the opportunity to practice Icelandic.

In addition to Icelandic skills, migrant workers in the reception were also able to learn about Iceland, including its nature and culture. Hotel guests would ask various questions about Iceland and it was necessary that receptionists would be able to answer the questions. One of the participants said:

> I think it's always a good start because you can understand the culture more, I think, if you live in a hotel, because there are a lot of people coming in to experience Iceland so you need to know some stuff about it, at least. It's always good to start from that.

The participants claimed that they learned about Iceland from their co-workers. In their spare time, many of them travelled in Iceland which further helped them to get to know the country. Little by little they acquired knowledge about Iceland, which was not only "the tourism product" sold to hotel guests but also their new environment. What they learned thus proved to be useful both at work as well as in their spare time.

Housekeepers

Similar to the workers in the reception, workers in housekeeping appreciated the international environment of the hotel although it was considerably more limited. They also described that they enjoyed learning about their co-workers' home countries and that building relations with people from all over the world provided the opportunity to travel to new countries in the future. Having co-workers from the same home country as oneself provided "that taste of home". Since the communication between all co-workers took place in English, the participants welcomed the opportunity to improve their English skills. In contrast to those working in the reception, those employed in housekeeping were not required to have good English skills and therefore some of them spoke little or even no English at the time they moved to Iceland. Being able to speak to their manager as well as co-workers in English, provided a motivation to learn English and their work environment similarly provided them with the opportunity to practice on a daily basis.

Whereas the migrant workers in housekeeping had frequent interaction with co-workers, communication with guests was limited. The participants working in housekeeping explained that they rarely spoke to guests due to the nature of their work. One of the housekeepers said:

> I don't really have that much time, and you know also because of my position I don't think I have that much opportunity to talk to them [guests]. Sometimes, but it's just "Hello" and that's pretty much it.

If they engaged with guests, the communication would most often be about practical matters, such as where guests could get additional towels or pillows. The participants felt that the conversations were rather shallow. Unlike those working in the reception, housekeepers would not talk to guests about Iceland and thus they were not required to learn about Iceland in the same way as their co-workers in the reception.

Moreover, the participants in housekeeping did not often encounter the Icelandic language or Icelandic people. They worked together with fellow migrant workers so that communication took place in English. There were no Icelanders working in housekeeping and the daily tasks did not require Icelandic skills. Most of the housekeepers, however, expressed an interest in learning Icelandic. They either had been or were planning to go to an Icelandic course, but their work environment did not provide a base for them to practice what they learned in the course. It was therefore difficult to acquire Icelandic skills. This in turn made it difficult for the migrant workers to develop careers in Iceland. As one of them described:

> I'm trying [to be an engineer] but I don't know Icelandic and that's a problem so no way I can be an engineer in Iceland. […] I was trying. I am here two years, so I am trying every month to get a let's say normal job. Not housekeeping or something. But it's very hard because I'm from abroad. I don't know Icelandic, so it's a big problem.

The participants agreed that Icelandic skills were necessary, if they wanted to find a job other than their current one. Without knowledge of Icelandic, they perceived their employment options in Iceland to be limited.

Discussion

The case study highlights how working in tourism enacts diverse socio-spatial configurations of mobility and immobility depending on which job people have. Receptionists and housekeepers have access to and are able to enact different kinds of networks that affect their opportunities for acquiring skills and develop careers within and beyond the sector. It is clear that regardless of their occupation migrant workers are able to improve their cultural knowledge and skills through interaction with people from various countries. However, the work environment of the reception proves to be an important ground for learning about Iceland as well as for practicing Icelandic. A lack of contact with guests as well as Icelanders restrains opportunities for housekeepers to learn about Iceland's nature, culture and language. This proves to have important implications for the migrant workers' personal as well as professional lives.

Implications for Professional Lives

With regard to their professional lives, migrant workers in the reception were able to develop and improve skills which are valuable for tourism employment. Through interaction with hotel guests and co-workers, the migrant workers were able to increase their cultural skills, their knowledge about Iceland as well as their English and Icelandic skills. These skills and an understanding of Iceland are necessary for communicating with hotel guests, including foreign as well as domestic travellers, and for delivering the tourism product to them. Reception staff must be able to share information about Iceland, its touristic attractions, culture and nature, with guests from various countries and the ability to do so increases the chances for migrant workers to find further employment in the tourism industry. It can therefore be argued that working in a hotel reception offers migrant workers the opportunity to sustain or increase their occupational mobility within the tourism industry.

In addition, their occupational mobility is also supported through the acquisition of Icelandic skills. Icelandic has historically played a prominent role in forming Iceland's identity and migrants in Iceland are encouraged to learn the national language in order to integrate into Icelandic society. In the years before 2007, migrants were able to find diverse employment in Iceland, but during a time of high unemployment following the years after the financial crash of 2008 knowledge of Icelandic became more important than before for finding employment (Skaptadóttir and Innes 2017). On a more general note, Baum (2012, p. 40) argues that "the acquisition of language skills by migrant workers in the hotel industry is seen as crucial to their progression beyond low skills and menial work". Employment in the reception and the consequent possible improvement of Icelandic skills can therefore assist migrant workers with career development beyond the tourism industry, increasing their occupational mobility not only within the tourism sector but also within the larger Icelandic labour market.

In contrast, participants in housekeeping described that they would be able to improve their English skills as well as their cultural skills, but knowledge of Iceland and its language remained very limited. Their work environment did therefore not contribute to a significant improvement of qualifications for tourism employment in Iceland. In addition, their work environment did not actively support them in acquiring Icelandic skills, limiting employment options for them in Iceland in general. Therefore, migrant workers in housekeeping had difficulties pursuing career options other than tourism employment in Iceland.

Finally, the importance of networks manifests itself also in providing migrant workers with tourism employment. This study found that knowing someone, who had lived and worked in Iceland, was often a key to employment in Iceland, in particular since hotel managers valued applicants who were recommended by other hotel employees. This helped migrant workers find a job and prepare for what life in Iceland was going to be like.

Implications for Personal Lives

Similar to Devine et al. (2007), we argue that learning about the host country, in this case Iceland, is not only useful for migrant workers' professional lives but also for their personal lives, in particular their ability to participate in wider Icelandic society. Working in a hotel reception can assist migrant workers in learning about the host country and its language, whereas those working in housekeeping need to take initiative outside of the work environment, should they wish to get to know their host country. By working in a hotel reception and thus learning about Iceland and its language, migrant workers gain a better understanding of their new environment which positively influences their opportunity for settling in.

While there were clear differences with regard to benefits and challenges associated with working in either the reception or housekeeping of a hotel, hotel employment had one clear advantage which was accessible to all migrant workers in a greater or lesser degree dependent on which department they worked in. Being able to meet and talk to people from different countries, guests and/or co-workers, stood out as a key benefit of hotel employment. Hotels are an international environment with culturally diverse individuals staying long-term or short-term. This was perceived as positive by the participants, as it offered them the chance to learn about other countries and cultures. Having co-workers from different countries also opened up the possibility of visiting these co-workers in their home countries. The migrant workers engaged in social networks with co-workers and hotel guests, which they could later utilise for travelling to a new country. In addition, they could improve their English skills which proved useful for international travel and work elsewhere. It can therefore be argued that the networks created through hotel employment have the potential to increase the migrant workers' geographical mobility.

In some cases, tourism employment also provides migrant workers with the chance to perform connections to their home country. This became evident as participants told how encounters with people from their home country, either guests or co-workers, provided them with a taste of home. Migrant workers also became connected to their home country when other guests or co-workers showed an interest in the migrant worker's home country. The importance of social networks thus goes beyond the creation of new travel possibilities, as it also outlines the ties that migrant workers have to their home countries and how they engage with these ties away from home.

Conclusion

Based on a case study of migrant workers at three hotels in rural Iceland we have outlined how the meaning and implications of working in tourism is by no means uniform. Moreover, the relational affordances of different jobs they perform produce different kinds of socio-spatial configuration of mobility and immobility. Migrant workers account for a diverse group of individuals, whose opportunities to make use of, perform and shape their mobilities vary greatly. In the case of hotel employment, it was found that receptionists and housekeepers engage in distinct networks, creating different opportunities and limitations for migrant workers with regard to their occupational and geographical mobility.

Housekeepers experience limitations to their mobility. Hotel employment does not assist them in acquiring knowledge of Iceland's culture, nature and language, at least not in any significant way. This in turn makes it more difficult for them to fully participate in Icelandic society. A lack of Icelandic skills further limits their occupational mobility. In contrast, receptionists actively learn about Iceland which helps them to get to know their new environment. Should they wish to settle down in the country, their work may assist them in doing so. Receptionists can also increase their occupational mobility, both within Iceland in general as well as within the tourism industry. Therefore, receptionists' ability to control and shape their occupational as well as geographical ability is comparably greater.

What housekeepers and receptionists have in common is that they can both utilise their social networks to increase their geographical mobility. Social relations are essential for securing initial employment in Icelandic hotels. Once in Iceland, migrant workers engage with people from various countries and the acquired knowledge of other countries and cultures is of value for being able to travel or move to other destinations.

It is evident that communication and the creation of social relations, which stem not only from interaction with hotel guests but also co-workers, are a key benefit of hotel employment. Employers as well as the overall tourism industry must therefore recognise the importance of creating a culture of learning within hotels. There is a need for understanding the value which is inherent in employing a group of culturally diverse individuals. The diverse backgrounds of migrant workers together with local workers offer the chance to learn from each other, which all workers, employers and the tourism industry can and should use to their advantage. It is, however, important to recognise the relational affordances of different jobs within tourism as it has profound implications for opportunities and challenges migrant workers face.

References

Baum, Tom. 2007. Human resources in tourism: Still waiting for change. *Tourism Management* 28: 1383–1399.

Baum, Tom. 2012. *Migrant Workers in the International Hotel Industry.* Geneva: International Labour Organization.

Baum, Tom, Eli Dutton, Shamim Karimi, Jithendran Kokkranikal, Frances Devine, and Niamh Hearns. 2007. Cultural diversity in hospitality work. *Cross Cultural Management: An International Journal* 14: 229–239.

Bauman, Zygmunt. 2000. *Liquid Modernity.* Cambridge: Polity Press.

Burns, Peter M. 1997. Hard-skills, soft-skills: Undervaluing hospitality's 'service with a smile'. *Progress in Tourism and Hospitality Research* 3: 239–248.

Devine, Frances, Tom Baum, Niamh Hearns, and Adrian Devine. 2007. Cultural diversity in hospitality work: The Northern Ireland experience. *The International Journal of Human Resource Management* 18: 333–349.

Duncan, Tara, David G. Scott, and Tom Baum. 2013. The mobilities of hospitality work: An exploration of issues and debates. *Annals of Tourism Research* 41: 1–19.

Ferðamálastofa. (2019). Heildarfjöldi erlendra gesta 1949–2018. Retrieved from Ferðamálastofa July 22. https://www.ferdamalastofa.is/is/tolur-og-utgafur/fjoldiferdamanna/heildarfjoldi-erlendra-ferdamanna.

Franklin, Adrian. 2003. The tourist syndrome: An interview with Zygmunt Bauman. *Tourist Studies* 3: 205–217. https://doi.org/10.1177/1468797603041632.

Hannam, Kevin, Mimi Sheller, and John Urry. 2006. Editorial: Mobilities, immobilities and moorings. *Mobilities* 1: 1–22.

Heimtun, Bente. 2016. Emotions and affects at work on Northern Lights tours. *Hospitality and Society* 6: 223–241. https://doi.org/10.1386/hosp.6.3.223_1.

Janta, Hania, Lorraine Brown, Peter Lugosi, and Adele Ladkin. 2011a. Migrant relationships and tourism employment. *Annals of Tourism Research* 38: 1322–1343.

Janta, Hania, Adele Ladkin, Lorraine Brown, and Peter Lugosi. 2011b. Employment experience of Polish migrant workers in the UK hospitality sector. *Tourism Management* 32: 1006–1019.

Janta, Hania, Peter Lugosi, Lorraine Brown, and Adele Ladkin. 2012. Migrant networks, language learning and tourism employment. *Tourism Management* 33: 431–439.

Jóhannesson, Gunnar Thór, Carina Ren, and René van der Duim (eds.). 2015. *Tourism Encounters and Controversies: Ontological Politics of Tourism Development*. Farnham: Ashgate.

Joppe, Marion. 2012. Migrant workers: Challenges and opportunities in addressing tourism labour shortages. *Tourism Management* 33: 662–671.

Ladkin, Adele. 2011. Exploring tourism labor. *Annals of Tourism Research* 38: 1135–1155.

Latour, Bruno. 2005. *Reassembling the Social: An Introduction to Actor-Network-Theory*. Oxford: Oxford University Press.

Law, John. 2004. *After Method: Mess in Social Science Research*. London and New York: Routledge.

Linge, T. T., T. Furunes, T. Baum, and T. Duncan. (2020). Hospitality through Hospitableness: Offering a Welcome to Migrants through Employment in the Hospitality Industry. In *Tourism Employment in Nordic Countries: Trends, Practices and Opportunities*, ed. A. Walmsley, K. Åberg, P. Blinnikka, and G. T. Jóhannesson. Cham: Palgrave.

Law, John, and John Urry. 2004. Enacting the social. *Economy and Society* 33: 390–410.

Piso, Annemarie. 2014. Migrant labour in rural tourism: Continuity and change. *International Journal of Tourism Research* 18: 10–18.

Ren, Carina, Gunnar Thór Jóhannesson, and René van der Duim (eds.). 2018. *Co-creating Tourism Research: Towards Collaborative Ways of Knowing*. London: Routledge.

Sheller, Mimi, and John Urry. 2006. The new mobilities paradigm. *Environment and Planning A* 38: 207–226.

Skaptadóttir, Unnur Dís, and Pamela Innes. 2017. Immigrant experiences of learning Icelandic and connecting with the speaking community. *Nordic Journal of Migration Research* 7: 20–27.

Skaptadóttir, Unnur Dís, and Kristín Loftsdóttir. 2019. Sérhefti Íslenskur vinnumarkaður - erlent starfsfólk. Special issue, *Íslenska þjóðfélagið* 10: 3–13.

Statistics Iceland. n.d.a. Register based employment by economic activity, years, sex, age group and origin 2008–2018. https://px.hagstofa.is/pxis/pxweb/is/Samfelag/Samfelag__vinnumarkadur__vinnuaflskraargogn/VIN10020.px. Accessed on July 5, 2019.

Statistics Iceland. n.d.b. Register based employment by years, sex, age and origin 2005–2018. https://px.hagstofa.is/pxen/pxweb/en/Samfelag/Samfelag__vinnumarkadur__vinnuaflskraargogn/VIN10003.px. Accessed on October 1, 2019.

UNWTO. 2018. *UNWTO Tourism Highlights*, 2018th ed. Madrid: UNWTO.

Uriely, Natan. 2001. 'Travelling workers' and 'working tourists': Variations across the interaction between work and tourism. *International Journal of Tourism Research* 3: 1–8.

Urry, John. 2000. *Sociology Beyond Societies: Mobilities for the Twenty-first Century*. London: Routledge.

van der Duim, René, Carina Ren, and Gunnar Thór Jóhannesson (eds.). 2012. *Actor-Network Theory and Tourism: Ordering, Materiality and Multiplicity*. London and New York: Routledge.

Wojtynska, Anna. 2012. Polish workers in the capital area of Iceland. In *Rannsóknir í félagsvísindum XIII: félags- og mannvísindadeild*, ed. Sveinn Eggertsson and Ása G. Ásgeirsdóttir, 1–12. Reykjavík: Félagsvísindastofnun Háskóla Íslands.

8

"You Need to Consider How It Looks in the Eyes of the Guest": The Work of Teenage Girls in Tourism in Iceland

Anna Vilborg Einarsdóttir and Laufey Haraldsdóttir

Introduction

In June 2019 the number of tourists in Iceland reached 2.3 million (Icelandic Tourist Board 2019a), in a country with a population of just 357,000 people (Statistics Iceland 2019a). Although the number of tourists outside high season has increased from 2010, (Icelandic Tourist Board 2019b) the tourism industry is impacted by seasonality that affects the hospitality industry leading to seasonal employment and underemployment (Jolliffe and Farnsworth 2003). Especially, areas outside of the capital city, Reykjavík, are economically dependent on tourism in the summer. Tourism is an industry that offers various jobs and skills. As a service industry, tourism relies on human resources, often a small number of permanent staff with reliance on peripheral labour

A. V. Einarsdóttir (✉) · L. Haraldsdóttir
Department of Rural Tourism, Hólar University, Hólar, Iceland
e-mail: annavilborg@holar.is

L. Haraldsdóttir
e-mail: laufey@holar.is

A. Walmsley et al. (eds.), *Tourism Employment in Nordic Countries*,
https://doi.org/10.1007/978-3-030-47813-1_8

(Zopiatis et al. 2014). Much of the work is low-skilled, even though the employees reflect job satisfaction (Walmsley 2015, 2017). Jobs in tourism in Iceland have increased considerably with the average number of total employees in the key tourism sectors estimated at 29,300 in April–March 2018–2019 (Statistics Iceland 2019b). The industry has responded to the situation by, e.g. recruiting adolescents as well as foreign workers (Statistics Iceland 2019c) and according to Walmsley (2015) the tourism industry depends on and is characterised by high level of youth employment, with the financial crisis resulting in youth becoming a new social class.

Employees under 30 years seem to sustain the tourism industry (Fig. 8.1), with adolescents contributing to just over 15% of the total number in 2016–2017.

In this study teenage girls between the ages of 16 and 17 are the focus. The aim of the research is to get a clearer picture of the circumstances of this age group in the tourism sector in Iceland. To facilitate this, we investigated the following: (a) the significance of youth employment in tourism in the socio-economic context of Iceland, (b) the girls' perspectives on working in the tourism sector and (c) their job satisfaction.

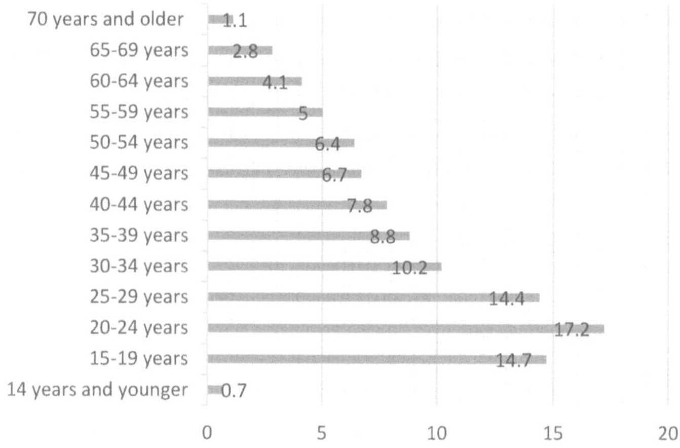

Fig. 8.1 Percentages of employees in tourism by age in 2016–2017

Literature Review and Theoretical Framework

Official information about the workforce in tourism in Iceland is scarce (Halldórsdóttir 2019) and research on the subject mainly focuses on workers over 18 years old. This study is the first one which focuses on the work of teenage girls in tourism.

Child and Adolescents' Environment and Contribution to the Labour Market in Iceland

Child and adolescent labour has historically been viewed in a positive way in Iceland (Garðarsdóttir 2009). Although expansion of education and legislation on childhood work in the latter half of the twentieth century (Garðarsdóttir 1997), along with urbanization in western societies, has prohibited or reduced children's work, an early start on the labour market is still valued in a positive way. In this subchapter we shed some light on the legal background of, as well the share of teenage girls within the tourism workforce as an illustrative background for the interview results.

The Convention on the rights of the child (United Nations 2019) state that a child means every human being below the age of 18 years as do the Statistics Iceland (2018a). Around 80% of 17-year-old adolescents in Iceland have already gained experience in the labour market, mostly during the summer months. Analysis of children's participation in the Icelandic labour market indicates that children, to some extent, start working before 13 years of age. Examples of that, are 10–11 years old working in guesthouses and restaurants Statistics Iceland (2018a).

In the Icelandic *Regulation on the work of children and adolescents no. 426/1999* 4th article the following is stressed:

> When considering the type and organization of work for youth under the age of 18, emphasis should be placed on ensuring safety as well as mental and physical health of young people, and that work does not interfere with their education or development. (Ministry of Social Affairs 2019)

According to *the Act on Working Environment, Health and Safety at work no. 46* (Althingi 2019), children under 18 years of age are entitled to 12–14 hours rest every 24 hours. Despite of that it is known that children and adolescents in Iceland work long hours and take responsibilities beyond their age and maturity. The Administration of Occupational Safety and Health (2019) oversees the above-mentioned laws and considers reports of possible violations. The role of the Ombudsman for Children (2019) is to "further the wellbeing of children and to look after their interests, rights and vis-à-vis public as well as private parties in all walks of life".

According to a study from 2018 a quarter of all children in 8, 9 and 10th class, that are 13–15 years old are in paid work during the school year, more common in rural areas (Icelandic Centre for Social Research and Analysis 2018) and not unreasonable to assume that a large part of this group works in tourism-related businesses.

Seasonality suits many adolescents as the peak summer season coincides with school leave. Figure 8.2 shows teenage girls contribution to tourism. Their work is temporary, both part time and full time jobs.

Figure 8.3 shows the teenage girls' diverse types of jobs. Most of their work is in restaurants and lodging. Other jobs are within recreation, museums and tourism offices.

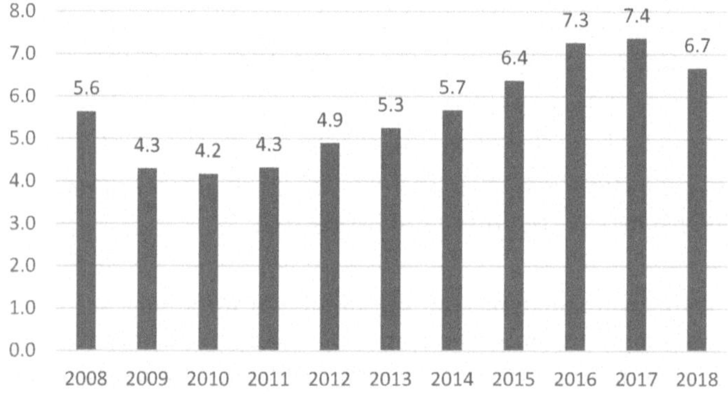

Fig. 8.2 Percentage of 13–17 year old girls in the tourism and hospitality industry 2008–2018 (*Source* Statistics Iceland 2019d)

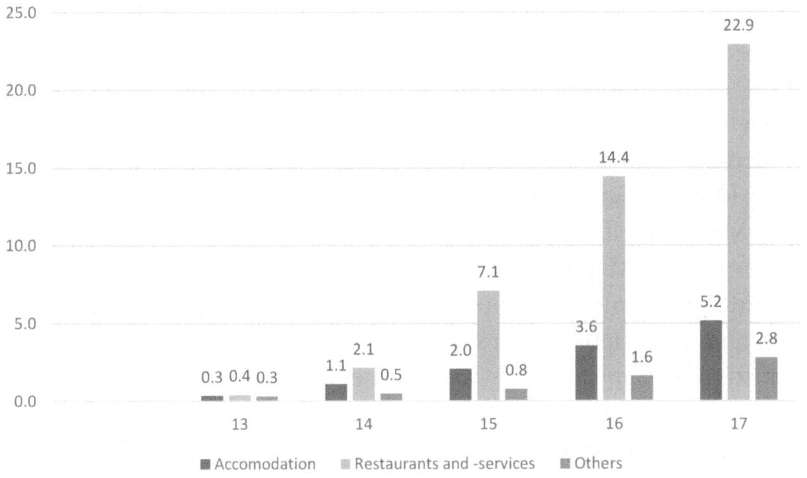

Fig. 8.3 Percentage of girls 13–17 years old in diverse occupations within the tourism industry 2008–2018 (*Source* Statistics Iceland 2019e)

The teenage girls, participating in this study, were first employed in the tourism industry at the age of 13–15. Whether defined as children or adolescent, it is obvious from the data above that girls under 18 years old make a significant contribution to the tourism industry in the socio-economic context of Iceland. The fact that they are paid less than adults (Efling 2019) means that they impact the economic situation of the tourism firms and even though they may lack experience, the majority of their work within the firms are low-skilled, but still highly demanding at least during high season.

Teenage Girls as Human Resources in the Tourism Industry

Above we have gathered data on child and adolescent contribution, as well as their legal environment on the Icelandic labour market. The findings from the empirical part of this study, enlighten how the teenage girls are recruited, trained, supervised, valued and rewarded with analytical perspective on human resource management and job satisfaction theories.

One of the most important assets of tourism and hospitality are human resources. It's considered critical for service quality, customer satisfaction, loyalty, competitive advantage and organizational performance (Nadiri and Tanova 2010; Kusluvan et al. 2010; Sigurðardóttir 2011; Madera et al. 2017; Neshybová 2019). Nevertheless, studies show that formal HRM practices in tourism and hospitality are rare. Even unprofessional, underdeveloped and inferior in bigger tourism businesses and not practiced in a way that generates employee commitment, satisfaction and motivation (Kusluvan et al. 2010). Neshybová (2019) argues that HRM practices in Icelandic hotels are taking their first steps regarding front-desk employees. Madeira et al. (2017) believe HRM practices have a positive relationship with the employee's feelings of job satisfaction and influence motivation to better performance. Managers must be aware of the condition of the employees, the importance of human needs that affect the behaviour and attitudes of employees and what factors will continue to inspire and satisfy them, at least teenage girls (Grétarsdóttir 2013).

Adolescents, possibly starting a career in tourism may be motivated to accept a low paid job with strange working hours. Maslow's (Maslow 1954; Kay 2003) key conclusion, however, was that only through our needs can motivation be handled on purpose because people try to satisfy their different and basic needs at work. Safe working environment is important, as well as social needs; to feel kindness and compassion at work. Herzberg's et al. (1993) motivator and hygiene theory focuses on two influencing factors (a) The motivation factors reflect internal incentives related to the nature of the job and its challenges lead to job satisfaction, e.g. how interesting the job is, the individual's independency at work, responsibility, opportunity for personal development and maturity, being valued and acknowledged. (b) The hygiene factors are external incentives who do not directly affect job satisfaction but are linked to the working environment, wages, job security, communication in the workplace and training. Herzberg stressed how work should be designed and redesigned to incorporate internal incentives.

Training and Work Performance

The teenage girls' work in the tourism and hospitality industry is mainly the tangible and intangible services performed on site. The intangible services are produced and consumed simultaneously through personal interactions between the service provider and the customer (Nadiri and Tanova 2010). Employee personality is gaining attention as a selection criterion for tourism and hospitality businesses because of its role in employee performance. Services become tangible in the personality, appearance, attitudes and behaviour of the service provider (Kusluvan et al. 2010). They become part of the product, represent the tourism businesses and take part in forming its image (Bitner et al. 1990; Nadiri and Tanova 2010; Robinson et al. 2019). Due to the nature of the work, comprehensive training of employees is extremely important.

The Road Map of Tourism in Iceland (Ministry of Industry and Innovation and the Icelandic Travel Industry Association 2015) reflects increased emphasis on employees' qualification and training. It is reasoned that training will result in higher overall quality of job performances in the tourism industry. The management of human resources gets fairly good coverage, and issues to attract qualified employees are addressed. Employees under 18 years, however, are not mentioned, in spite of their obvious and essential contribution to the workforce in the industry.

The girls first real job is a challenge. They expect fair treatment and are willing to perform their duties well. Vroom's theory (Vroom 1964; Kay 2003; Chiang and Jang 2008) based on expectancy, instrumentality and valence asserts that motivation is high when employees believe that if they put much effort into the job it will be rewarded.

The teenage girls, as front-line employees, communicate in a foreign language with tourists and even their colleagues. They are in a role of grown ups and bear a big responsibility because the success of the business depends on the quality and performance of its employees. The girls are required to look, sound and provide service that is compatible with the customers expectations i.e. from diverse cultures and increasingly demanding customers.

Studies show that personality traits of service employees are key variables in predicting service quality. Extrovert people represent self-confidence, conscientious employees work hard and the open-minded are described as broad-minded and imaginative (Cheng 2011). According to Kusluvan et al. (2010) service-oriented people are helpful, dependable and cooperative, have social skills and willingness to follow rules. Service employees are required to manage their emotions for the benefit of customers. They shall show enjoyment at work even though their working conditions are not conducive. Curbing emotions requires skills and training in order to be maintained (Baum 2007).

Still the employees often get minimum training (Jolliffe and Farnsworth 2003; Kusluvan et al. 2010). Usually a colleague with short experience trains the new employee, yet the need for clear information about the job is crucial (Chiang and Jang 2008). Supervisors, motivating young employees, should be trained to show them goodwill and collaboration (Gibson et al. 2013). Better knowledge creates higher motivation, better performance and more likely job satisfaction.

Job Satisfaction

Job satisfaction is a prominent concept in studies about workforce in general. Aziri (2011) states that job satisfaction is one of the most important challenges facing managers as job satisfaction represents a combination of positive or negative feelings that the employees have towards their work and it can be considered as one of the main factors when it comes to efficiency and effectiveness of firm. He states that strong connection is between job satisfaction and performance as job satisfaction influences employees' commitment, affects quality, productivity and profitability (Chiang and Jang 2008; Aziri 2011; Lillo-Bañuls et al. 2018; Neshybová 2019). It seems obvious that job satisfaction has to do with feelings, and as Spector (1997) argues, to some extent, job satisfaction reflects good treatment of employees, as well as how people feel about their job and their perspectives towards it. The employees' sense of achievement and success at work is commonly perceived to be directly linked to personal well-being and productivity. Job satisfaction implies doing a job one enjoys, doing it well and being rewarded for one's efforts (Aziri 2011). Job satisfaction is significantly lower among tourism employees and differences exist between the factors influencing job satisfaction compared with the service sector (Lillo-Bañuls et al. 2018). Poor working conditions are still frequently reported despite claims of the importance of employees to organizational success (Walmsley 2017). Zopiatis et al. (2014) argue that organizations concerned about employee commitment should stress information sharing, decision-making and empowerment initiatives which can win "hearts and minds". Job satisfaction has particularly challenged the hotel industry (Chiang and Jang 2008; Neshybová 2019; Zopiatis et al. 2014). Table 8.1 shows some determinants that influence job satisfaction, according to Lillo-Bañuls et al. (2018). The bold ones of those listed were mentioned by the interviewees (see methodology below).

According to Grétarsdóttir (2013) young employees, generally, seem to be very satisfied with their opportunities, motivation and challenges they get at work. Where and how employees find meaning in work is fundamental to how they approach and experience their work and workplaces (Syukrina et al. 2014). Some employees choose demanding and responsible jobs while others don't. The psychological needs of people determine whether people make use of all the incentives that the job offers. When employees experience undemanding jobs as tiresome they often develop it further with new solutions both in terms of performance and quality of service. In order to make jobs interesting, inspiring and increase job satisfaction Hackman and Oldham (1976, 1980) stress that job characteristics need to involve autonomy, feedback,

Table 8.1 Determinants of job satisfaction. Lillo-Bañuls et al. (2018)

Part time job	**Wages**	Job development
Temporary job	Workplace image	Education
Shift systems	Management	Over-qualification
Weekend work	**Next manager**	Number of years at job
Working schedule	**Colleagues**	**Challenging work**
Hourly pay	Willingness to cooperation	**Monotonous job**
Lower employee turnover	Job and job conditions	**Physically hard work**

diverse skills, task identity and significance. Employees that feel a sense of belonging, experience their work important and meaningful, their feeling of responsibility towards the job increases (Hackman and Oldham 1976, 1980; Grétarsdóttir 2013; Madera et al. 2017; Zopiatis et al. 2014; Neshybová 2019).

Methodology

The research was conducted during the period of October 2018 to July 2019. As the aim of the study is to increase an understanding rather than to generalize and the subjects relatively few in number a qualitative approach was used to examine issues and attitudes related to teenage girls working in tourism in Iceland. The study is based on in depth, informal, semi-structured, interviews (Creswell and Poth 2017) where the respondents can themselves indicate areas of satisfaction or dissatisfaction (Spector 1997).

Based on the theoretical framework a checklist was designed. Interviews were conducted with eleven girls. The interviewees were chosen via snowball sampling (Creswell and Poth 2017). The researchers contacted friends and colleagues who suggested potential interviewees. The girls received a text message with short information about the research, an offer to participate, and an information that their parents had to agree to the participation and sign an informed consent because of the interviewees young age.

The interviewees were mainly from the Northern and Southern parts of the country, one in Reykjavik. All have some experience in tourism companies, housekeeping in hotels and service jobs in cafes and restaurants. Interviews, that varied between 25 and 55 minutes, were conducted face-to-face or took place through the online platform Zoom. The interviews were recorded and transcribed verbatim and content analysis used to identify themes. Although 11 interviews might seem a small sample, the researchers identified a certain saturation already before finishing 10 interviews, with a high consistency between the interviewees.

Information and official documents about this group were gathered in the government sector as well as offices responsible for their working conditions and rights. Legislation on labour and official statistics on the contribution of children and youth to the employment market were examined as well.

Findings

Challenges When Beginning Work

According to the interviewees, the job opportunities for their age group are mostly limited to the tourism sector or the municipalities' youth employment programs (Vinnuskóli Reykjavíkur 2019) and according to them the chances of getting a job within tourism in rural areas is good. The girls don't have any formal or informal qualifications in the field, but they all emphasize that they have good skills in the English language, which is partly due to their jobs in tourism.

> My English has improved, co-workers and the guests talk English. One becomes more independent and secure. (Interviewee 4)

The first challenge the girls meet is applying for a job. In most cases they approached the tourism businesses to ask for a job, or their parents did. No formal application form was needed in most cases, but in some the girls needed to send in their CV. Most of the girls attended a face-to-face interview, while some were conducted on the telephone. One girl was asked to do a trial, where she should try out the work she was supposed to do in the job.

The findings suggest that, when it comes to this age group, employment-contracts are not the rule but rather an exception. In one case the interviewee did not know what a contract of employment was at all, in spite of having worked in the same company for two years. One of the girls, with some experience in the field, took the initiative to ask for a contract in her job interview and the employer then pulled it out from a drawer in his desk.

The first day at work was a challenge for most of the girls. Their line-manager usually welcomed them and the reception was informal. They were generally introduced to their co-workers on their first day at work. One of the girls commented that she (had first) got to know the other employees at a staff party.

A handbook for employees was available at two of the places the interviewees worked at. In both cases the handbook needed an update as the information no longer applied to the present situation in the firms. Some of the girls had never heard of an employee handbook and reacted with „Uh…what is that "? (Interviewee 7).

None of the girls had a written job-description and no formal courses were offered for new staff at their working places. In most cases, an existing staff member introduced the work procedure for the girls.

> I started in March and then it wasn´t very busy. A girl who was at work when I started taught me everything. (Interviewee 4)

Most of the girls were positive about the first day at work while some had negative experiences that day, mostly because they were not taught or trained sufficiently. Those who didn't get sufficient training also reported a negative interaction with their boss already on the first day.

> A girl younger than I was told to teach me. Soon I got criticism for what I was doing. It was awful, being scolded in my first job for something I didn't know. I saw that I wasn't properly taught. In another place I was taught, in detail, how to take off the bed and fold. (Interviewee 2)

All of the girls had experienced challenging customers. None of them were taught how to deal with this type of customer but were given some guidelines on how to behave towards customers in general.

> I was told how to talk to the customers. I was told to always take the initiative to greet them and to be polite. […] We weren't taught how to handle our difficult customers, we just had to manage that ourselves. (Interviewee 2)

Remuneration and Benefits

The interviewees are in general satisfied with their wages, provided that they get paid on time and for the hours they work. All the girls get paid by the hour and according to an agreed rate. Some of them keep a record of their working hours themselves, others don't. In general, the girls trust that the wages are correct, even though they don't always understand the payslip. One of the girls was an exception and didn't trust her employee.

> I was in the highest salary grade there, but still it was so low. See... the payslip never showed that I was working full time, even though I worked 100% work. [...] It happened very often that they didn´t count all the hours I worked. [...] I never got paid for two days when I was sick once. I didn´t know until a year into that job, that I was supposed to get those paid. (Interviewee 2)

Most of the girls are aware that their employers are obligated to pay to their union, but they knew less about their pension funds.

> I think it is paid to the union, but I don´t know about the pension fund. He must do it. [...] most of those that work there are not thinking about this. (Interviewee 1)

Perks, if any, include free meals when working and working clothes. Some mentioned not being provided with enough working clothes, while others got more than enough. All except one interviewee had to wash their working clothes themselves, labelled or not.

Working Environment

In general, the working environment seems to be good. The girls enjoy having real work and overall seem to like what they are doing. However, the working days can be long and the number of employees on a shift insufficient. Long hours, mostly running around or in a standing position can also be challenging.

> Now I am mostly doing 12 hour shifts which is rather long. I look forward to take 8 hour shifts again. (Interviewee 6)

If a staff member is absent, the others share the tasks and responsibilities. To call for an extra help doesn't seem to be the rule. All the girls expressed commitment to and independence in their work.

> When I sense that things are going out of control, I often take charge and tell the others what they need to do, so that we can finish on time. (Interviewee 3)

Communication and team spirit in the working place appear to be the most important factor when it comes to the girls' attitude towards work and job satisfaction. Their line-manager and co-workers are the people they communicate mostly with and they play a big role in the girls' everyday work conditions. A couple of interviewees reported that language can be a challenge when it comes to work culture;

> The atmosphere in the working place was a bit uncomfortable, many couldn´t speak much English and needed to use manual communication. Only Polish was spoken in the breaks and then I sit alone. (Interviewee 5)

Being young and in some cases doing their first proper job, a praise for a good job done is important for our interviewees. They never received direct rewards for doing a great job, but the interviews show that their employers did sometimes acknowledge a job well done. The girls appreciated that.

> My boss always thanked us for the day, for good collaboration and that we bothered to take so many shifts. (Interviewee 3)

They liked being trusted and to take increased responsibility;

> I started recently and have been asked to lock up and to check if everything is alright. (Interviewee 1)

Nevertheless, compliment and encouragement seem to be rather rare. One of the girls explained that she would get compliment occasionally from her bosses (Interviewee 6). Responding to a question about getting compliments for a good job, another one replied;

> No, I don´t recall that. The tourists gave us compliments. Once the whole group got a credit [from the boss] for responding quickly when a big group came unexpected. (Interviewee 4)

The superior's attitude isn't always ideal or likely to result in a higher performance among the young girls;

> [We had a] high turnover rate because of the shift arrangement. The supervisor criticised everything, never gave a compliment. That was hard. (Interviewee 2)

The young employees could also get undeserved reprehensions.

> Two of us were doing the housekeeping and sometimes I got blamed for poor cleaning. One day we were asked to do one room each without us knowing what was happening. On inspection it turned out that it wasn't always me that didn´t clean as expected. It was good to know and then I was taught to do it right. (Interviewee 2)

According to this, the interviewee didn't get sufficient training before starting. Only when it turned out that she wasn't doing it well enough, she was taught how to.

The Best and Worst Aspect of Working in Tourism

In general, the interviewees like their jobs, they look forward to the working day and feel they were listened to by their employers. The satisfaction of their customers also mattered when it came to their own attitude towards their job.

Most of the girls considered having good co-worker a major impact on whether they stayed at the job or not, and most think it's best to work with people their own age:

> Most [co-workers] are outgoing, at a similar age, good music at work. Nice work. [I] feel good at work and sense that I am counted upon. (Interviewee 1)

Good training in one place can be useful in another and some of the girls mentioned that getting sufficient training in English was one of the best things about it:

> I have improved my skills in English. I have learned how to approach the foreigners. It was emphasised to us - the staff, that we were in charge and should not allow customers to bully us, and that we should act in a polite manner. (Interviewee 3)

Our interviewees reported that the work could also be boring, repetitious and lacking variability. Washing up, cleaning and housekeeping, difficult customers and challenging bosses are mentioned as the worst things about working in the tourism industry.

> To be stressed out because of a difficult boss…I would get stressed as soon as she arrived. In another place I soon figured the boss out and learned how to be around him. (Interviewee 2)

Discussion

Continued growth in global tourism requires a growing workforce, and as this study shows, the work of young people and children are of significant importance for the industry in Iceland. There is a strong legislation on youngsters' rights, regarding working hours and conditions, even though evidence shows that those rules can be disregarded (see interviewee 2, 3 and 6) and only few of the girls seem to have any knowledge about their rights as employees. Apart from one, all of the girls in the study, worked without a job contract, the job descriptions were in many cases unclear and even though exceptionally, even the wages. This shows that the girls' recruitment is not treated formally and only in exceptional cases they are informed about their rights by their employees. This raises questions about their working conditions, as well as whether their work contributions are taken seriously, despite of their decisive share on the labour market.

Aligned with the above-mentioned positive public attitude towards child and adolescent work, most of the participants in this study started working as early as 13 or 14 years old, largely in tourism. The first paid job is for most youngsters important and memorable. The interviewees were largely positive towards their jobs in tourism and committed to it. However, none of them saw tourism definite as their future career. That could of course be explained with their young age and that most of them hadn't decided much about their future jobs. Nevertheless, it is an indication of a lack of interest towards that field of profession, something that the tourism sector in Iceland should take seriously and is worth further investigations.

Bitner et al. (1990) as well as Aziri (2011) stress the importance of employees in the context of companies' productivity and images. This applies not least to service-driven industry such as tourism, where productivity depends largely on good utilization of workforce (Sigurðardóttir 2011) and the image on successful communications between the employees and the guests, contributing to a positive experience of the customers. In rural Iceland, tourism companies' productivity depends obviously also on the condition of infrastructure such as the transport system. The infrastructure influences seasonality, which again affects both access to competence, and utilization of staff in the sector. It is in this context that young people's work becomes most relevant because they are willing to work on weekends and nights, and the peak season in tourism coincides with the school leave.

For a positive image, it is important that staff are properly trained, self-confident and motivated to do well. Like Kusluvan et al. (2010) and Joliffe and Farnsworth (2003) point out, lack of supervision and training is often associated with seasonality and high turnover of employees in the tourism industry. However, lack of training, knowledge and skills among employers can affect the level of effectiveness and productivity of the firms (Sigurðardóttir 2007; Mulyanto et al. 2018). Additionally, seasonality does not exempt employers from the responsibility to follow Icelandic law, which requires them to look after the welfare of children and adolescents they employ.

The adolescent girls interviewed in this study, often lacked clear instructions and support from their employers and this affects their job satisfaction, motivations and performance. In certain instances, the girls had responsibilities that seemed beyond their maturity, while in others their jobs were repetitious and "boring" and not likely to provoke interest or ambition. Vroom (1964) argues that to encourage motivation, staff needs to be rewarded or at least get some acknowledgement for a job well done. Employees also need to be able to make the most of their abilities in order to be satisfied in their jobs (Chiang and Jang 2008). Allocating work and responsibility according to age and experience, as well as alternating monotonous jobs, would be one way to address these issues.

In Table 8.1 above, Lillo-Bañuls et al. (2018) illustrate factors influencing job satisfaction. Although the majority of the girls in the study say that they are satisfied in their jobs, many of the issues in Lillo-Bañuls et al.'s (2018) paper were reported as problematic in the interviews. This includes long shifts, physical hard work, requirements beyond their capacity, lack of training, challenging costumers and difficult line-managers. According to the girls they often have to find ways to manage those things themselves, or they learned by experience. Other factors listed in Lillo-Bañuls et al.'s (2018) paper were reported to add significantly to our interviewees' job satisfaction. Of those, the most important ones were good co-workers at similar age, good atmosphere in the workplace and supportive managers.

The core features in Hackman and Oldham (1980) job characteristics model also appeared in the girls' narratives. When they experience their work as important and meaningful, their feeling of responsibility towards the job increases. They become "part of the product" (Bitner et al. 1990) and have a stronger sense of themselves as representatives for their region and country. Making the guest feel welcome becomes important to them when they know that they are being trusted, when they feel motivated by the various tasks at work and when they get the chance to develop their skills and self-confidence.

Like this study shows, as well as former research in the context of Icelandic tourism and hospitality (Sigurdardottir 2007, 2011; Neshybová 2019), HRM practices are quite rare, despite of obvious necessity, not the least for youngsters working in the sector. Without reducing the importance of mentorship from fellow workers and peers, managers have an important role and it is significant that children and adolescents are not treated and managed as adult employees. We agree with Gibson et al. (2013) which points out that managers and supervisors of teenagers should be trained, to motivate them and better understand their needs, as that will affect the behaviour and attitudes of the young employees.

Conclusion

This research sheds light on the significant part teenage girls play in the socio-economic context of tourism in Iceland. Although the employment of youth is widely recognized in tourism (e.g. Walmsley 2015), very few studies have actually listened to the youth's voice or focused on their special needs as employees. It is, we argue, urgent to pay more attention to this group and include them in authorities' information and data.

It also seems obvious that the industry as a whole needs to pay more attention to this age group among their employees. They need more supervision and support appropriate for their age. We propose a recommendation, of a procedure within the industry, which manages better the start of employment, of children and adolescents. This needs to include informing them about their rights and obligations, as well as salary and terms of employment, and not the least to meet their needs for training and support. Training supervisors in managing teenagers in order to strengthen their working skills should be part of the procedure. With this approach, the industry will gain secure and self-confident young employees from the first day at work.

The aspects touched up on in this chapter are issues of challenges regarding competence, labour shortages and high employee turnover rates in the tourism sector. Acknowledging this group of workers, both in official numbers and by the industry, can be a tool for providing a future workforce in tourism. Good training and support are more likely to motivate an interest, build self-confidence and ambitions towards the job, than learning by experience—often the hard way. It is also more likely that the firms will retain those employees after the initial period, and ultimately, that the young people will see opportunities for future career in the tourism and hospitality sector.

References

Administration of Occupational Safety and Health. 2019 Vinna barna og unglinga [Childrens and adolescents work]. https://www.vinnueftirlit.is/vinnuvernd/vinna-barna-og-unglinga. Accessed on July 15, 2019.

Althingi. 2019. Lög um aðbúnað, hollustuhætti og öryggi á vinnustöðum. nr. 46/1980 [Act on Working Environment, Health and Safety in Workplaces, no. 46/1980]. Last modified September 20. https://www.althingi.is/lagas/122a/1980046.html. Accessed on June 15, 2019.

Aziri, Brikend. 2011. Job satisfaction: A literature review. *Management Research and Practice* 3 (4): 77–86. http://mrp.ase.ro/no34/f7.pdf. Accessed on June 3, 2019.

Baum, Tom. 2007. Human resources in tourism: Still waiting for change. *Tourism Management* 28: 1383–1399. https://doi.org/10.1016/j.tourman.2007.04.005. Accessed on September 19, 2019.

Bitner, Mary Jo, Bernhard H. Booms, and Mary Stanfield Tetreault. 1990. The service encounter: Diagnosing favorable and unfavorable incidents. *Journal of Marketing* 54: 71–84. https://www.ida.liu.se/~steho87/und/htdd01/9602131056.pdf. Accessed on June 19, 2019.

Cheng, Chun-Hao. 2011. A study of critical factors of the job involvement for hotel employees—Taking hotels in Taiwan as an example. *International Journal of Hospitality Management* 30: 990–996. https://doi.org/10.1016/j.ijhm.2011.03.001. Accessed on September 13, 2019.

Chiang, Chun-Fang, and SooCheong (Shawn) Jang. 2008. A expectancy theory model for hotel employee motivation. *International Journal of Hospitality Management* 27: 313–322. https://doi.org/10.1016/j.ijhm.2007.07.017. Accessed on June 19, 2019.

Creswell, John W., and Cheryl N. Poth. 2017. *Qualitative Inquiry and Research Design: Choosing Among Five Traditions*. Thousand Oaks: Sage.

Efling. 2019. Laun ungmenna skv. samningi Eflingar og SA [Wages of young people according to the agreement between Efling and SA (Confederation of Icelandic Employers)]. https://efling.is/laun-ungmenna-skv-samningi-eflingar-og-sa/?lang=en. Accessed on November 15, 2019.

Garðarsdóttir, Ólöf. 1997. Working children in urban Iceland 1930–1990: Ideology of work, work-schools and gender relations in modern Iceland. In *Industrious Children: Work and Childhood in the Nordic Countries 1850–1990*, ed. Ning de Coninck Smith and Bengt Sandin, 160–185. Odense: Odense University Press.

Garðarsdóttir, Ólöf. 2009. Working or going to school: Childhood experiences in post-war Reykjavík. *Barn* 27 (3–4): 173–185.

Gibson, Jane Whitney, Wei Chen, Erin Henry, John Humphrey, and Yunshan Lian. 2013. Mary Parker Follett through the lens of critical biography. *Journal of Management History* 19: 441–458. https://doi.org/10.1108/jmh-05-2012-0044. Accessed on May 8, 2019.

Grétarsdóttir, Vilborg. 2013. Áhrifaþættir starfsánægju hjá starfsfólki Hjallastefnunnar ehf [Effective factors of job satisfaction among the employees of Hjallastefnan ehf.]. MS thesis, University of Iceland.

Hackman, Richard J., and Greg R. Oldham. 1976. Motivation through the design of work: Test of a theory. *Organizational Behavior and Human Performance* 16 (2): 250–279. https://doi.org/10.1016/0030-5073(76)90016-7. Accessed on May 8, 2019.

Hackman, Richard J., and Greg R. Oldham. 1980. *Work Redesign*. Reading, MA: Addison-Wesley.

Halldórsdóttir, Íris Hrund. 2019. Workforce in tourism: What shall be examined, by what means and how is the situation in Iceland? http://www.rmf.is/is/um-rmf/frettir/vinnuafl-i-ferdathjonustu. Accessed on July 15, 2019.

Herzberg, Frederick, Bernard Mausner, and Barbara Bloch Snydermann. 1993. *The Motivation to Work*. New Brunswick: Transactions Publishers.

Icelandic Centre for Social Research and Analysis. 2018. *Ungt fólk 2018. Lýðheilsa ungs fólks í Reykjavík. Niðurstöður rannsókna meðal nemenda í 8., 9. og 10. bekk árið 2018* [Young people 2018. Young people´s public health in Reykjavík. Results of research among pupils in class, 8, 9 and 10, year 2018]. https://reykjavik.is/sites/default/files/ymis_skjol/skjol_utgefid_efni/reykjavik_heild_loka.pdf. Accessed on June 13, 2019.

Icelandic Tourist Board. 2019a. Tourism in Iceland in figures—June 2019. https://www.ferdamalastofa.is/static/files/ferdamalastofa/talnaefni/ferdatjonusta-i-tolum/2019/juni-2019-en2.pdf. Accessed on July 27, 2019.

Icelandic Tourist Board. 2019b. Hærra hlutfall ferðamanna komi til landsins utan háannar [A higher proportion of tourists visit the country outside high season]. https://www.maelabordferdathjonustunnar.is/is/hagstaerdir-1/maelikvardar. Accessed on June 20, 2019.

Jolliffe, Lee, and Regena Farnsworth. 2003. Seasonality in tourism employment: Human resource challenges. *International Journal of Contemporary Hospitality Management* 15 (6): 312–316. https://doi.org/10.1108/09596110310488140. Accessed on May 9, 2019.

Kay, Pandora. 2003. Consumer motivation in a tourism context: Continuing the work of Maslow, Rokeach, Vroom, Deci, Haley and others in ANSMAC 2003: A celebrations of Ehrenberg and Bass; marketing discoveries, knowledge and contribution. In *Proceedings of the 2003 Australian and New Zealand Marketing Academy conference ANSMAC*, December 1–3, Adelaide, SA. http://dro.deakin.edu.au/view/DU:30036658. Accessed on May 22, 2019.

Kusluvan, Salih, Zeynep Kusluvan, Ibrahim Ilhan, and Lutfi Buyruk. 2010. The human dimension: A review of human resources management issues in the tourism and hospitality industry. *Cornell Hospitality Quarterly* 51 (2): 171–214. https://doi.org/10.1177/1938965510362871. Accessed on June 15, 2019.

Lillo-Bañuls, Adelaida, José M. Casado-Díaz, and Hípólito Simón. 2018. Examining the determinants of job satisfaction among tourism workers. *Tourism Economics* 24 (8): 980–997. https://doi.org/10.1177/1354816618785541. Accessed on July 1, 2019.

Madera, Juan M., Mary Dawson, Priyanko Guchait, and Amana Mapel Belarmino. 2017. Strategic human resources management research in hospitality and tourism: A review of current literature and suggestions for the future. *International Journal of Contemporary Hospitality Management* 29: 48–67. https://doi.org/10.1108/ijchm-02-2016-0051. Accessed on September 5, 2019.

Maslow, Abraham. 1954. *Motivation and Personality*. New York: Harper.

Ministry of Industry and Innovation and Icelandic Travel Industry Association. 2015. Road map for tourism in Iceland. https://www.stjornarradid.is/media/atvinnuvegaraduneyti-media/media/Acrobat/Road-Map-for-Tourism-in-Iceland.pdf. Accessed on July 16, 2019.

Ministry of Social Affairs. 2019. Regulation on the work of children and adolescents no. 426/1999. Last modified November 1. https://www.reglugerd.is/reglugerdir/allar/nr/426-1999. Accessed on November 4, 2019.

Mulyanto, Mulyanto, I Gusti Putu Diva Awatara, and Ardi Gunardi. 2018. Enhancing competence, environmental management system, job satisfaction and employee performance. *Journal of Environmental Management & Tourism* 9: 40–45. https://doi.org/10.14505/jemt.v9.1(25).06. Accessed on November 5, 2019.

Nadiri, Halil, and Cem Tanova. 2010. An investigation of the role of justice in turnover intentions, job satisfaction, and organizational citizenship behaviour in hospitality industry. *International Journal of Hospitality Management* 29: 33–41. https://doi.org/10.1016/j.ijhm.2009.05.001. Accessed on September 25, 2019.

Neshybová, Paulína. 2019. Employee motivation and satisfaction practices—The case of Iceland. MA thesis, University of Iceland.

Ombudsman for Children. 2019. What is the role of Ombudsman for children. https://www.barn.is/um-embaettid/upplysingar-a-erlendum-tungumalum/enska. Accessed on July 21, 2019.

Robinson, Richard, N.S., Tom Baum, Maria Golubovskaya, David J. Solnet, and Victor Callan. 2019. Applying endosymbiosis theory: Tourism and its young workers. *Annals of Tourism Research* 78: 102751. https://doi.org/10.1016/j.annals.2019.102751. Accessed on November 19, 2019.

Sigurðardóttir, Ingibjörg. 2007. Þjálfun og fræðsla í ferðaþjónustu [Training and education in tourism]. In *Research in Social Sciences VIII*, ed. I. Hannibalsson, 321–329. Reykjavík: Háskólaútgáfan. http://www.rmf.is/static/research/files/2007-ispdf. Accessed on October 7, 2019.

Sigurðardóttir, Ingibjörg. 2011. Framleiðnihugtakið; þekking og viðhorf stjórnenda í íslenskum afþreyingarfyrirtækjum [The term productivity; knowledge and attitudes of managers in Icelandic recreation companies]. Paper presented at the spring conference of the Institute of Business, April 13, Reykjavík, Iceland. https://skemman.is/bitstream/1946/8527/1/8.Framleidnihugtakid_Ingibjorg.pdf. Accessed on November 10, 2019.

Spector, Paul E. 1997. *Job Satisfaction: Application, Assessment Causes and Consequences*. Thousand Oaks, CA: Sage.

Statistics Iceland. 2018a. Childrens participation in the Icelandic labour market. Statistical Series, Working papers. http://hagstofan.s3.amazonaws.com/media/public/2018/431f4ec3-d3b7-4cb6-a530-75d94b099829.pdf. Accessed on July 15, 2019.

Statistics Iceland. 2018b. Staða barna á Íslandi [Childrens situation in Iceland]. https://hagstofa.is/utgafur/frettasafn/born/hagskyrslur-um-born. Accessed on July 22, 2019.

Statistics Iceland. 2019a. Population—Key figures 1703–2019. Population 1 January 2019. Last-modified March 20. https://px.hagstofa.is/pxen/pxweb/en/Ibuar/Ibuar__mannfjoldi__1_yfirlit__yfirlit_mannfjolda/MAN00000.px/table/tableViewLayout1/?rxid=74ceb7a0-a560-4de8-9b6d-9373ea99c20a. Accessed on July 15, 2019.

Statistics Iceland. 2019b. Tourism short-term indicators in June. Last modified June 13. https://statice.is/publications/news-archive/tourism/tourism-short-term-indicators-in-june. Accessed on June 14 2019.

Statistics Iceland. 2019c. Register based employment by economic activity by months, sex, age groups and origin 2008–2019. Last modified August 23. https://px.hagstofa.is/pxen/pxweb/en/Samfelag/Samfelag__vinnumarkadur__vinnuaflskraargogn/VIN10022.px/table/tableViewLayout1/?rxid=699ea262-eb2b-48d3-bac8-4178046bb832. Accessed on September 17, 2019.

Statistics Iceland. 2019d. Percentages of 13–17 years old girls in the tourism and hospitality industry 2008–2018. Specific processing.

Statistics Iceland. 2019e. Percentages of girls 13–17 years in diverse occupations within the tourism industry 2008–2018. Specific processing.

Syukrina, Alini Mat Ali, Noor Azzah Said, Noor'ain Mohamed Yunus, Sri Fatiany Abd Kader, Dilla Syadia Ab Latif, and Rudzi Munap. 2014. Hackman and Oldham's job characteristics model to job satisfaction. *Procedia – Social and Behavioral Sciences* 129: 46–52. https://doi.org/10.1016/j.sbspro.2014.03.646. Accessed on August 9, 2019.

United Nations. 2019. Conventions on the Rights of a Child. For every child, every right. Last modified October 31, 2018. https://www.unicef.org/child-rights-convention/what-is-the-convention. Accessed on May 19, 2019.

Vinnuskóli Reykjavíkur. 2019. Hlutverk Vinnuskólans [The role of the school of work]. Last modified May 24. https://vinnuskoli.is/um-vinnuskolann. Accessed on June 3, 2019.

Vroom, Victor H. 1964. *Work and Motivation*. New York: Wiley.

Walmsley, Andreas. 2015. *Youth Employment in Tourism and Hospitality: A Critical Review.* Oxford: Goodfellow Publishers Ltd.

Walmsley, Andreas. 2017. Overtourism and underemployment: A modern labour market dilemma. Paper presented at Responsible Tourism in Destinations 13—Tackling Overtourism—Local Responses, September 29–30, Reykjavík, Iceland.

Zopiatis, Anastasios, Panayiotis Constanti, and Antonis L. Theocharous. 2014. Job involvement commitment, satisfaction and turnover: Evidence from hotel employees in Cyprus. *Tourism Management* 41: 129–140. https://doi.org/10.1016/j.tourman.2013.09.013. Accessed on September 15, 2019.

9

Migrant Workers in Tourism: Challenges of Unions and Workers in the Icelandic Tourism Boom

Magnfríður Júlíusdóttir and Íris H. Halldórsdóttir

Introduction

In the last decade, the booming tourism industry in Iceland has increasingly recruited international migrant workers. These developments pose new challenges for labour unions as migrant workers in tourism are often on temporary contracts, highly mobile and differentiated by sociocultural identities and motives (McDowell et al. 2009; Duncan et al. 2013; Carson et al. 2014). In the wider context of economic globalisation, neoliberal ideas of free and flexible markets (Harvey 2005) and increased immigration in the last thirty years, the Nordic model of

M. Júlíusdóttir (✉)
Geography, Faculty of Life and Environmental Sciences,
University of Iceland, Reykjavík, Iceland
e-mail: mj@hi.is

Í. H. Halldórsdóttir
Tourism Studies, Faculty of Life and Environmental Sciences/Icelandic
Tourism Research Centre, University of Iceland, Reykjavík, Iceland
e-mail: irishh@hi.is

A. Walmsley et al. (eds.), *Tourism Employment in Nordic Countries*,
https://doi.org/10.1007/978-3-030-47813-1_9

industrial relations and public social safety net has been under pressure. The outcome, seen in increased income inequality and weakening of unions varies between the Nordic countries (Høgedahl and Kongshøj 2017; Pareliussen et al. 2018).

Industrial relations in Iceland have much resemblance with the Nordic model, but neoliberal ideas had a stronger hold in ruling parties in government than in the other Nordic countries in the 1990s and until the bank crash in 2008 (Ólafsson 2011; Mjøset 2011). Despite of this, Iceland has now the highest union density and collective bargaining coverage in Europe (Vandaele 2019), but increasingly both media and labour movement have paid attention to breaches of collective agreements and substandard treatment of migrant workers in construction and tourism (ASÍ 2019; RÚV 2018).

In the chapter our focus is on the position of migrant workers in the expanding tourism sector in Iceland, in times of high mobility and neoliberal influences on labour market relations. In our analysis we combine insights from studies in labour geography, industrial relations and critical tourism and migration studies. In this short narrative, only a handful of research in these broad fields has been selected, attempting to understand the drivers of change in labour market relations and migrants' trajectories in this case. Among them are different drivers of precarisation (Alberti et al. 2018) and the structural hierarchy of migrant workers in tourism (Underthun and Jordhus-Lier 2018; McDowell et al. 2009). The interlinking of geographical mobility, job mobility and mobility power are highly relevant in these processes (Vickers et al. 2019), as well as the responses of labour unions to differentiated migrant workers arriving at their territory (Connolly et al. 2014; Tapia and Alberti 2019). These changes also raise important issues for tourism companies striving to reach goals of sustainable and responsible tourism (Robinson et al. 2019) as well as Nordic governments celebrating the inclusiveness and progress gained from the Nordic model.

We examine these developments mainly through interviews with representatives of labour unions outside the capital area, statistical data, as well as analysing the responses of the social partners to the open market context of labour after Iceland joined the European Economic Area

(EEA) in 1994. In 2018 and 2019 we interviewed 16 union representatives in six unions outside the capital area, where some areas have become whole year destinations while others mainly have tourist in the summer season. We also interviewed 16 migrant workers in tourism and had informal discussions with workers and employers in the two main study sites, the vicinity of Keflavík international airport and the tourist town of Vík on the South coast.

Tourism Growth and Migrant Workers

From 2011 tourist arrivals and growth of the industry took off, with tourist numbers climbing from 565 thousand in 2011 to 2.3 million in 2018 (Icelandic Tourist Board, n.d.). Between 2011 and 2017, yearly increase in tourist arrivals was 19–39%. The Icelandic tourism industry faced many challenges with the rapid development which involved a doubling of employees in the sector between 2011 and 2018 (Statistics Iceland, n.d.a).

The tourism industry had no other option than turning to international labour markets for workers due to the small size of the national labour force and phenomenal growth in tourism. In 2018, 16% of employees in the labour market in Iceland worked in tourism-related industries, 33,423 people. Almost one third were immigrants, up from 14.6% in 2011, most of them foreign citizens. From 2011 to 2018 the number of immigrants working in tourism in Iceland increased by 306% (Statistics Iceland, n.d.a). The majority from Central and Eastern Europe (CEE) and by far the largest group is from Poland, 37% in 2017 (Statistics Iceland, n.d.b). People from Spain, Portugal, Czech Republic, Croatia, Romania and Hungary are newer immigrant groups increasingly seen in tourism employment in Iceland, according to our study. This composition of workers in tourism, deviates from countries with a longer history of either minority groups from former colonies or refugees from outside EEA. In the Nordic context Sweden has received most refugees and Denmark and Norway have big immigrant groups with refugee background, compared to both Iceland and Finland (Heleniak 2018).

Most of the foreign workers in tourism in Iceland worked in accommodation and catering in 2017, 76.4%, followed by travel agencies and other tourism-related services, for example car rentals (Statistics Iceland, n.d.c) (see Fig. 9.1). Only 1.4% were registered in aviation, which prior to October 2018 didn't include workers servicing air transport. Today they are included with aviation in official figures on tourism workers, but excluded in Fig. 9.1.

Economic restructuring in the last decades and increased tourism has led to more reliance on tourism for providing employment around the country. The largest share of foreign citizens working in tourism is found in South (41%) and East (37%) part of the country (Fig. 9.2). (Statistics Iceland, n.d.d). The South and the Southeast are the most visited destinations of tourists outside the capital area (Statistics Iceland, n.d.e), with tourism now a whole year industry in some places. In the Southwest, where Keflavík international airport is located, over one third of workers in tourism are foreign citizens. Their lower share in the Northeast might be due to a shorter tourist season or more availability of locals for work in tourism.

More Icelanders saw opportunities in tourism for economic support, following the financial crash 2008. Operational tourism companies

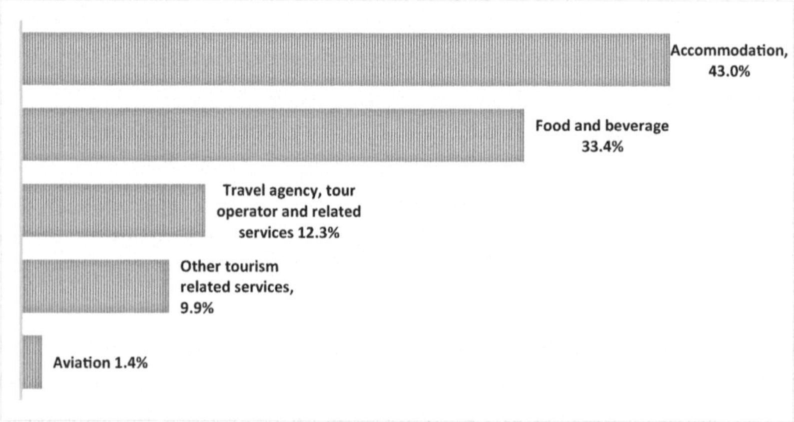

Fig. 9.1 Proportion of foreign employees in tourism subsectors, 2017 (*Source* Of underlying data: Statistics Iceland, n.d.c)

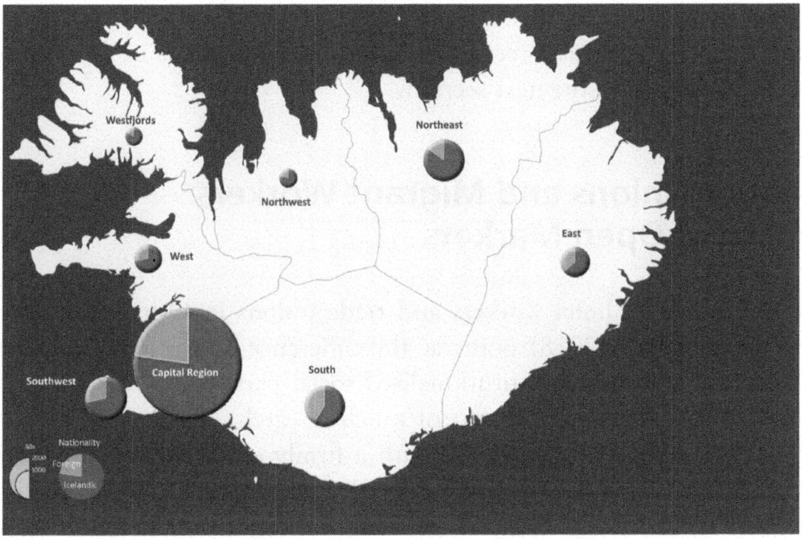

Fig. 9.2 Employees by foreign or Icelandic citizenship and region in 2017 (*Source* Statistics Iceland. *Map* Íris H. Halldórsdóttir & Benjamin David Hennig)

increased by 84% from 2008 to 2017 (Statistics Iceland, n.d.f.) and the tourism sector grew fastest in the economy. "*Gold rush*" was used to describe the situation by union representatives interviewed for this research. In 2017, 78% of 3.620 companies in tourism in Iceland had 0–4 employees, with almost one thousand micro and small companies entering the market since 2011 (Statistics Iceland, n.d.f). The chairman of a big union was critical on how easy it is to start a new company:

> In Iceland, anyone can start a business and I'm just going to say that any moron can start a business. He will maybe be offering boat trips but has never operated a boat [...] anyone can start a business without showing any knowledge in the field, no education or anything.

It is noteworthy that only 380 active businesses in tourism were in 2019 registered members of the *Icelandic Travel Industry Association* [Samtök ferðaþjónustunnar—SAF] (Communication with SAF representative, September 2019). This shows that only a fragment of all operating tourism businesses are members of SAF. On the other hand, the labour unions have managed to recruit most of the new workers in tourism, as will be explained in the next section.

Labour Unions and Migrant Workers in Era of Open Markets

In a research on hotel workers and trade unions in Oslo, Underthun and Jordhus-Lier (2018) point at the difference between the national context of Norwegian institutionalised social partnership and the neo-liberal Anglo-American context of much research on hospitality workers. This is also true for Iceland with a firmly established negotiations on collective agreements between social partners in the labour market and the government as third partner facilitating agreements. These include various acts passed to regulate the new labour market reality of free movement of labour and firms after Iceland joined the EEA in 1994. In the 1990's cases of substandard working conditions and wages of construction workers of European firms, subcontracted by the public energy company, caught the attention of unions and media. Unions interventions at that time met with disinterest from both government and public institutions like the Directorate of Labour (Ísleifsson 2013).

After these cases, exposing the precarious situation of foreign workers in a borderless EEA market, the *Icelandic Confederation of Labour* [Alþýðusamband Íslands—ASÍ] has emphasised defending the Nordic model of universal coverage of collective agreement to hinder social dumping. One of the first agreement of the social partners, on foreign workers in the Icelandic labour market, was signed in March 2004 in advance of the EU enlargement. A positive stance was taken on free movement and the social partners should jointly uphold the rules of collective agreements in all companies using foreign workers (Vinnustaðaskírteini, n.d.).

In the battle against social dumping in last 25 years, article one of the *Act on Working Terms and Pension Rights Insurance*, from 1980, is frequently referred to:

> Wages, and other working terms agreed between the social partners shall be considered minimum terms, independent of sex, nationality or term of appointment, for all wage earners in the relevant occupation within the area covered by the collective agreement. Contracts made between individual wage earners and employers on poorer working terms than those specified in the general collective agreement shall be void. (Ministry of Social Affairs 2002)

The addition "*independent of sex, nationality or term of appointment*" came into the act when Icelandic laws were made compatible to EU treaties before joining EEA.

In the 1990s ASÍ managed to defend payment of fees to unions for the service they provide, which include working on collective agreements guaranteeing minimum wages and other benefits to all workers (Ísleifsson 2013). The disputes were formally about interpretation of freedom of association for workers and employers in the European Social Charter, involving different interpretations from the Council of Europe committee of experts and ILO specialist. According to Ísleifsson (2013, pp. 54–55), writing from the perspective of the Icelandic labour movement, this questioning of membership and fees to unions were signs of the growing strength of neoliberal ideology, "*Although the discussions were formally about human rights, no one doubted that at the core of it was an attempt to decrease peoples' membership in unions and thereby weaken their foundations*". Unlike in many countries where unions find it difficult to recruit migrant workers in tourism, especially temporary workers (Underthun and Jordhus-Lier 2018; Jordhus-Lier 2014; Alberti 2014), the migrant workers in Iceland are automatically registered through employers' deduction of service fees to local unions (ASÍ, n.d. labour law), if they follow labour market rules.

This partly explains the high union density in Iceland. In a comparative European study of 32 countries, the average union density was 28% in 2017, ranging from 4.2% in Estonia to 90.4% in Iceland, in 2016 (Vandaele 2019). The other Nordic countries had union density ranging from 51–67% and Nordic countries "*continue to top the unionisation league … largely due to … a relatively benevolent institutional settings*". (Vandaele 2019, p. 17). The average collective bargaining coverage 2010–2016 in Iceland was 92%, compared to 17% in Poland. In the other Nordic countries the coverage ranged from 73% in Norway to 90% in Sweden (Vandaele 2019). For the predominantly young migrant workers from CEE, joining the tourism workforce, the unionised labour market in Iceland can be an unfamiliar situation.

During the economic boom years in banking and construction, 2005–2008, attention of unions and media continued to focus on working conditions of foreign workers in construction as well as growing presence of temporary staffing agencies and posting of workers by foreign companies. New acts were passed on both, to regulate these new actors in the Icelandic labour market (Ísleifsson 2013). The bank crash in 2008 wiped out these new actors (Friberg et al. 2014), but with the tourism led economic upswing in recent year both types of mediation of foreign workers have returned on a bigger scale than before 2008 (Magnússon et al. 2018; Directorate of Labour 2018).

In 2010 a new act on workplace inspections came to force, initiated by the social partners who were mandated to carry out the inspections, should bear all the cost and inform public institutions if suspecting non-compliance to laws (Alþingi 2010). At first the sectors of construction, accommodation and catering were to be inspected, but three years later tourism in general was added (Vinnustaðaskírteini, n.d.). According to our informants attempts to coordinate workplace inspections among social partners and public institutions only lasted a short period and in recent years the local labour unions have taken on this task. One reason given was the lack of interest of past governments to strengthen the public institutions with mandate to monitor the labour market. This is in accordance with the neoliberal ideas seen in government agreements in Iceland 1995–2003, emphasising that inspections will not unnecessarily burden companies (Stjórnarráð Íslands, n.d.).

Despite of better legal protection of migrant workers at the start of the tourism boom, in recent years media coverage of substandard treatment of migrant workers in tourism has grown, with a programme on national TV in the autumn 2018 shocking many with stories of ill treatment of migrant workers in both construction and tourism (RÚV 2018). The Minister of Social Affairs responded by forming a group from ministries, public institutions and the social partners, which came with recommendations on actions to deal with the situation (Ministry of Social Affairs 2019). A follow up study by ASÍ (2019), on foreign workers and criminal activities in the labour market, speaks of substantial wage theft mainly affecting young people and migrant workers in the Icelandic labour market. The study showed that 45% of the 487 cases in legal dept collection process in 2018 were on behalf of employees in tourism, the majority in hospitality work. Over half (54%) were on behalf of migrant workers, although they were only 26% of members of unions covered by the study (ASÍ 2019). This study only shows the tip of the iceberg for many reasons. It is based on data from four unions, with just over half of all union members in Iceland; it is only cases who have sufficient data to be sent by unions to formal dept collection and perhaps most importantly, it is only cases from people who seek assistance. The *Icelandic Travel Industry Association* (SAF 2019) also responded with a short report on illegal activities in the labour market. Although all criminal activities in tourism were mentioned as distorting market competitiveness, the main focus was on foreign companies and the need to strengthen inspections of their activities. In our study a union representative came with a critical reflection on Icelanders tendency to see themselves in a better light than foreign companies: "*This is just a very unpleasant situation, because we accuse foreigners of being greedy and dishonest ... but when we start to look at the situation here, then we are not at all better than they*".

Unions' Concerns on Migrant Workers in Tourism

The chairmen/-women and other staff we interviewed at the offices of six unions are the people in daily contact with union members and carry out workplace inspections. These unions cover large part of Iceland outside of the capital area. In areas experiencing the fastest growth in tourism, majority of employees in tourism paying fees to the unions were foreign citizens according to the information we obtained. We had problems getting comparable information as the unions don't have a uniform system of registration. In the Keflavík area 52% of union members were foreign citizens in 2018, the majority working in tourism-related activities at the airport, in hotels or car rentals. In the East we got the most detailed information, showing that foreign citizens were 62% of workers in tourism, July 2018–June 2019, of which 55% were women. We now discuss some of the key themes that emerged in the interviews, beginning with breaches of collective agreements and placing migrant tourist-workers at the bottom of the wage scale.

Non-compliance with Collective Agreements

Monitoring compliance with collective agreements in tourism takes a large share of unions' time. The most common area of non-compliance related to too low wages, as they were not correctly calculated. Paying a flat rate instead of making a division between daytime and overtime wage rates was often mentioned. According to one union representative, employees always lost wages by this arrangement, as the flat rate increase to daytime rates was not high enough to compensate for no payment of overtime. Split shifts added to this issue, including not paying for the in-between time of three to four hours. Other main issues were not respecting rules of 11 hours of rest or weekly holidays. Lack of work schedules, payslips and contracts can be added to the long list provided. There were also cases of people being hired full time but getting less work and pay if fewer tourist than anticipated used the service.

A union representative claimed that the tourism sector in Iceland is notorious for paying the lowest possible wage rates. Even though tourism companies pay according to collective agreement the tendency is to pay the migrant workers the stipulated minimum rate. This was also found to be the case for Polish construction workers in Iceland in 2010, in one of the few comparative Nordic studies on migrant workers including Iceland (Friberg et al. 2014). In the first Icelandic study on wage differences between immigrants and locals, covering the period 2008–2017 and companies with ten or more employees, immigrants had on average 8% lower wages, with the largest pay gap in construction, 16%. Information on pay gap by background in tourism is still lacking, due to limited data (Statistics Iceland 2019).

The union representatives noted a common fear among migrant workers about seeking their rights while still employed. They would not confront their employer or asked to be anonymous if asking the union to intervene. The fear of losing the job was cited as the main reason why most migrant workers only seek assistance after they have quitted the job and then ask unions to make salary claims that can extend many years back. On the employer's side SA has emphasised flexibility in the labour market making it easy for firms to fire employees, which as historically been easy in Iceland in European comparison (Ólafsdóttir 2008). In beginning of the twentieth-century SA hindered the ratification of the ILO convention on *Termination of Employment*, from 1982 (NFS 2017). Our interviews suggest this ease of dismissing employees was seen as misused by some employers and making it more difficult for migrant workers in tourism to seek compliance to collective agreements.

Concerns were also raised about people working in tourism for free, as they were recruited as volunteers or trainees. From the union's standpoint it was clear that "*it is illegal to use volunteers in an economic activity*" as one interviewed union representative put it. The social partners have agreed that companies competing for profit in the market cannot use volunteers. In our interviews the practice was said to be widespread among horse rental companies, where the lack of insurance tied to formal registration in the labour market was found especially worrying for the volunteers or trainees. The practice of using volunteers in tourism in Iceland is also addressed by Einarsdóttir et al. in this volume.

Among the union representatives there was frustration that employers repeatedly found guilty of not paying according to collective agreements were not sanctioned apart from having to make up the shortfall in salary for the individual worker seeking unions assistance. We asked if some type of tourism companies stood out as more problematic regarding migrant workers, but variations were seen in all types of firms. One union representative divided them into three categories based on communication between the union and tourism companies in the area. In one group were "*relatively honest companies*" and those were usually the bigger or older ones. They had experienced people handling wages and all matters were solved professionally. The there was a number of companies where the problem was seen as lack of knowledge on collective agreements or rashness. Usually the breaches were corrected after the union intervened on behalf of the employee. The third group was described as "*just criminals*" that avoid making formal contracts with employees or have them obscure, never promising anything. Those are the ones that take up most time of the unions.

In the Keflavík area a clear distinction was made between the big companies at the airport and car rentals and hotels. The big airport service firms paid all employees by the same agreements, irrespective of being "*Polish or French or Icelandic*". On the other hand, many car rentals and hotels were described as "*the wild market*", some of them intentionally saving on wages. As soon as the union had claimed corrections for one worker "*they get rid of him .... and hire a new one instead and start at the same practice*".

Problematic Issues of Housing and Social Protection

Shortage of housing has been a bottleneck in some of the popular tourist location in Iceland and some bigger tourism companies have either built or bought houses for their mostly migrant staff. In 2010s at Keflavik international airport, the largest two employers in ground handling bought apartment buildings, left by the U.S. army 2006, to house their foreign employees. When the tourism boom started full fledge in 2010 many of these houses were still available (Hagfræðistofnun Háskóla Íslands 2010). The ground handling companies recruited numerous workers, directly in Poland by agents from the companies, over 100 people in one trip (Páll Ketilsson and Hilmar Bragi Bárðarson 2016; Jóhann Bjarni Kolbeinsson 2015). In 2016, 1490 people were working for these two big companies (ISAVIA 2018 p. 31).

The smaller companies in popular tourist locations need to improvise to find solutions for their employees. One of the migrant workers interviewed in the South described cohabiting with the father of her employer:

> I live with her father; the grandpa of my boss. [There are] really, really small rooms. Everything a little bit improvised so it's like I have two beds in my room. I was asking to remove one so he was like "It's difficult". So, I don't really feel at home [...]. I have been there now for seven months.

The common interlinking of housing and employment among migrant workers in tourism was frequently regarded as problematic. The main issues were substandard housing, crowding, high rent with dubious mixing of salary and rent, and not least the precarious situation of workers dependant on employers for housing. These cases are difficult for the unions to handle for two main reasons: If they make any fuss the employees could face the risk of losing their "*home*" and secondly the unions have no jurisdiction over housing of employees.

There were cases where rent was paid under the table to avoid income tax and examples of employers getting manifold the normal market price from renting to their own employees, like in this case of an expensive hotel in the South:

> A nice and respectable place that advertises a lot and is expensive [...]. They had 18 people in a small summerhouse, in bunk beds and there each slept on top of the other. For this, each was charged 38 thousand kronas per month. So let's multiply 38 thousand with 18 to find what they are getting. [...] they are getting at least worth over two monthly salaries just out of the rent.

Workers depending on their employer for housing, fears losing not only the job but also housing if raising issues on wages or work conditions. In crisis situations the unions often have to intervene to assist migrant workers:

> Workers are dependent on their employer with housing. They do not know anyone, and they are in some sort of shackles because otherwise they will end up on the street. There is a recent example where an employer threw an employee out of a restaurant here nearby and the person was here crying at four a clock on Friday when we were closing before the weekend.

The lack of official strategy of social assistance and protection regarding cases like these where migrant workers are "*thrown out*", came up repeatedly in our interviews. One union representative claimed there was a need for a "*shelter for foreigners, like women's shelters*". In the case cited above, the only solution the union could find was a bed in a shelter for women victims of domestic violence. In workplace visits the union inspectors leave contact information of the local union, which is often the only contact migrant workers have with institutions and locals outside the workplace. The unions are therefore often in the situation of acting like social authorities and providers of information about Icelandic society, that they thought, municipalities or government institutions should provide.

The housing issues are evidently a thorn in the eyes of the labour unions and one representative said these issues needed to be taken up in the next negotiations on collective agreements. There is a resistance from the Confederation of Icelandic Enterprise (SA), as seen in their refusal of addressing changes of laws governing leasing of housing, in the ministers group discussing criminal activities in the labour market. In that forum ASÍ wanted to address the problematic interlinking of housing and work for migrant workers but SA did not find it relevant (Ministry of Social Affairs 2019).

Diversity in Vulnerability Among Migrant Workers

When discussing variation among groups of migrant workers the Polish migrants were thought of as seeking their rights in the labour market and having a good network of sharing information. One informant claimed that for an Icelandic company today having a Polish worker was "*as much hassle as employing Icelanders*". This was considered a positive trend from the perspective of the unions. In the South the Romanians and Bulgarians working in tourism were thought of as a more vulnerable group, recently coming into tourism in the area. A workplace inspector experienced them as more isolated, not having similar support network as the Polish migrants. Approaching some of the Romanians in workplace inspections he used the metaphor of trying to win the trust of a "*scared kitten*". He claimed employers were increasingly recruiting Romanians through Irish temporary staffing agencies, where labour market rules provided limited protection for workers.

The "*enormous mobility of people*" coming to work in tourism, often for only three to six months was considered a barrier to reaching all with information on their labour market rights. Some of these people were claimed to be just not interested in the information and happy to work for free on a horse farm for a free horse trip in Iceland or in a hotel as part of their travel in Iceland. This group resembles the migrant tourist-workers, where boundaries between leisure and work are blurred (Bianchi 2000). They are not registered in the labour market and unclear how big this group is.

Employees in Tourism Industry Policy and Quality Systems

In our interviews some concerns were raised about little emphasis on staff in Icelandic tourism sectors promotion of sustainability and corporate social responsibility (CSR). Some felt the social dimension of sustainability was missing. One described CSR as just being interested in grants to local communities: "*Corporate social responsibility is to throw a hundred thousand Krona's into the football club or give a scan to the retirement home or the nursing home. But all matters of the staff are just left out.*"

There was also reservation surrounding unions' direct participation in qualifying the working conditions of workers in tourism, as was initiated with Vakinn, the official Icelandic tourism quality and environmental system. Around 100 companies participate in Vakinn, which is partly based on the Qualmark quality system used in New Zealand (Vakinn, n.d.). In 2017 the Icelandic Tourist Board, on behalf of Vakinn and the Icelandic Confederation of Labour (ASÍ) signed an agreement on collaboration (Snorri Már Skúlason 2017). It stated that labour unions would certify that tourism companies participating in Vakinn were operating according to collective agreements. It was clear that this kind of collaboration was controversial among the unions from the start. It was considered a double-edged sword; it could improve conditions for the employees but could also weaken trust towards the unions if approved companies were later found to mistreat their employees. Some criticised the execution of the audit and felt that this collaboration was based on criteria set out by the tourism companies with little influence from the unions. Some unions refused to participate unless they were able to make a more extensive audit, which the representatives of the tourism companies did not agree on. One of the union representatives said quite distinctively: "*I cannot be held responsible for any company I have not entered*".

In an open meeting in December 2018 titled "*Prosperity based on abuse?*", held by one of the biggest unions in the country, a question was raised if a quality system like this could assist unions in reaching their goals. One union representative refused the idea and said that unions should not be a "*tool for capitalistic businesses to be able to whitewash themselves*".

With a growing media coverage on bad treatment of migrant workers in tourism, some business owners feel that there is a need for a special quality system on treatment of staff. Systems in line with OK Forhold in Denmark, Schyssta Vilkor in Sweden and Fair Hotels in Ireland (Bergene et al. 2014). A company owner in the South approached the local union with this idea and in our interview expressed frustration of being suspected of treating the foreign migrants unfairly: "*It is quite miserable to always be under this suspicion of being in an underground operation, that there is tax evasion and other things*". In his view the negative media image affects all companies, irrespective of how they treat their employees.

A campaign to promote responsible tourism was recently initiated, by the Iceland Tourism Cluster and Festa, an NGO promoting CSR. Early 2017 roughly 300 tourism companies signed a declaration on responsible tourism, committing themselves to publish their goals towards responsible tourism, before end of the year (Iceland Tourism Cluster, n.d.). A survey in 2018 showed however that only 28% of the companies had published some goals, mainly relating to environmental protection and safety issues. Main reasons given for limited activities was lack of time or knowledge (Iceland Tourism Cluster 2019). In this responsible tourism initiative, respecting the rights of employees is one of four themes and the topic is more prominent than in the Vakinn quality system. Despite this, a recent draft for a "road map" guiding public policy in tourism to 2030 (Ministry of Industries and Innovation 2019), seems to have missed the increased spotlight on working conditions of migrant workers in tourism, in both Icelandic society and critical tourism academy discussing the SDGs (Robinson et al. 2019). Although the "road map" picks up many goals from the 2030 Agenda for Sustainable Development, the target set for decent work in goal 8 (United Nations, n.d.), are left out.

Conclusion

In this chapter we have provided insights into the hitherto largely neglected mapping of the international migrant tourism workers in Iceland and their working conditions, mainly viewed from the standpoint of labour union representatives. Due to unions' big involvement in carrying out workplace inspections and high unionisation rate among migrant workers, we claim they have the best insight into many aspects of the situation of migrant tourism workers. The context of rapid growth in tourism, increasingly depending on migrant workers, poses new challenges to unions, not least in the gold-rush-like situation of new companies popping up. For the unions this is a new reality compared with earlier groups of labour migrants, mainly recruited by well-established firms in the fishing industry and construction. The geographical spread of tourism activities, variations in type of firms and identities and motivations of people seeking jobs in tourism in Iceland add to the complexity faced.

In the context of growing neoliberalism at the end of the twentieth century and open market of the EEA, our results show that some migrant workers in tourism experience inequality and fall outside the notion of inclusiveness and protection, despite the Icelandic labour movement managing to defend many aspects of the Nordic model of industrial relations. Some of the main results are signs of precarisation in the Icelandic context that in our view need to be given more attention by public institutions, tourism industry actors and labour unions to enhance equality, decent working conditions and responsible tourism.

A common theme in our interviews with union representatives, as well as in other studies (ASÍ 2019; Skaptadóttir and Wojtyńska 2019) is the lack of coordinated governance structures to deal with the situation of growing presence of migrant workers, that often are recruited for temporary work in tourism. Many of them are in precarious situation, depending on employers for housing, with little social network in Iceland and at the bottom of the wage scale. Repeated cases of non-compliance to labour market agreements without sanctions, stand out as especially problematic. The unions' critique is not directed at all tourism industry actors, but to many companies that show little respect for established rules in the labour market and wellbeing of the migrant workers. The strongest critique is directed at the lack of enforcement of existing laws and the labour movement having almost sole responsibility for both labour inspections and providing social welfare service to migrant workers in distress, due to a lack of other providers. Both government institutions and municipalities need to be more involved in preventing abuse and social marginalisation of this group.

From the unions' standpoint the tourism industry and policy makers in tourism could also put more focus on decent work conditions in tourism. Lack of knowledge on collective agreement and rights of employees, is seen as a common problem among tourism companies. The extent of cases of all kinds of breaches of collective agreements, especially by many smaller companies that have joined the gold rush in tourism in the last decade, supports the unions' critique of lack of focus on labour issues in the official quality and environmental system used in Icelandic tourism. A low share of tourism companies in the Icelandic Travel Industry Association (SAF) adds to this problem, as majority of companies are not receiving support from the industry's social partner.

Apart from the many cases of wage theft mentioned in our interviews and the ASÍ (2019) study, other drivers of precarisation (Alberti et al.

2018) can be noticed in developments in work conditions of migrant workers in tourism in Iceland, like transfer of risk to workers by unclear contracts or no contracts at all, as in the case of volunteers and trainees, and the flexibility of employers to dismiss workers. Theorising the dynamics of precarity through mobility (Vickers et al. 2019) draws attention to the interlinking of geographical mobility, job mobility and mobility power. A constant threat of job mobility is a feature of precarity, and the workers' mobility power can be regarded as weak in the current situation of temporary contracts and flexibility to dismiss people. One of our results is that the perceived threat of dismissal often deters migrant workers in tourism to seek the assistance of unions they are entitled to. On the other hand the mobility power of migrant workers in tourism in Iceland can be claimed to have been strong in the years of rapid growth with ever-growing demand for workers. It enables people to seek another job in the sector in different parts of the country, like some of the migrants interviewed had done after a bad experience. This example of workers' mobility power is likely to be less now, with downturn in tourism arrivals in 2019. In subsectors of tourism the decrease in number of employees in tourism has been highest in air transport, 21% between August 2018 and 2019 (Statistics Iceland, n.d.g.) and in 2019 unemployment has risen most among foreign citizens and in the vicinity of Keflavík international airport (Directorate of Labour 2019). As shown, the majority of migrant workers in tourism in Iceland come from eastern or southern part of Europe and can move freely inside EEA. Because of this migration status and earned unemployment benefits they have the choice of staying in Iceland, which can also be interpreted as mobility power in the institutional setting of the Nordic model.

Unlike the "stuck workers" of ethnic minority and often refugee background in the lowest ranked jobs in the structural hierarchy of city hotels in Norway (Underthun and Jordhus-Lier 2018), the stuck workers in our study were rather overqualified Europeans intending to stay in Iceland. They experienced being stuck in the lowest paying work in the tourist sector, irrespective of work experience, education and proficiency in Icelandic and other languages. This indicates little job mobility outside the liminal group space of migrant workers in tourism

(Underthun and Jordhus-Lier 2018). Instead of the tourism industry and authorities often regarding migrant workers in tourism as, cheap, temporary and easily replaceable labour, this group of migrant workers wanting to settle should rather be regarded as a future resource in need of more support to enhance their upward job mobility in the sector.

The Icelandic unions could reflect on how they can best represent the growing number of migrant members working in tourism, this heterogeneous group of people with various motives, sociocultural identities and economic in/security and responsibilities, providing a bigger share of unions income in the fastest growing tourist areas. A coherent registration is needed by the unions to enable analysis of the heterogeneity, to enable unions to serve migrant workers as a differentiated group, engaging more with problems and discrimination migrant workers encounter as migrants (Alberti 2014). Some poorly staffed unions outside the capital area would need assistance from national federations or ASÍ and the research community to enable a more nuanced knowledge on this group. More attention to migrant intersectionality at different levels is suggested for research and unions strategies by Tapia and Alberti (2019). In the Icelandic context few studies have attempted to analyse developments in arrivals of migrant workers using an intersectional approach (Júlíusdótir et al. 2013) and for the recent developments of migration for work in tourism such an approach is needed to give more information on the heterogeneity in sociocultural identities and experiences of this group. The current emphasis of unions in Iceland on adherence to collective agreement and calls for more security in housing and safety net from authorities, places the logic of action of unions increasingly on social rights, instead of only class based struggles, according to the frame of analysis used by Connolly et al. (2014). The need for more information on different aspects of rights and duties in Icelandic society for stronger embeddedness in society outside the workplace is an important result of our study. Instead of treating migrant workers in tourism like replaceable guest workers at the bottom of the labour market, their important contribution to postcrisis recovery in Iceland needs to be acknowledged and steps taken to secure good working conditions for all, as important part of responsible and sustainable future in tourism.

References

Alberti, Gabriella. 2014. Multi-scalar organising in London's hotels: The challenges of engaging transient workers through labor and community alliances. In *A Hospitable World? Organising Work and Workers in Hotels and Tourist Resorts*, ed. David Jordhus-Lier and Andres Underthun, 156–174. London: Routledge.

Alberti, Gabriella, Ioulia Bessa, Kate Hardy, Vera Trappmann, and Charles Umney. 2018. In, against and beyond precarity: Work in insecure times. *Work, Employment & Society* 32 (3): 447–457.

Alþingi [Parliament]. 2010. Law nr. 42/2010 Lög um vinnustaðaskírteini og eftirlit á vinnustöðum. https://www.althingi.is/altext/stjt/2010.042.html.

ASÍ [Icelandic Confederation of Labour]. n.d. Icelandic Labour Law. https://www.asi.is/vinnurettarvefur/icelandic-labour-law/labour-market-and-trade-unions/union-funds/. Accessed on July 20, 2019.

ASÍ. 2019. Íslenskur vinnumarkaður 2019. Erlent launafólk og brotastarfsemi á vinnumarkaði. https://www.asi.is/media/315797/islenskur_vinnumarkadur_2019_brotastarfsemi_130819_2.pdf. Accessed on August 13, 2019.

Bergene, Ann C., Karla Boluk, and Ethel Buckley. 2014. Examining the opportunities and challenges of union organization within the hospitality industry: The fair hotels Ireland strategy. In *A Hospitable World? Organising Work and Workers in Hotels and Tourist Resorts*, ed. David C. Jordhus-Lier and Andres Underthun, 195–212. London: Routledge.

Bianchi, Raoul V. 2000. Migrant tourist-workers: Exploring the ones of post-industrial tourism. *Current Issues in Tourism* 3 (2): 107–137.

Carson, Doris. A., Dean B. Carson, and Linda Lundmark. 2014. Tourism and mobilities in sparsely populated areas: Towards a framework and research agenda. *Scandinavian Journal of Hospitality and Tourism* 14 (4): 353–366.

Connolly, Heather, Stefania Marion, and Miguel Martínez Lucio. 2014. Trade union renewal and the challenges of representation: Strategies towards migrant end ethnic minority workers in the Netherlands, Spain and the United Kingdom. *European Journal of Industrial Relations* 20 (1): 5–20.

Directorate of Labour [Vinnumálastofnun]. 2018. Erlendir ríkisborgarar á íslenskum vinnumarkaði á árunum 2012–2017. https://www.vinnumalastofnun.is/maelabord-og-tolulegar-upplysingar/erlendir-rikisborgarar-a-islenskum-vinnumarkadi. Accessed on February 20, 2018.

Directorate of Labour. 2019. Vinnumarkaðurinn á Íslandi. Yfirlit, horfur og þjónusta Vinnumálastofnunar, November. https://www.vinnumalastofnun.is/media/2342/nov2019-skyrsla.pdf. Accessed on December 20, 2019.

Duncan, Tara, David G. Scott, and Tom Baum. 2013. The mobilities of hospitality work: An exploration of issues and debates. *Annals of Tourism Research* 41: 1–19.

Friberg, Jon H., Jens Arnholtz, Line Eldring, Nana W. Hansen, and Frida Thorarins. 2014. Nordic labour market institutions and new migrant workers: Polish migrants in Oslo, Copenhagen and Reykjavik. *European Journal of Industrial Relations* 20 (1): 37–53.

Hagfræðistofnun Háskóla Íslands. 2010. *Áhrif Keflavíkurflugvallar og Ásbrúar á atvinnuástand á Suðurnesjum*. Report nr. C10:03. Reykjavík: Hagfræðistofnun Háskóla Íslands. http://www.ioes.hi.is/sites/hhi.hi.is/files/C-Series/2010/C10_03.pdf. Accessed on September 13, 2019.

Harvey, David. 2005. *A Brief History of Neoliberalism*. Oxford: Oxford University Press.

Heleniak, Thimothy. 2018. Migration: The wary welcome of newcomers to the Nordic region. In *State of the Nordic Region 2018*, ed. Julian Grunfelder, Linus Rispling, and Gustaf Norlén, 48–55. Copenhagen: Nordic Councils of Ministers.

Høgedahl, Laust, and Kristian Kongshøj. 2017. New trajectories of unionization in the Nordic Ghent countries: Changing labour market and welfare institutions. *European Journal of Industrial Relations* 23 (4): 365–380.

Icelandic Tourist Board. n.d. Number of foreign visitors. https://www.ferdamalastofa.is/en/recearch-and-statistics/numbers-of-foreign-visitors. Accessed on July 1, 2019.

Iceland Tourism Cluster. n.d. Ábyrg Ferðaþjónusta. http://www.icelandtourism.is/verkefni/abyrg-ferdathjonusta/. Accessed on September 1, 2019.

Iceland Tourism Cluster Initiative. 2019. *Ábyrg ferðaþjónusta: niðurstaða könnunar meðal þátttakenda vorið 2018*. Reykjavík: Icelandic Tourism Initiative and Festa. http://www.icelandtourism.is. Accessed on September 12, 2019.

ISAVIA. 2018. *Keflavíkurflugvöllur — Stóriðja í stöðugum vexti*. https://www.isavia.is/media/1/storidja-i-stodugum-vexti_skyrsla_isavia_aton.pdf. Accessed on July 1, 2019.

Ísleifsson, Sumarliði R. 2013. *Til velferðar. Saga Alþýðusambands Íslands*, 2. Reykjavík: Alþýðusamband Íslands and Forlagið.

Jordhus-Lier, David C. 2014. Extinguishing fires: Coping with outsourcing in Norwegian hotel workplace. In *A Hospitable World? Organising Work and*

Workers in Hotels and Tourist Resorts, ed. David C. Jordhus-Lier and Anders Underthun, 137–155. London: Routledge.
Júlíusdóttir, Magnfríður, Unnur D. Skaptadóttir, and Anna Karlsdóttir. 2013. Gendered migration in turbulent times in Iceland. *Norwegian Journal of Geography* 67 (5): 266–275.
Ketilsson, Páll, and Hilmar Bragi Bárðarson. 2016. Airport Associates stækkar um 75–80% á þessu ári. *Víkurfréttir*, June 1. vf.is/vidskipti/airport-associates-staekkar-um-75-80-a-thessu-ari. Accessed on June 15, 2019.
Kolbeinsson, Jóhann Bjarni. 2015. Fjölga um 1.500 á Keflavíkurflugvelli. *Ruv.is*, November 25. https://www.ruv.is/frett/fjolga-um-1500-a-keflavikurflugvelli. Accessed on June 10, 2019.
Magnússon, Gylfi, Inga Minegaite, Erla Kristjánsdóttir, and Þóra H. Christiansen. 2018. Here to stay? The rapid evolution of the temporary staffing market in Iceland. *Stjórnmál Og Stjórnsýsla* 14 (2): 135–158.
McDowell, Linda, Adina Batnitzky, and Sarah Dyer. 2009. Precarious work and economic migration: Emerging immigrant divisions of labour in Greater London's service sector. *International Journal of Urban and Regional Research* 33 (1): 3–25.
Ministry of Industries and Innovation. 2019. *Framtíðarsýn og leiðarljós íslenskrar ferðaþjónustu til 2030*. June. https://samradsgatt.island.is/oll-mal/$-Cases/Details/?id=1408. Accessed on July 1, 2019.
Ministry of Social Affairs. 2002. Act on Working Terms and Pension Rights Insurance, No. 55/1980. https://www.government.is/Publications/Legislation/Lex/?newsid=89000204-fbd6-11e7-9423-005056bc4d74. Accessed on July 21, 2019.
Ministry of Social Affairs. 2019. Skýrsla samstarfshóps Félags- og barnamálaráðherra um félagsleg undirboð og brotastarfsemi á vinnumarkaði. https://www.stjornarradid.is/lisalib/getfile.aspx?itemid=c6a38acb-2567-11e9-942f-005056bc530c. Accessed on June 3, 2019.
Mjøset, Lars. 2011. Nordic political economy after financial deregulation: Banking crises, economic experts, and the role of neoliberalism. In *The Nordic Varieties of Capitalism, Comparative Social Research*, ed. L. Mjøset, vol. 28, 1–51. Bingley: Emerald Group Publishing Limited.
NFS [Council of Nordic Trade Unions]. 2017. Nordic ILO report 2017. ILO conventions ratified by the Nordic Countries in the period 1980–2016. http://www.nfs.net/aktuellt/ilo/ilo-conventions-ratified-by-the-nordic-countries-36446164. Accessed on June 20, 2019.

Ólafsdóttir, Katrín. 2008. Er íslenskur vinnumarkaður sveigjanlegur? http://www.ru.is/media/publications/SoB/Er-islenskur-vinnumarkadur-sveigjanlegur_2008.pdf. Accessed on July 22, 2019.

Ólafsson, Stefán. 2011. Icelandic capitalism: From statism to neoliberalism and financial collapse. In *The Nordic Varieties of Capitalism*, ed. Lars Mjøset, 1–51. Bingley: Emerald Group Publishing Limited.

Pareliussen, Jon K., Mikkel Hermansen, Christophe André, and Orsetta Causa. 2018. Income inequality in the Nordics from an OECD perspective. In *Increasing Income Inequality in the Nordics: Nordic Economic Policy Review 2018*, ed. Rolf Aaberg, et al., 19–62. Copenhagen: Nordic Council of Ministers.

Robinson, Richard N.S., Antje Martins, David Solnet, and Tom Baum. 2019. Sustaining precarity: Critically examining tourism and employment. *Journal of Sustainable Tourism* 27 (7): 1008–1025.

RÚV. 2018. Svarta hliðin á íslenskum vinnumarkaði, October 2. https://www.ruv.is/kveikur/svarta-hlidin-a-islenskum-vinnumarkadi-2/. Accessed on October 5, 2019.

SAF. 2019. Tillögur til skilvirkara eftirlits með erlendri og ólöglegri starfsemi í ferðaþjónustu. https://www.saf.is/2019/03/05/tillogur-saf-ad-skilvirkara-eftirliti-med-erlendri-og-ologlegri-starfsemi/. Accessed on April 2019.

Skaptadóttir, Unnur D., and Anna Wojtynska. 2019. Sveigjanlegur vinnumarkaður, hrakvinna oginnflytjendur. *Íslenska þjóðfélagið* 10 (2): 14–28.

Skúlason, Snorri Már. 2017. *Samstarf við ASÍ um ábyrgð fyrirtækja innan Vakans*. ASÍ, March 1. https://www.asi.is/frettir-og-utgafa/frettir/almennar-frettir/samstarf-vid-asi-um-abyrgd-fyrirtaekja-innan-vakans/.

Statistics Iceland. n.d.a. Register based employment by economic activity by years sex, age groups and origin 2008–2018. https://statice.is/statistics/society/labour-market/labour-force-register-data/. Accessed on July 4, 2019.

Statistics Iceland. n.d.b. Nationality_Tourismjobs. Statistics Iceland special processing, received on February 13, 2019.

Statistics Iceland n.d.c. Proportion of foreign employees in tourism divided by subsectors, 2017. Statistics Iceland special processing, received on June 22, 2018.

Statistics Iceland n.d.d. Employees by foreign or Icelandic citizenship and region in 2017. Statistics Iceland special processing, received on June 22, 2018.

Statistics Iceland. n.d.e. Overnight stays and arrivals in all types of accommodation by municipalities 2008–2018. https://statice.is/statistics/business-sectors/tourism/accommodation/. Accessed on September 9, 2019.

Statistics Iceland. n.d.f. Number of enterprises and operational information by industry and size 2008–2017. https://statice.is/statistics/business-sectors/enterprises/structural-business-statistics/. Accessed on August 15, 2019.

Statistics Iceland n.d.g. Number of employees in activities related to tourism 2008. https://px.hagstofa.is/pxen/pxweb/en/Atvinnuvegir/Atvinnuvegir__ferdathjonusta__ferdaidnadurhagvisar/SAM08051.px. Accessed on December 20, 2019.

Statistics Iceland. 2019. Analysis on pay gap by background 2008–2017. *Hagtíðindi* 104 (6). https://hagstofa.is/utgafur/utgafa/laun-og-tekjur/rannsokn-a-launamun-eftir-bakgrunni-2008-2017/. Accessed on August 15, 2019.

Stjórnarráð Íslands [Government of Iceland]. n.d. Stefnuyfirlýsingar fyrri ríkissjórna frá 1995. https://www.stjornarradid.is/rikisstjorn/sogulegt-efni/stefnuyfirlysingar-fyrri-rikisstjorna/. Accessed on July 22, 2019.

Tapia, Maite, and Gabriella Alberti. 2019. Unpacking the category of migrant workers in trade union research: A multi-level approach to migrant intersectionalities. *Work, Employment & Society* 33 (2): 314–325.

Underthun, Anders, and David C. Jordhus-Lier. 2018. Liminality at work in Norwegian hotels. *Tourism Geographies* 20 (1): 11–28.

United Nations. n.d. Sustainable development goals: Knowledge platform. https://sustainabledevelopment.un.org/?menu=1300. Accessed on December 20, 2019.

Vakinn. n.d. About Vakinn. https://www.vakinn.is/en/about-vakinn. Accessed on July 1, 2019.

Vandaele, Kurt. 2019. Bleak prospects: Mapping trade union membership in Europe since 2000. https://www.etui.org/Publications2/Books/Bleak-prospects-mapping-trade-union-membership-in-Europe-since-2000. Accessed on June 20, 2019.

Vickers, Tom, John Clayton, Hilary Davison, Lucinda Hudson, Maria A. Caãadas, Paul Biddle, and Sara Lilley. 2019. Dynamics of precarity among 'new migrants': Exploring the worker-capital relation through mobilities and mobility power. *Mobilities* 14 (5). https://doi.org/10.1080/17450101.2019.1611028.

Vinnustaðaskírteini. n.d. Samkomulag ASÍ og SA um útlendinga. http://www.skirteini.is/samkomulag-asi-og-sa/. Accessed on July 20, 2019.

10

Employee Motivation and Satisfaction Practices: A Case from Iceland

Magnus Asgeirsson, Paulína Neshybová, Brynjar Thor Thorsteinsson and Ester Gústavsdóttir

Introduction

In this chapter, the importance of employee motivation and satisfaction is discussed in the context of human resource management (HRM) strategy and how internal marketing (IM) practices can be utilised as tool to support and implement such strategy in the hospitality industry.

M. Asgeirsson (✉)
Faculty of Life and Environmental Sciences,
University of Iceland, Reykjavik, Iceland
e-mail: mha@hi.is

P. Neshybová
University of Iceland, Reykjavik, Iceland

B. T. Thorsteinsson
University of Bifrost, Borgarnes, Iceland
e-mail: brynjar@bifrost.is

E. Gústavsdóttir
Reykjavik University, Reykjavik, Iceland
e-mail: esterg@ru.is

A. Walmsley et al. (eds.), *Tourism Employment in Nordic Countries*,
https://doi.org/10.1007/978-3-030-47813-1_10

The tourism industry is one of the fastest developing in the world, is highly labour-intensive and constantly faces difficulties in retaining its employees (Santa Cruz et al. 2014). An example is provided from the hotel sector in Iceland, where in-depth interviews were conducted with managers and front desk employees of the same hotel chain to gather their perceptions and perspectives on employee motivation and job satisfaction practices. Iceland presents an interesting case for this exploration, as its tourism industry has experienced unprecedented year-on-year growth, in terms of tourist arrivals, since the eruption of Eyjafjallajökull in 2010. As a result, due to the lack of time and resources, the already labour-intensive industry has faced increasing problems in hiring, training and retaining their employees.

HRM in the Hospitality Industry

HRM has evolved from personnel administration through personnel management to a more holistic view of human resources (HR) as an asset for organisations to develop and use to gain a competitive advantage. HRM has been defined by Price (2011) as follows:

> A philosophy of people management based on the belief that human resources are uniquely important in sustained business success. An organisation gains a competitive advantage by using its people effectively, drawing on their expertise and ingenuity to meet clearly defined objectives. HRM is aimed at recruiting capable, flexible and committed people, managing and rewarding their performance and developing key competencies. (Price 2011 p. 29)

HRM is therefore not a single action performed by a single person or department within an organisation, but rather a continuum throughout the organisation, dealing with recruitment procedures, management and development of employees (Storey 2007; Legge 1995; Bratton and Gold 2017). Many different models and practices are related to HR, but which models or practices apply or are applied is highly dependent on the structure of the organisation and department, or the nature of the industry (Sherafat and Elahi 2018). However, strategic decisions, the selection tools and the implementation of practices—the *what, how and why*—should be under the control of HRM professionals (Sherafat and Elahi 2018; Swart et al. 2012).

Implementing a successful HR strategy is essential for any business, since employees who are well motivated, trained and satisfied in their roles are generally more loyal and demonstrate better performance with regard to service and quality of work, which in turn leads to more satisfied and loyal customers and better overall organisational performance (Mu Yeh 2013; Yee et al. 2011; Bratton and Gold 2017; Heskett et al. 1994; Sekhar et al. 2013; Hazra et al. 2015). With this in mind, implementation of HR strategy and HRM should be of primary importance for managers in the hospitality industry, and especially in the hotel sector, since it is a highly labour-intensive and service-orientated sector, with high staff turnover (Thompson and Heron 2005; Santa Cruz et al. 2014).

HR strategy and implementation of such strategy varies between organisations, especially those of different sizes. Small organisations rarely have the financial resources to employ HR specialists or to deploy an HR department, and the role of HR is therefore often performed by managers or owners. Large firms, however, often have extensive resources to utilise within their structure. The role of the HR department and its position in the hierarchy is also dependent on the organisation's size and/or number of service locations. In firms with few service locations, the HR department is situated in the uppermost levels of the hierarchy, and participates in strategy planning with a hands-on approach, supporting managers in implementing that strategy. In companies with numerous service locations, however, the HR department tends to be situated near to or in the head office or main hotel offices, and its role is to create policies and guidelines, while the actual implementation is outsourced to the service locations or their managers (de Guzman et al. 2011; Davidson and Wang 2011; Orfin-Tomaszewska et al. 2015). Organisations that have several service locations have a greater need for a functional HR department at a strategic level. This applies, for example, to hotel chains, where uniformity in quality and service is the augmented product. Here, HR should be proactive at a strategic level, creating methods and delegating tasks and functions to on-site managers in order to promptly and effectively react to various personnel situations and improve basic HR functions. These principles also ensure the standardisation of employee behaviour, thus resulting in service quality that can be guaranteed across all hotels that belong to the chain (Orfin-Tomaszewska et al. 2015). Building a strong organisational culture, promotion, appreciation and appropriate treatment of employees are also important (Hazra et al. 2015; Sekhar et al. 2013).

Despite the importance of HR and HRM strategies in the sector, perhaps due to high staff turnover and short-term employment, managers consider investment in training, knowledge development or HR strategies that help the empowerment of their employees to be a risk. It is difficult for them to achieve a balance between holistic long-term HR commitments and managing operational costs. They find that they lack know-how in resolving various HR issues, which if done correctly, can contribute to job satisfaction and motivation among employees, who then take pride in their work, and thus staff turnover is decreased (Davidson and Wang 2011; Solnet et al. 2015; Cheng et al. 2013; Hechanova et al. 2006).

With specific regard to the hotel industry, various factors influence overall job satisfaction. The most common ones relate to salary (Fisher and McPhail 2011; Jung and Yoon 2015), possibilities for promotion (Kong 2013) or working conditions (Gallardo et al. 2010). Empowerment (Zhao et al. 2016), workplace fairness (Nadiri and Tanova 2010), ethical leadership and leadership styles (Cheng et al. 2013) as well as job recognition and appraisals (Ogbonnikan 2012) also have a significant effect on the satisfaction of hotel employees. As work-related variables, both motivation and satisfaction are influenced by the same or similar factors; in other words, the factors that influence motivation correlate with those that affect satisfaction (Roos and Van Eeden 2008). These influencing factors lead not only to improved customer satisfaction, high morale and productivity, but to motivated employees who remain loyal and committed to the organisation. Employee motivation has thus become a major consideration in HRM strategies. Generally, the more that an organisation is committed to advancing its HRM, the greater the variety of incentives that it is able to offer (Naile and Selesho 2014; Teitler-Regev et al. 2014).

One of the most effective ways to improve employee motivation and job satisfaction is to combine marketing and HR together. This strategy called Internal marketing puts company employees forward as its main customers, as opposed to only external customers, focusing especially on their engagement and retention rate and thus influencing their well-being, job satisfaction and loyalty that are perceived as main goals. According to Kotler (2000) and Kotler et al. (2015) IM consists mainly of three attributes, namely hiring employees who are suited to the organisational culture, training employees correctly, and most importantly, motivating employees.

In order to support and implement these three IM attributes into everyday reality various commonly used practices or components need to be looked at and applied, such as (1) role clarity, (2) work environment, (3) evaluation of managers and (4) reward systems with promotional and career development opportunities (Arnett et al. 2002; Gjurašić and Lončarić 2018). The first key practice, role clarity, means that employees are fully aware of their duties in their jobs and feel comfortable performing these. The roles are suitable regarding employees' previous experience and knowledge, and adequate training is provided to build employee confidence (Hanaysha and Tahir 2016). Within the hotel sector, three training stages are usually applied: introduction to the organisation, general training in company practice and finally, on-the-job training (Crick and Spencer 2011). While introduction and general training, such as a first aid course, is usually performed by HR, the on-the-job training should be performed by management employees who know best what each job requires (Cheng and Brown 1998). The second component refers to work environment that should include all tangibles, processes and support that enable employees to perform, and should enhance their well-being (Wahlberg et al. 2017). Third, organisational culture, managerial styles and feedback (formal and informal) and their evaluation play a significant role in creating a positive work environment (Saleem 2015; Schein 2017). The last component, crucial for supporting and implementing the IM attributes, is that reward systems should be in place that are based on team and individual performance and adhere to the goals and objectives of either the organisation as a whole or individual departments. The organisation's goals and objectives must be clear, simple and achievable, as must the reward system itself. Career paths and promotional opportunities also play a role in rewarding effective performance (Nadiri and Tanova 2010; Güngör 2011).

All of the aforementioned factors can lead employees to take pride in their work, which can benefit organisations by lowering staff turnover, increasing the quality of service and improving employee work-related motivation, satisfaction and performance (Santa Cruz et al. 2014; Marques et al. 2018; Arnett et al. 2002; Hanaysha and Tahir 2016).

Methodology

This study mainly followed the literature connected to work-related motivation and satisfaction, with the aim of gaining an understanding of the organisation and its HR department, for which the functions and usage of practices related to employee motivation and job satisfaction were investigated. For a methodological approach, qualitative research has been chosen mainly because it helps to investigate a problem or specific area that has not been researched before by delving deeper into experiences and knowledge of individuals. Especially in hospitality industry that heavily relies on human factor, various qualitative methods are being used in order to explore different circumstances and issues that cannot be measured by numbers (Arendt et al. 2012).

In this case, semi-structured interviews were chosen as a data gathering method to acquire sensitive information about work-related issues and processes. Four hotel or reception managers and eight front desk employees of the same hotel chain in Iceland were interviewed with a set of open-ended questions about HRM, motivation and job satisfaction at their workplace. The hotel chain used as a case study in this chapter is a well-established organisation in the Icelandic context and has been operating in the market for several years. Even though the sample size cannot clearly represent most of the hotels in Iceland or management personnel and front desk employees, it was big enough to sufficiently answer the research question, as the qualitative research is usually built on non-representative sampling (O'Reilly and Parker

2013). Also, given the structure of the hotel sector in Iceland, unlike in most larger countries, it is not possible to expand in detail on the nature of the individual hotels for fear of revealing their identity.

Findings and Discussions

In this chapter findings from the interviews conducted are provided and compared to theory discussed in the previous chapters. The findings extracted, reflect the perspectives, experiences and perceptions of the interviewees on how things are done within their specific place of work and the hotel chain. Some discrepancies between theory and practice are pointed out but also practises that are somewhat in line with the theory.

All interviewed participants were aware of the existence of the HR department for the whole hotel chain and that it is located in the main offices in one of the hotels. However, neither the managers nor the front desk employees were able to define the HR department's responsibilities or how many people work within it. None of the front desk employees had any personal experience with the HR department while working for the company and had neither encountered HR in relation to recruiting nor in other work-related issues. For this reason, they automatically assigned various functions that usually belong to HR to their direct supervisors. The managers who were interviewed also had little information to contribute with regard to this department. They discussed its responsibility for the legal paperwork of employees of the company, or for addressing certain urgent issues that might arise across all hotels, but in general, the managers did not have any further knowledge about the roles and functions of the HR department. Based on the theoretical background, HRM should not be a single action-based administrative function, but rather a continuum that deals with recruitment, management and development of employees (Bratton and Gold 2017; Sherafat and Elahi 2018). Some managers also mentioned the strategies and goals of HRM regarding strengthening teamwork and improving guest satisfaction, while others seemed to be unfamiliar with such things. However, all managers admitted that these general hotel strategies do not include any details about employee motivation or satisfaction practices. As the literature has shown, the HR department usually has a complicated position in the hotel industry, especially with regard to forming and applying suitable systems across businesses that own several hotels and employ hundreds of people (de Guzman et al. 2011).

Since most of the motivation and satisfaction practices are dependent on a company's structure and strategy, HRM or managerial style (Naile and Selesho 2014), managers were expected to have at least some knowledge about what keeps front desk employees satisfied and how to achieve it. All the managers who were interviewed were informed about the topic of the interviews, but most of them did not seem to have much knowledge of the practices aimed at employee motivation and satisfaction, nor did they express particular interest in them. None of them saw motivation and satisfaction in the workplace as a complex issue or something that should be regularly pursued or even be made a part of the company's HR strategy. This result is in alignment with Solnet et al. (2015), who claimed that one of the issues facing the hotel industry is the absence of any practices or strategies, mainly due a to lack of interest or know-how from management, or their focus on flexible management. The only motivation and satisfaction factors of which the managers were aware were fair salaries, positive working environment and regular appraisal, which is in accordance with several studies (Fisher and McPhail 2011; Kong 2013; Gallardo et al. 2010). These answers were based only on the manager´s opinions and guesses and were not related to knowledge of the hotel business or any specific practices or strategies within their hotel or chain.

In contrast, front desk employees were more aware of various motivation and satisfaction factors, mainly due to their own experience in this field. The most important satisfaction factors for them appeared to be appraisal, regular feedback, responsibility, freedom in decision-making and opportunities to learn new things, thus corresponding with the outcomes of various motivation theories and studies (Zhao et al. 2016; Ogbonnikan 2012). The front desk employees did not perceive money to be one of the most important factors in influencing their motivation and satisfaction, which is in contrast with many of the studies that document the effects of monetary incentives on employee motivation and satisfaction (Jung and Yoon 2015; Fisher and McPhail 2011).

In terms of the methods that are being used in the implementation of HR strategies, participants were asked to elaborate on practices related to IM. Findings reveal that there is a general policy in place within the chain for the initial introduction of the company and their hotels, that also includes the signing of employment contracts. This policy is delegated to managers to implement and use in introductory training. However, some managers were unaware of its existence, and therefore not every hotel follows the introduction guidelines set by the HR department. Regarding general training, both groups of participants mentioned that there is first aid training for all employees, and that it is carried out in cooperation with the Red Cross. However, neither the introductory training nor the first aid course was available to everyone at the time at which the interviews were conducted. All managers claimed the on-the-job training to be the most important kind of training and stated that it is performed by them or other colleagues in a similar position. The front desk employees' answers showed that such training is either non-existent or too brief. It also seems to be chaotic and is usually left up to the front desk personnel themselves to organise. This means that neither the HR department nor the hotel management prepares training for their employees, as no training programme or strategy is applied across all hotels within the hotel chain. According to the literature, very few hotels pay attention to training due to the additional costs involved and high staff turnover within the industry, which accords with the case of this hotel chain (Cheng et al. 2013).

In terms of reward systems and acknowledgement, various non-financial tangible rewards are applied, such as vouchers for restaurants, free trips with travel companies or various events for employees. The initiative for this often comes from the tour operators and/or restaurants in order to introduce their products and services, rather than being planned by management. According to employees this seems to be given out somewhat randomly and not necessarily linked to job performance. Tangible financial rewards are also given by the hotel for selling various items, such as trips or up-selling breakfast (where it was not included). The rules for receiving rewards, from commission-based sales differ between hotels; some allocate them based on personal sales performances, but others divide part of the commission equally among all front desk employees. According to employees, salaries are kept to a minimum and the possibility for a pay raise comes only through Union contracts. Incentives in the form of appraisal and recognition by management seem to be common, according to the interviewees. However, the system in place is neither clear nor simple and employees are uncertain of what they can achieve based on evaluation (Nadiri and Tanova 2010; Güngör 2011). According to P.-Y. Cheng et al. (2013), career development is one of the most important factors that contributes to employee motivation and satisfaction, but this seems to be underdeveloped within the hotels. Although career development opportunities are clear to everyone within the reception department, once an employee wants to transfer to another department or another hotel within the same hotel chain, there is no structure to be followed. Whenever a position becomes available, an email is sent out, but the hiring process is unclear and existing employees are seemingly not encouraged to apply. The hotel managers can promote employees only within their area, and for changes between departments, they require the approval of higher management or sometimes even the HR department.

The findings demonstrate that neither the HR department nor the management had developed any kind of practice or general strategy towards mutual understanding and communicating the company's vision for the future. Most of the front desk employees did not know what is happening in the company and they lacked information regarding its goals and plans. The reason behind this might be that neither the HR department nor the hotel management is aware of various positive outcomes, that emerge by informing employees about a company's strategies and projects, such as the ability to deal with different work-related issues, opportunities for further growth or effective employee performance (Arnett et al. 2002). The literature also suggests that the more organisational culture is communicated, and employees are involved in decision-making, the more committed the employees feel to working towards company's objectives (Naile and Selesho 2014). Most of the front desk personnel were highly satisfied with their tasks and enjoyed the freedom that they had in decision-making. However, their responsibilities differed between hotels, based on managerial style. This correlates with many studies and theories see for example

Hechanova et al. (2006) and Zhao et al. (2016) that have documented that the effect of the proper delegation of responsibilities is being able to see the results of one's work. Also, that responsibilities given to employees makes them feel trusted and empowered.

The importance of interpersonal relationships also proved to be a recurring point in all interviews. Managers were described as possessing good interpersonal communication skills; they are supportive and open, which positively influences the working relationship. The front desk employees stated that they genuinely like their managers and the way in which they are treated, and it seems to be one of the most important factors that keeps them satisfied in their workplace. This is in line with Wahlberg et al. (2017) finding that having skilled and supportive management creates a pleasant working environment and strong working relationships, which can have a significant influence on employee satisfaction. At the company in question, there seems to be no system in place for employees to evaluate or give feedback regarding their work or work environment, other than directly to their supervisor, to their managers or even to the HR department. The department has not established any specific tools for this purpose, except for an anonymous questionnaire that was supposed to be used to analyse teamwork, customer service and provide feedback about management. However, this study found that not everyone was aware of the questionnaire's existence, and even those who had the chance to contribute to it had never heard of its results. This conflicts with advice given by Saleem (2015) and Arnett et al. (2002) on the importance of applying a system that allows employees to provide regular feedback to the management, in order to directly influence employee satisfaction and their performance.

Organisational performance in the chosen hotel chain was not given much attention. Some managers did not seem to care how their employees felt about the company if they were happy to work, while others claimed that the front desk employees are very proud to be working there. The front desk employees felt the same way, and most of them did not have strong opinions on this topic or had never previously thought about it. In addition, expressed by the interviewees, neither the management nor the HR department did anything specific to make their employees aware of the company's performance on the market for example by communicating key performance indicators. This study therefore did not confirm the importance of this factor as argued in the literature.

Conclusion

The findings of this study reveal that an HR department in this specific case exists, but that its presence and role are somewhat unclear to both managers and front desk employees. In interviews, managers stated that the role of HR was at an administrative functions level—for example, processing legal paperwork—and not all were aware of the training materials and other support that the department could provide. Similarly, employees had no knowledge of the department, as it did not interact with them through the process of hiring and training. They had only minimal knowledge about the hierarchy of the company, or about who is responsible for various issues related to employee management, including practices aimed at motivation and satisfaction among the front desk staff. Some knew about a job survey that was apparently performed by the HR department, but those who had participated had not been shown any of its results or findings. It seems that many of the functions usually assigned to HR are transferred to hotel management, without a clear strategy or guidelines. A better structured organisation hierarchy as well as cooperation and communication between managers and the HR department is needed in order to influence employee motivation and satisfaction.

Uniformity and consistency seem to be absent in terms of training initiatives within the chain. Managers are left to decide how training is performed and what is emphasised at all stages, i.e. introductory, general and on-the-job training, which seem to be either missing, too short or inadequate. Managers seem to be reluctant to consider employees as a long-term workforce or investment, and that is perhaps one of the reasons why training is kept to a bare minimum, but lack of know-how could also play a role. However, employees do not necessarily see their employment as a short-term relationship and would like the structural approach to improve from a chaotic state to one that is more organised and proactive. They would also like to be more involved and have more knowledge about the company and the opportunities that can arise within it, for example, in terms of promotions and changing venues. A long-term investment in training and career building is therefore needed, in addition to the devotion of more attention to employees as internal customers.

Although employees are rewarded for their work and given incentives, no clear and uniform system is in place throughout the chain. Instead, it is left up to individual hotels or managers to reward employees as they see fit. This means that employees are unsure whether their efforts will be rewarded due to unpredictability or the random appearance of rewards regardless of achievements or efforts. An interesting point that has not been mentioned in any of the previous research cited in this chapter is that most of the non-financial rewards that are used in this hotel chain come from external sources. This could be emphasised by management and be included in their reward strategy aimed at enhancing employee motivation and satisfaction. The employees interviewed did not consider this a part of a strategy, but rather as random outside incentive. Promotion opportunities and career paths within the chain seem unclear to employees, and they have little or no knowledge of such things outside of their department or hotel. This will inevitably affect their relationship with work and long-term prospects, and perhaps reduce their job motivation and satisfaction. The vision and performance of the organisation must be communicated more effectively to managers and employees alike to ensure their loyalty, which can result in better organisational performance.

Although the company lacks an IM strategy and as well, according to the interviewees, a systematic approach from the HR department or management. Several tools and practices that can enhance employee motivation and satisfaction were identified in this case, especially in relation to the work environment. For example, interviewees reported to have experienced freedom in decision-making, delegation of tasks, open communication and good interpersonal relationships, personal appraisal and general support and rewards. Perhaps for that reason, front desk employees were not dissatisfied, although they did not see themselves working for the company in the distant future. As previously mentioned, employees were willing to stay and develop their careers, but found themselves without a cause or opportunity.

It is therefore recommended, given the nature of the industry and sector, that employee motivation and satisfaction should be put at the forefront of HR strategies and aligned with managerial practices. HR specialists must have a holistic view and understand the linkage between employee satisfaction and motivation, service performance, customer satisfaction and profits, all of which affect the sustainability of the organisation. This means that the HR personnel and managers need to create and nurture a strong organisational culture that emphasises effective internal communication, the well-being of its employees, systematic rewards and the constant development of its employees' key competencies.

On that note it is important to recognise, as this case shows in some instances, that theory and practice sometimes seem to be worlds apart and may in fact view things very differently. Even more importantly realising that it does not have to be, or perhaps shouldn't be that way. In many instances, theory could and should be viewed as best practice and utilised by managers as such. This notion opens to further research on the topic in terms of knowledge transfer from theory (best practice) to practitioners within the field and what role scholars should have in that. Perhaps more cooperation is needed in order to combine the two worlds to enhance continuous learning and knowledge transfer.

References

Arendt, Susan W., Kevin R. Roberts, Catherine Strohbehn, Jason Ellis, Paola Paez, and Janell Meyer. 2012. Use of qualitative research in foodservice organizations. *International Journal of Contemporary Hospitality Management* 24 (6): 820–837. https://doi.org/10.1108/09596111211247182.

Arnett, D.B., D.A. Laverie, and C. McLane. 2002. Using job satisfaction and pride as internal marketing tools. *Cornell Hotel and Restaurant Administration Quarterly* 42 (2): 87–96.

Bratton, John, and Jeff Gold. 2017. *Human Resource Management: Theory and Practice*, 6th ed. London: Palgrave Macmillan.

Cheng, Angeline, and Alan Brown. 1998. HRM strategies and labour turnover in the hotel industry: A comparative study of Australia and Singapore. *The International Journal of Human Resource Management* 9 (1): 136–154. https://doi.org/10.1080/095851998341233.

Cheng, Pi-Yueh, Jen-Te Yang, Chin-Sheng Wan, and mei-chin Chu. 2013. Ethical contexts and employee job responses in the hotel industry: The roles of work values and perceived organizational support. *International Journal of Hospitality Management* 34: 108–115. https://doi.org/10.1016/j.ijhm.2013.03.007.

Crick, Anne P., and Andrew Spencer. 2011. Hospitality quality: New directions and new challenges. *International Journal of Contemporary Hospitality Management* 23 (4): 463–478. https://doi.org/10.1108/09596111111129986.

Davidson, Michael C.G., and Ying Wang. 2011. Sustainable labor practices? Hotel human resource managers views on turnover and skill shortages. *Journal of Human Resources in Hospitality & Tourism* 10 (3): 235–253. https://doi.org/10.1080/15332845.2011.555731.

de Guzman, Gloria M., James P. Neelankavil, and Kaushik Sengupta. 2011. Human resources roles: Ideal versus practiced—A cross-country comparison among organizations in Asia. *The International Journal of Human Resource Management* 22 (13): 2665–2682. https://doi.org/10.1080/09585192.2011.560865.

Fisher, Ron, and Ruth McPhail. 2011. Internal labour markets as a strategic tool: A comparative study of UK and Chinese hotels. *The Service Industries Journal* 31 (2): 137–152. https://doi.org/10.1080/02642060802644942.

Gallardo, Eva, Maria Margarida Nascimento Jesus, Sandra-M Sánchez-Cañizares, and Tomás López-Guzmán. 2010. Employee satisfaction in the Iberian hotel industry: The case of Andalusia (Spain) and the Algarve (Portugal). *International Journal of Contemporary Hospitality Management* 22 (3): 321–334. https://doi.org/10.1108/09596111011035936.

Gjurašić, Matina mag oec, and Dina Lončarić. 2018. International market orientation and organizational citizenship behaviour in the hospitality industry—A literature review. *Ekonomska Misao i Praksa* 27 (1): 157–176.

Güngör, Pınar. 2011. The relationship between reward management system and employee performance with the mediating role of motivation: A quantitative study on global banks. *Procedia—Social and Behavioral Sciences* 24: 1510–1520. https://doi.org/10.1016/j.sbspro.2011.09.029.

Hanaysha, Jalal, and Putri Rozita Tahir. 2016. Examining the effects of employee empowerment, teamwork, and employee training on job satisfaction. *Procedia—Social and Behavioral Sciences* 219: 272–282. https://doi.org/10.1016/j.sbspro.2016.05.016.

Hazra, Krishnendu, Prosenjit Ghosh, and Partha Pratim Sengupta. 2015. Motivation and performance linkages: A comparative study between two sectors of hospitality industry. *Journal of Business Studies Quarterly* 7 (1): 77.

Hechanova, Ma Regina M., Ramon Benedicto A. Alampay, and Edna P. Franco. 2006. Psychological empowerment, job satisfaction and performance among Filipino service workers. *Asian Journal of Social Psychology* 9 (1): 72–78. https://doi.org/10.1111/j.1467-839X.2006.00177.x.

Heskett, James L., Thomas O. Jones, Gary W. Loveman, W. Earl Sasser Jr., and Leonard A. Schlesinger. 1994. Putting the service-profit chain to work. *Harvard Business Review* 72 (2): 164–170.

Jung, HyoSun, and HyeHyun Yoon. 2015. Understanding pay satisfaction: The impacts of pay satisfaction on employees' job engagement and withdrawal in deluxe hotel. *International Journal of Hospitality Management* 48: 22–26. https://doi.org/10.1016/j.ijhm.2015.04.004.

Kong, HaiYan. 2013. Relationships among work-family supportive supervisors, career competencies, and job involvement. *International Journal of Hospitality Management* 33: 304–309. https://doi.org/10.1016/j.ijhm.2012.09.006.

Kotler, Philip. 2000. Marketing management: The millennium edition. *Marketing Management* 23 (6): 188–193.

Kotler, Philip, Suzan Burton, Kenneth Deans, Linen Brown, and Gary Armstrong. 2015. *Marketing*. Australia: Pearson Higher Education AU.

Legge, Karen. 1995. What is human resource management?" In *Human Resource Management: Rhetorics and Realities*, 62–95. London: Macmillan Education UK.

Marques, Carlos Peixeira, Carmem Teresa Leal, Carla Susana Marques, and Breno Tinoco Cabral. 2018. Internal marketing and job satisfaction in hotels in Via Costeira, Natal, Brazil. *Tourism & Management Studies* 14 (1): 36–42.

Nadiri, Halil, and Cem Tanova. 2010. An investigation of the role of justice in Turnover intentions, job satisfaction, and organizational citizenship behavior in hospitality industry. *International Journal of Hospitality Management* 29 (1): 33–41. https://doi.org/10.1016/j.ijhm.2009.05.001.

Naile, Idah, and Jacob M. Selesho. 2014. The role of leadership in employee motivation. *Mediterranean Journal of Social Science* 5 (3): 175–182. https://doi.org/10.5901/mjss.2014.v5n3p175.

O'Reilly, Michelle, and Nicola Parker. 2013. 'Unsatisfactory Saturation': A critical exploration of the notion of saturated sample sizes in qualitative research. *Qualitative Research* 13 (2): 190–197. https://doi.org/10.1177/1468794112446106.

Ogbonnikan, Esther. 2012. Job satisfaction and efficiency among hotel workers: A case study Giant Star Hotel Nigeria. Theseus. http://www.theseus.fi/handle/10024/41439.

Orfin-Tomaszewska, Katarzyna, Marta Sidorkiewicz, and Anna Tokarz-Kocik. 2015. Human resource management in chain hotels on the example of Radisson Blu hotel in Szczecin. *Ekonomiczne Problemy Turystyki* 31. https://doi.org/10.18276/ept.2015.3.31-15.

Price, Alan. 2011. *Human resource management*. Andover: Cengage Learning EMEA.

Roos, Wanda, and René Van Eeden. 2008. The relationship between employee motivation, job satisfaction and corporate culture. *SA Journal of Industrial Psychology* 34: 54–63.

Saleem, Hina. 2015. The impact of leadership styles on job satisfaction and mediating role of perceived organizational politics. *Procedia—Social and Behavioral Sciences* 172: 563–569. https://doi.org/10.1016/j.sbspro.2015.01.403.

Santa Cruz, Francisco González, Tomás López-Guzmán, and Sandra Ma Sánchez Cañizares. 2014. Analysis of job satisfaction in the hotel industry: A study of hotels in Spain. *Journal of Human Resources in Hospitality & Tourism* 13 (1): 63–80. https://doi.org/10.1080/15332845.2013.807394.

Schein, E.H. 2017. *Organizational Culture and Leadership*. Edited by 5th ed. Hoboken, NJ: Wiley.

Sekhar, Chandra, Manoj Patwardhan, and Rohit K. Singh. 2013. A literature review on motivation. *Global Business Perspectives* 1 (4): 471–487. https://doi.org/10.1007/s40196-013-0028-1.

Sherafat, Shapoor, and Ahmad Ali Khaef Elahi. 2018. Developing a model for alignment between individual and organizational performance management and organizational objectives. *Journal of Human Resource Management* 21 (2): 49–57.

Solnet, David, Anna Kralj, and Tom Baum. 2015. 360 degrees of pressure: The changing role of the HR professional in the hospitality industry. *Journal of Hospitality & Tourism Research* 39 (2): 271–292. https://doi.org/10.1177/1096348012471380.

Storey, John. 2007. *Human Resource Management: A Critical Text*, vol. 3. London: Thomson Learning.

Swart, J., C. Mann, S. Brown, and A. Price. 2012. *Human Resource Development: Strategy and Tactics*, vol. 2. Oxford: Elsevier.

Teitler-Regev, Sharon, Oksana Goziker, and Shosh Shahrabani. 2014. Factors affecting the decision to remain in the hospitality industry among hotel front office desk employees: The case of Israel. *Journal of Tourism and Hospitality Management* 2 (9): 364–375. https://doi.org/10.17265/2328-2169/2014.09.003.

Thompson, Marc, and Paul Heron. 2005. Management capability and high performance work organization. *International Journal of Human Resource Management* 16 (6): 1029–1048. https://doi.org/10.1080/09585190500120806.

Wahlberg, Tiia Annika, Nelson Ramalho, and Ana Brochado. 2017. Quality of working life and engagement in hostels. *Tourism Review of AIEST—International Association of Scientific Experts in Tourism* 72 (4): 411–428.

Yee, Rachel W.Y., Andy C.L. Yeung, and T.C.E. Cheng. 2011. The service-profit chain: An empirical analysis in high-contact service industries. *International Journal of Production Economics* 130 (2): 236–245.

Yeh, Chien Mu. 2013. Tourism involvement, work engagement and job satisfaction among frontline hotel employees. *Annals of Tourism Research* 42: 214–239. https://doi.org/10.1016/j.annals.2013.02.002.

Zhao, XinYuan, R. Ghiselli, R. Law, and Jing Ma. 2016. Motivating frontline employees: Role of job characteristics in work and life satisfaction. *Journal of Hospitality and Tourism Management* 27: 27–38.

11

Managerial Practices of Co-creation and Psychosocial Work Outcomes

Olga Gjerald and Trude Furunes

Introduction

Managing a service firm is a multifaceted task that requires handling customers', employees', and other stakeholders' interests simultaneously. One of the main goals of service providers is to satisfy customers and ensure that the service they pay for is of value to them so that they are willing to repurchase and also recommend the service to others. To provide high quality and stable service over time, it is crucial to have a stable, well-trained workforce. Thus, the second goal is to make sure that employees are satisfied with their job and intend to stay with the company. In most service companies, some employees are working closely with customers (also called front stage or front office) whereas

O. Gjerald (✉) · T. Furunes
Norwegian School of Hotel Management,
University of Stavanger, Stavanger, Norway
e-mail: olga.gjerald@uis.no

T. Furunes
e-mail: trude.furunes@uis.no

A. Walmsley et al. (eds.), *Tourism Employment in Nordic Countries*,
https://doi.org/10.1007/978-3-030-47813-1_11

others are handling back-office jobs (backstage). Consequently, one of the challenging management tasks is to set up a team of employees who understand how their job tasks contribute to the end service, i.e., making employees in all parts of the service chain responsible for customer satisfaction and value. The idea is that the job will be more meaningful to the individual if he or she comprehends its entirety (i.e., the service they are a part of).

The literature on service management is vast. In the last decade, research on co-creation (i.e., the customer or guest is involved in the making of the service they are buying) has evolved with an aim to create value for customers and firms (Neghina et al. 2017). The concept remains at the root of both service-dominant logic (SDL) (Vargo and Lusch 2016) and service logic (SL) (Grönroos 2012). However, in their recent paper titled "Service-dominant logic 2025", Vargo and Lusch (2017) called for more empirical research on providers' experiences of the co-creation. There is also a need to connect the empirical research on managerial practices of co-creation to the working environment of service employees (Grissemann and Stokburger-Sauer 2012). In this chapter, we aim to give an account of service management practices that are targeted to improve the co-creation along with the psychosocial work outcomes. We exemplify how managers in service occupations can apply different practices to improve co-creation and how these practices intersect with managing psychosocial risk factors at work. Our approach is both theoretical and empirical. The goals in the theoretical analysis are to (a) contribute with knowledge about provider practices of co-creation in service, as called for by Vargo and Lusch (2017), (b) highlight the interplay between psychosocial work outcomes and managerial practices of co-creation, and (c) discuss how managers in hospitality, tourism, and service occupations can work to minimise psychosocial risks, as called for by Grissemann and Stokburger-Sauer (2012) and Jaakkola and Alexander (2014). The empirical material comes from a study conducted in Norway. Based on our empirical material, we propose six particular practices that managers in service occupations apply to their work and the specific psychosocial factors these practices intersect. Before presenting the managerial practices, we briefly summarise some aspects of the literature on co-creation. The body of literature on this topic is rapidly growing, and some recent attempts have been made to integrate the different approaches (Oertzen et al. 2018; Harrington et al. 2019). Even though it is not possible to study providers' co-creation practices exhaustively in the present chapter, we aim to take a small step in this direction. In the third section, we present our method and in the subsequent two sections, we illustrate managerial practices of co-creation on the basis of our empirical material. We focus particularly on the different ways in which managers in tourism and hospitality can work to prevent psychosocial risks at work. Finally, we link our empirical results with the theoretical discussion on Nordic leadership and meaning of work.

Background

The literature offers different perspectives on how service providers and service customers create value, and on the role the service provider and the customer have in the value creation process.

The service-dominant logic (SDL) (Vargo and Lusch 2008a, b) stresses that *value* is no longer delivered but *co-created*. SDL prioritises the interaction between the company and the customer and implies that value is created in the interaction process itself rather than exclusively in the provision of the service (Etgar 2008, Grönroos 1997). Therefore, involving customers in the service creation helps tailor the service to the customers' particular needs and hence assists in creating a unique experience (Grissemann and Stokburger-Sauer 2012). The customer realises value by considering and using the service provider's offers and resources. The SDL views co-creation in terms of participatory, interactive activities that involve different actors, while value (sense of being better-off) is defined as "value-in-use", that is, "the value for customers, created by them during their usage of resources" (Grönroos and Gummerus 2014, p. 209). Co-creating customers are conceptualised as resource integrators who possess certain skills, knowledge, and capabilities (termed "operant resources") and use these to "act upon" generally tangible, static things ("operand resources") to create an effect (Vargo, Maglio, and Akaka 2008). For SDL, the number of studies on consumer involvement in SDL value co-creation in tourism and hospitality has grown substantially. Shaw et al. (2011) were among the few to empirically assess the concept of SDL and its implications for tourism management in a hospitality setting from the providers' side. Recent studies have investigated how different technologies facilitate co-creation (e.g., Morosan 2018). Both SDL and Service Logic by Gronroos continues the Nordic tradition of viewing the outcomes of the service and the process of service delivery as a forming part of the total experience (Grönroos 2011).

Heinonen and colleagues criticised the SDL for being provider-dominant (Heinonen et al. 2013), and proposed a customer-dominant logic of service (Heinonen and Strandvik 2015, Heinonen et al. 2010). CDL assumes that value extends beyond the co-creation interactions and consumption, and rather than being restricted to co-creation in interactions between the company and the customer, value emerges solely in the customer's sphere (Heinonen et al. 2010). The internal logic of CDL is based on positioning customer insight in the foreground in place of the type of offering or the system of providers (Heinonen and Strandvik 2015). The CDL relocates value-creating agency into customers' own life contexts and beyond specific visible service interactions that are normally in service providers' sphere of influence, rendering customers' value-forming processes "invisible" to service organisations (Heinonen et al. 2010). This is somewhat opposite to SDL which views consumer value (value-in-use) as the final step of a process that begins with and is always facilitated by the provider (Grönroos and Gummerus 2014).

Recent studies of customer-dominant logic have looked at complaining practices on social media in tourism (Dolan et al. 2019) (i.e., three practices of tourist complaints and their potential for both value co-creation and co-destruction), customer-to-customer co-creation practices in tourism (i.e., tourists' social practices of co-creation at festivals) (Rihova et al. 2018), or tourists' creation of their own events (i.e., how off-track travellers use their real-life experience to turn into successful events) (Ben Gamra Zinelabidine et al. 2018).

The evolution of business logics from goods-dominant to customer-dominant has been fuelled by three streams of research literature (Harrington et al. 2019). First, the marketing literature which considers the ideas of SDL from a perspective of shared responsibilities by both customers and service providers (Vargo and Lusch 2004) and revisits the customer experience concept to include a co-created experience (Boswijk et al. 2012). Second, the innovation literature which considers co-creation among firms to develop relationships and interact to gain new knowledge (Roberts and Darler 2017). Third, the customer engagement literature which overlaps with the former two (e.g., Chathoth et al. 2016). While goods-dominant logic's primary value proposition is based on value-in-exchange (transactional value), SDL is associated with value-in-use, and customers perceive the value during the service consumption process in terms of reliability, quality, consistency, and responsiveness to customers' needs (Sandström et al. 2008). Customer-dominant logic's value proposition is a value-in-experience concept and is thus tied to characteristics of the experience—aesthetics, entertainment, education, and escapism (Pine and Gilmore 1999). A further extension of the range of business logics is memory-dominant logic, a value-in-memory approach, and a firm strategy (Harrington et al. 2019). Studies on MDL are yet scarce since this is a rather novel concept, and so far they attempt to get insight into customers' perceptions of memorable experiences (e.g., Sthapit and Coudounaris 2018).

Co-creation is a process in which both customers and providers participate, so we can assume that both service employees and customers get something out of the co-creation. Six co-creation dimensions might influence the overall co-creation experience for providers as well as customers: (1) personal benefits ("gaining a better status and recognition"), (2) social benefits ("being able to connect with other people"), (3) hedonic benefits ("having pleasurable experiences"), (4) cognitive benefits ("acquiring new knowledge/skills"), (5) economic benefits ("compensation in line with effort made"), and (6) pragmatic benefits ("solutions better meeting personal needs") (Verleye 2015, pp. 323–324). Outcomes of the co-creation process do not necessarily have to be beneficial, they can also be counterproductive for both customers and service providers, according to Oertzen et al. (2018).

There are still few empirical studies investigating beneficial as well as counterproductive service provider outcomes associated with the co-creation experience according to Oertzen et al. (2018) supporting earlier reviews of Mustak et al. (2013, 2016). For instance, the literature rarely discusses hedonic benefits for service providers, although a case study of customer engagement behaviour (Jaakkola and Alexander 2014) showed how co-creation leads to an improved working environment. Previous research suggests that the co-creation of service can lead to personal benefits for the service provider through improved decision support (Pinho et al. 2014) and a better provider image (Toivonen and Tuominen 2009). Social benefits of co-creation for service providers in tourism and hospitality include better service innovation synergy (Hsieh et al. 2013), contract retention with customers (Burdon et al. 2015), and cross-functional integration and teamwork within the service provider (Alam 2011). The counterproductive outcomes of co-creation for providers include personal co-creation outcomes, such as need for service protection and increased uncertainty (Hurmelinna-Laukkanen and Ritala 2010), and job stress (Hsieh and Yen 2005). The counterproductive social co-creation outcomes include service failures attributed to provider and role conflict (Hsieh et al. 2004, Hsieh and Yen 2005). The economic co-creation outcomes are for example declining market performance (Carbonell et al. 2012) and perceived workload (Hsieh et al. 2004). And finally, the pragmatic co-creation outcomes which include less productive ideas (Magnusson et al. 2003).

Regarding service jobs, psychological and social relations with both colleagues and guests are of particular importance for employees' job satisfaction and well-being at work. Typically, in Nordic countries, Work Environment Acts regulate how employers should work to systematically improve the psychosocial work environment by continuously surveying, judging potential risks, and conducting follow-up (Arbetsmiljöverket 2015). Norwegian studies show that 15% of sickness leaves are related to the negative psychosocial work environment (Aagestad et al. 2014a, b). The most important risk factors across studies are high demands and low control, combined with low autonomy, low social support, low job security, high role conflicts, and high emotional demands (STAMI 2018).

Although every occupation has its psychosocial risk factors, job demands and job resources are two general categories identified in the literature. The "Job-demands-resources model" (JD-R), introduced by

Bakker and Demerouti (2007), explains that occupational stress occurs due to imbalance between the job demands (e.g., work pressure or emotional demands) placed on the employee, and the individual resources (e.g., career opportunities, autonomy, role clarity, performance feedback, social support, or self-efficacy) that this employee has available in order to deal with the demands. If a manager can recognise specific job demands and job resources that each employee faces, health impairments due to strain can be reduced, and job resources can be increased. Previous studies have shown that increasing job resources can lead to increased motivation, engagement, and performance. Increased job resources can also buffer the effect of job demands. For instance, an employee who can decide how he or she is going to solve a job task (high autonomy) can handle higher job demands, unlike someone who is forced to do a task in a certain way.

How do service providers manage co-creation in order to increase beneficial and decrease counterproductive psychosocial outcomes? In this chapter, we give an account of different managerial practices associated with service-dominant and customer-dominant logics mainly, and the ways in which these practices are related to psychosocial provider outcomes. To date, few studies have explored managerial challenges and empirical practices of co-creation in labour-intensive services, such as hospitality, tourism, or business consulting. Using the JD-R model as a theoretical point of departure, we take a closer look at what service managers do to handle the co-creation, how these managerial practices reflect different business logics and which psychosocial outcomes they are associated with. By doing so, we respond to recent calls for empirical research on SDL (Vargo and Lusch 2017), linking these to management challenges of building sustainable organisations and sound work environments (Jaakkola and Alexander 2014).

Method

Empirical data that we present in this chapter come from a qualitative, exploratory study of psychosocial risk factors in service jobs. We conducted four focus-group interviews with 18 health and safety representatives from 15 companies in Rogaland, Norway, in 2018. The sample comprised employees from service occupations in the broad sense, from hotel, cleaning, food, and catering to service companies that work closely with the oil industry. The interviews lasted just over an hour and the two authors participated in all of them. We recorded the interviews and then transcribed them verbatim. The transcribed interviews formed the basis for the qualitative content analysis carried out in NVivo 12 (QSR). Data were analysed through a stepwise content analysis as a deductive–inductive approach using the a priori research questions as a guide to systematise the data (Graneheim and Lundman 2004). In Step (1), the transcribed interviews, which were the units of analysis, were read through several times to obtain a sense of the whole. In Step (2), initial coding was conducted using the software program. In Step (3), the authors reflected on the content and discussed the themes which formed a group of content with commonality (Graneheim and Lundman 2004). In the following section, we use some of our data material (in italics, quotation marks) to illustrate the practices.

Our interpretation should be seen from the perspective of Nordic management traditions and working life regulations. Nordic countries have flatter organisational structure (less hierarchy), regulated employee rights and a higher degree of unionisation among employees. In Norway, for instance, the Work Environment Act (Arbeidsmiljøloven) states that the employer is responsible for promoting a healthy and meaningful working environment. These regulations may influence how the managers in the study handle challenges that occur.

Although there are differences between the Nordic countries, managers in Nordic countries are found to have some common ways to handle challenges that differ from their Southern-European peers. The Nordic leadership style has a background in the Nordic welfare state (Andreasson and Lundqvist 2018). The Nordic welfare model has a strong dimension of individuality. On a deeper level, individuality in this context means that the individual (employee) is not heavily dependent on either the employer or workplace. This is probably the driving force behind aspects such as the flat organisational structure. Nordic managers are found to rely on unwritten rules and to be more employee-oriented compared to other European managers. The

Nordic leader functions as a coach for their employees, rather than an authoritative autocrat. Aspects such as consensus and co-operation are important features for Nordic leadership (Andreasson and Lundqvist 2018). Studies have also shown that employees in Nordic countries prefer managers who are inspirational and of high integrity as well as team-oriented (Smith et al. 2003). Danish managers adopt informal management with bottom-up processes and report higher reliance on subordinates and lower reliance on superiors and formal rules (e.g., Staub-Bauer 2015). Norwegian managers have more collegial management with bottom-up processes, reporting higher reliance on subordinates and co-workers. Finnish managers report relatively strong reliance on subordinates and on their own experience and training but very low reliance on formal rules (autonomy with a bottom-up emphasis) (e.g., Nenonen and Lindahl 2017). Icelandic managers report relatively strong reliance on co-workers and on their own experience, but stronger reliance on superiors compared to the other Nordic nations (collegial hierarchy) (Smith et al. 2003).

Managerial Practices of Co-creation in the Nordic Context

We have identified six managerial practices of value co-creation, reflecting different business logics, as well as handling the psychosocial risks. In the following paragraphs, we present the identified practices and explain what business logics they reflect, as well as how they connect to psychosocial outcomes. We also consider our findings in relation to Nordic leadership style.

Conditioning for Flexibility in Co-creation

Service employees experience an increasing number of quantitative job demands (the amount of work that has to be done during a shift) as well as qualitative job demands (the quality of the service provided). This is due to the need for "faster delivery" and increased customer expectations. Even though the established routines specify how the work should be carried out, deviations in demand occur daily. In fact, informants report that "*deviation has become the norm*". Therefore, a large part of a manager's task is to handle deviations, be it fluctuations in demand or changed customer expectations. Another managerial task is to prepare the employees for the "*flexibility as a new job demand*". Customers expect customisation (customer-dominant logic) and tailor-made solutions combined with faster delivery. For service providers to respond to customer expectations, there is a need to increase flexibility in tailoring the service. Thus, flexibility has become a new job demand for service employees. Especially service jobs have been known for high job demands, often in combination with low job control. In our study, informants revealed their concerns about flexibility, explaining that customers expect them to be highly flexible regardless of the circumstances.

This flexibility is conditioned by external factors (e.g., project deadlines, customer expectations, and local service providers) and is something service employees cannot directly control. Conditional flexibility can be related to perceptions of low role clarity, high job demands, and low job control. For some employees, it can also become a new job resource. Since flexibility implies fluctuation in job demands, employees who can influence decisions concerning the project deadlines or customer delivery may perceive conditioned flexibility as a source of job control. Conditioned flexibility appears to be the key factor in a value co-creation process, and as such, it is at the heart of SDL. Managers' task is to help employees to tackle this flexibility so that it would become a job resource rather than an ultimate job demand. It is also important to note that employees who wish to be more involved in the value co-creation process with customers want to exercise some control over the process and thus control the flexibility.

Customer-dominant logic is characterised by value-in-experience and different functions of the experience, such as education, entertainment, escapism, and aesthetics. Previous studies have shown that customers who show an increased level of involvement in co-creation often have some prerequisite cultural or physical capital and expertise (Shaw et al. 2011). They may, therefore, exhibit higher customisation demands. Customisation lies on the other side of flexibility (Harrington et al. 2019). As such, conditioning for flexibility in value co-creation is a managerial practice that intersects both service-dominant and customer-dominant business logics. Flexibility in a Nordic perspective is based on a high degree of autonomy. This means that the individual employee has significant power, influence, and responsibility for their own work, and that the organisation is decentralised and flat. Flexibility also assumes a set of rules that are neither restrictive nor binding (Andreasson and Lundqvist 2018).

Fostering Co-creation Competence Among Staff

In our data, we find several examples of how managers work on identifying the various competencies that are necessary to improve the co-creation process at all stages. For example, some managers conduct a gap analysis of current and future desired competencies. Other managers use job rotation to expose employees to different parts of the service job to make them more robust to future changes in job demands and roles and to make the service chain less dependent on the skills of one individual employee. Still, other managers increase employees' resources by boosting service employees' language skills or organising courses to allow their employees to "*adopt the role of the customer*". The service industry will experience changes, and the demand for new professions and occupations with new work tasks and capabilities will increase. Labour market analyses also suggest increased demands for education, even at advanced levels, and the need for continuing education to take place throughout one's working life (BFUF 2018). According to the SDL, all parties (e.g., service employees, customers, other actors) involved in a value generation process (i.e., "co-creation") contribute to customer's value-in-use. The value the customer obtains as an actor in the process depends on the actions taken by all the other actors who influence the process (Grönroos and Gummerus 2014). One critical dimension of the SDL is the co-creation process while another key element is effective engagement with employees with an understanding that they are a part of operant resources. Collaborative competence is the main determinant of competitive advantage in the experience. The managerial challenge is to create a working environment based on curiosity, collaboration, and connectedness. Psychosocial factors that managers can influence by fostering co-creation competence among their staff are role clarity, social relations, social support, and perceived job demands.

This practice touches upon the memory-dominant logic by appealing to value-in-memory and evocation of customers' engagement and emotions in the co-creation process. To contribute to a memorable experience (MDL), the industry needs competent service employees. One managerial challenge is thus to raise the awareness of co-creation competence while the managerial task is to make sure all employees (front stage and backstage) have co-creation competencies.

The evolution of tourism and hospitality work in experience—from production, to co-production, to co-creation, to co-experience and towards memory—will continue with the use of cutting-edge technologies. Artificial intelligence (AI), Internet of Things, service robots (chatbots), virtual reality and augmented reality and biometric identifications (e.g., facial recognition), will be adopted to enrich memorable tourism experiences (Leung 2019). Consumers' memory will be assisted by technology which will enable them to "save" the experience or even relive it again. Fostering competence among service providers is crucial in this context. Employees can be flexible thanks to further education, enabling them to take a lot of responsibility within the organisation. Education is also the key to the Nordic leadership style in many ways. The Nordic countries have small, open, and knowledge-intensive economies at the forefront of technological development. This has created a need for adaptability and it also means that employees will continuingly enhance their competencies (Andreasson and Lundqvist 2018). Fostering co-creation competences among service employee will enhance their perceptions of work security in the age when service workers' role is changing rapidly.

Mediating Social Support from Customers

In some service occupations where employees have few colleagues (e.g., tour guides) or rotate between shifts (e.g., hotel day and night shifts), employees do not necessarily get to establish social relations with colleagues over time. In our study, we discovered that service employees use guests and customers as a source of social support. Informants underlined the importance of "*making sure the staff gets recognized for doing a good job*" and of the immediate feedback from customers in order for service employees to believe their job is meaningful. In back-office jobs, it is important for employees to know their customer, and it becomes a managerial task to make sure that back-office employees also get positive feedback from customers. In order to improve social relations at work and enhance work engagement, managers mediate customer feedback (received either face-to-face or electronically) to the employees. Such social interactions positively affect the work environment, create additional job resources for the employees, and emphasise the meaning of service work.

The CDL perspective highlights the importance of a social context beyond single encounters with the service organisation. Value in the CDL is both an intra- and inter-subjectively determined, it stems from personal experiences as well as social context and processes that may involve several different actors (Heinonen and Strandvik 2015). Managers can acknowledge the importance of customers' experience for their businesses and employees by providing social support and appraisal to their staff. Value is created not solely during, but also before and especially after the interaction between employees and customers. Mediating customers' feedback and support to the back office is essential for improving the social relations at work and increasing work engagement.

Providing a Safe Space to Ventilate Emotions

Particularly for frontline employees, co-creation of service experiences may cause strain due to emotional demands. For instance, when welcoming guests to a hotel, receptionists are expected to behave in a certain polite and hospitable way even when guests complain. Over time, multiple situations like that can cause frustration or anger and thus create an emotional demand. The informants in our study recognised the employees' need to vent their negative emotions, or "*to be able to let off steam*"—frustration, anger, disgust, etc.,—somewhere, preferably backstage. Having supporting colleagues and leaders who understand their feelings can help buffer their stress.

How a firm applies its operant resources (i.e., service employees' knowledge and skills) to cater to the customers is crucial for creating a competitive advantage. Getting employees emotionally ready to participate in co-creation with customers and making sure they can do this over time without emotional burnout is, therefore, an important managerial task and an essential part of SDL.

Moderating Multicultural Dialogue

Many service occupations have a multicultural workforce. Due to low training requirements service jobs become entry jobs for immigrants. Thus, handling multicultural work environments becomes a managerial challenge. The informants in our study expressed that having a multicultural workforce is a major challenge that involves handling psychosocial risks at work. Workers with similar cultural backgrounds "*tend to stick together*" and form in-groups and out-groups. Several informants reported that these in-groups and out-groups sometimes coincide with frontline and backline jobs, which adds to interpersonal tension at work, as it reinforces the existing communication barriers and creates conflict. This affects individual employees' perceptions of social relations, social support, and role clarity.

Getting the multicultural workforce to nurture a culture based on curiosity, openness, and connectedness is one of the key elements of the SDL. Re-organising a firm to reflect the ideas of SDL requires a new leadership (Shaw et al. 2011), one that embraces the diversity of knowledge types and skills. This, in turn, may change the role of the team leader or manager–employee interactions (Lafley and Charan 2008).

"Getting Them to See the Whole [Service Chain]"

Informants discussed the importance of highlighting the contribution of backstage employees to the co-created service experience. One central managerial challenge is to make the invisible work visible, to emphasise that all tasks contribute to the whole service chain, and point out their influence on the end product or service. Not every staff member is in the position to engage in co-creation together with customers. It is, however, crucial to see and understand the whole value chain and each staff member's place in it and their contribution to the co-creation process from the psychosocial perspective (i.e., the meaning of work). "*You are not 'just picking up the supplies'—you are creating value!*" A managerial task is to make sure that back-office staff know the "end product" and understand the customers' contribution, being able to actually imagine themselves going through the service experience. It is important to make the back-office staff feel like an equal part of the co-creation process. Another challenge our informants have highlighted is for managers to ease the job rotation and help service employees rotate between front stage and backstage job roles. Job rotation can also contribute to improving social relations and social support among colleagues as well as to increasing job resources by providing a better understanding of other employees' perspectives and work roles. It can also clarify the employees' role in the value creation process and elucidate the whole and the parts that constitute the service. By analysing the co-creation value chain and communicating to the employees where they are and how they contribute to the co-creation of experience, managers can help employees increase perceived job security, enhance perceptions of role clarity, and contribute to employees' perceptions of doing meaningful work.

In our view, this practice reflects service-dominant logic in that SDL considers co-creation through the entire value chain, backstage, and front stage (Shaw et al. 2011). This practice also relates to the CDL as it highlights the centrality of the customer experience (Heinonen and Strandvik 2015); and the MDL which further underlines the importance of creating memorable experiences (Harrington et al. 2019). Meaningful and memorable experiences are personal, more or less social and cultural, and have to do with discovery, adventure, and new initiatives (Boswijk et al. 2012). Creating a memorable experience is, therefore, a collective undertaking of customers and employees along the entire service chain.

Discussion and Concluding Remarks

An important responsibility of service managers is to establish the conditions that enhance meaningfulness of work along the whole service chain. Our empirical findings indicate that managers do that by conditioning for flexibility, fostering competence, mediating social support, providing a safe space to ventilate emotion and a common ground for diverse workforce, and highlighting individual employees' contribution to the customers' experience. It is crucial for managers to help employees build the connection between day-to-day tasks, the company's goals, and the overall tourism, hospitality, or service experience that is being created. Carton (2018) uses the term "connection-building" about essential work that all managers need to do on reminding their employees how the job they do is important for the customer experience.

The quest for meaningful work is a central and defining feature of organisational life (Bunderson and Thompson 2009). The meaningfulness of work is the perception that daily responsibilities have broader significance (Rosso et al. 2010). When daily tasks are marked by a sense of significance, employees can weather the most unpleasant elements of employment, including challenging tasks and stigmatised work (Ashforth and Kreiner 1999). This is especially important for the future of Nordic leadership, as the issues of ethics, education, and stakeholder involvement will gain even more significance in the years to come. The core of the Nordic leadership style consists of several different ethical considerations, such as democracy, human dignity, responsibility, obligations, rights, and the individual's role in relation to the community (Andreasson and Lundqvist 2018). Baum et al. (2019) summarise some of the paradoxes of tourism and hospitality employment. One such paradox is that many of the conditions that underpinned work in the hospitality in the 1930s remain identifiable in today's industry environment in several parts of the world (precarity, low pay, poor working conditions, and jobs of low quality for women and migrants) (Baum 2019). How do we make sure that a better world of tourism work emerges in the next 75 years? Our findings show that for the tourism workforce to be able to co-create authentic and memorable experiences, tourism managers need to carefully craft co-creation practices that enhance employees' perceptions of job resources, create positive social relations at work, provide and mediate social support from different stakeholders, understand the need to let off steam, and provide a safe space for it. The role of service workers is changing (Bowen 2016). The demand for services with high customer interaction is increasing despite the rise of AI, and more and more companies predict that their employees will be facing higher levels of complexity in the interaction with customers in the upcoming years (Kleinschmidt et al. 2019). This is why it is essential to give priority to HRM and handling of psychosocial labour-force issues in service firms.

References

Aagestad, C., H.A. Johannessen, T. Tynes, H.M. Gravseth, and T. Sterud. 2014a. Work-related psychosocial risk factors for long-term sick leave: A prospective study of the general working population in Norway. *Journal of Occupational and Environmental Medicine* 56 (8): 787–793. https://doi.org/10.1097/JOM.0000000000000212.

Aagestad, C., R. Tyssen, H.A. Johannessen, H.M. Gravseth, T. Tynes, and T. Sterud. 2014b. Psychosocial and organizational risk factors for doctor-certified sick leave: A prospective study of female health and social workers in Norway. *BMC Public Health* 14: 1016.

Alam, I. 2011. Process of customer interaction during new service development in an emerging country. *The Service Industries Journal* 31 (16): 2741–2756.

Andreasson, Ulf, and Mikael Lundqvist. 2018. Nordic Leadership. Analysis No. 02/2019. Copenhagen: Nordic Council of Ministers.

Arbetsmiljöverket. 2015. *Den organisatoriska och sociala arbetsmiljön – viktiga pusselbitar i en god arbetsmiljö. Vägledning till Arbetsmiljöverkets föreskrifter om organisatorisk och social arbetsmiljö, AFS 2015:4.* Stockholm: Arbetsmiljöverket, 2016. https://www.av.se/arbetsmiljoarbete-och-inspektioner/publikationer/bocker/vagledning-organisatorisk-social-arbetsmiljo-h457/. Accessed on December 10, 2018.

Ashforth, Blake E., and Glen E. Kreiner. 1999. How can you do it?": Dirty work and the challenge of constructing a positive identity. *The Academy of Management Review* 24 (3): 413–434. https://doi.org/10.2307/259134.

Bakker, Arnold B., and Evangelia Demerouti. 2007. The job demands-resources model: State of the art. *Journal of Managerial Psychology* 22 (3): 309–328. https://doi.org/10.1108/02683940710733115.

Baum, T. 2019. Hospitality employment 2033: A backcasting perspective. *International Journal of Hospitality Management* 76: 45–52.

Baum, T., D. Solnet, R. Robinson, and S. Mooney. 2019. Tourism employment paradoxes, 1946–2095: A perspective article. *Tourism Review* ahead-of-print (ahead-of-print). https://doi.org/10.1108/tr-05-2019-0188.

Ben Gamra Zinelabidine, B., L. Touzani, N. Ben Dahmane, and M. Touzani. 2018. How off-track tourists create their own event: A customer-dominant logic perspective. *Qualitative Market Research* 21 (4): 549–566. https://doi.org/10.1108/QMR-01-2017-0037.

BFUF. 2018. A growing industry during times of change: Forecasts for the skills and competences needed in the hotel and restaurant sector up to the year 2030. *BFUF Report No. 08*. https://bfuf.se/wp-content/uploads/2018/09/BFUF_8_A_Growing_Industry_During_Times_of_Change.pdf. Accessed on October 29, 2019.

Boswijk, Albert, Thomas Thijssen, and Ed Peelen. 2012. *The experience economy: A new perspective*, 3rd ed. Amsterdam, The Netherlands: Pearson Education.

Bowen, D.E. 2016. The changing role of employees in service theory and practice: an interdisciplinary view. *Human Resource Management Review* 26 (1): 4–13.

Bunderson, J., and J. Thompson. 2009. The call of the wild: Zookeepers, callings, and the double-edged sword of deeply meaningful work. *Administrative Science Quarterly* 54: 32–57. https://doi.org/10.2189/asqu.2009.54.1.32.

Burdon, S., G.R. Mooney, and H. Al-Kilidar. 2015. Navigating service sector innovation using co-creation partnerships. *Journal of Service Theory and Practice* 25 (3): 285–303.

Carbonell, Pilar, Ana I. Rodriguez-Escudero, and Devashish Pujari. 2012. Performance effects of involving lead users and close customers in new service development. *Journal of Services Marketing* 26 (7): 497–509.

Carton, Andrew M. 2018. I'm not mopping the floors, I'm putting a man on the moon: How NASA leaders enhanced the meaningfulness of work by changing the meaning of work. *Administrative Science Quarterly* 63 (2): 323–369. https://doi.org/10.1177/0001839217713748.

Chathoth, P.K., G.R. Ungson, R.J. Harrington, and E.S.W. Chan. 2016. Co-creation and higher order customer engagement in hospitality and tourism services: A critical review. *International Journal of Contemporary Hospitality Management* 28 (2): 222–245. https://doi.org/10.1108/IJCHM-10-2014-0526.

Dolan, R., Y. Seo, and J. Kemper. 2019. Complaining practices on social media in tourism: A value co-creation and co-destruction perspective. *Tourism Management* 73: 35–45. https://doi.org/10.1016/j.tourman.2019.01.017.

Etgar, Michael. 2008. A descriptive model of the consumer co-production process. *Journal of the Academy of Marketing Science* 36 (1): 97–108.

Graneheim, U.H., and B. Lundman. 2004. Qualitative content analysis in nursing research: Concepts, procedures and measures to achieve trustworthiness. *Nurse Education Today* 24 (2): 105–112. https://doi.org/10.1016/j.nedt.2003.10.001.

Grissemann, Ursula S., and Nicola E. Stokburger-Sauer. 2012. Customer co-creation of travel services: The role of company support and customer satisfaction with the co-creation performance. *Tourism Management* 33 (6): 1483–1492. https://doi.org/10.1016/j.tourman.2012.02.002.

Grönroos, Christian. 1997. Value-driven relational marketing: From products to resources and competencies. *Journal of Marketing Management* 13 (5): 407–419.

Grönroos, Christian. 2011. Value co-creation in service logic: A critical analysis. *Marketing Theory* 11 (3): 279–301. https://doi.org/10.1177/1470593111408177.

Grönroos, Christian. 2012. Conceptualising value co-creation: A journey to the 1970s and back to the future. *Journal of Marketing Management* 28 (13/14): 1520–1534. https://doi.org/10.1080/0267257X.2012.737357.

Grönroos, Christian, and Johanna Gummerus. 2014. The service revolution and its marketing implications: Service logic vs service-dominant logic. *Managing Service Quality: An International Journal* 24 (3): 206–229. https://doi.org/10.1108/MSQ-03-2014-0042.

Harrington, R.J., R.K. Hammond, M.C. Ottenbacher, P.K. Chathoth, and B. Marlowe. 2019. From goods-service logic to a memory-dominant logic: Business logic evolution and application in hospitality. *International Journal of Hospitality Management* 76: 252–260. https://doi.org/10.1016/j.ijhm.2018.05.014.

Heinonen, Kristina, and Tore Strandvik. 2015. Customer-dominant logic: Foundations and implications. *Journal of Services Marketing* 29 (6/7): 472–484. https://doi.org/10.1108/JSM-02-2015-0096.

Heinonen, Kristina, Tore Strandvik, Karl-Jacob Mickelsson, Bo Edvardsson, Erik Sundström, and Per Andersson. 2010. A customer-dominant logic of service. *Journal of Service Management* 21 (4): 531–548. https://doi.org/10.1108/09564231011066088.

Heinonen, Kristina, Tore Strandvik, and Päivi Voima. 2013. Customer dominant value formation in service. *European Business Review* 25 (2): 104–123. https://doi.org/10.1108/09555341311302639.

Hsieh, An-Tien, C.H. Yen, and K.C. Chin. 2004. Participative customers as partial employees and service provider workload. *International Journal of Service Industry Management* 15 (2): 187–199.

Hsieh, An-Tien, and Chang-Hua Yen. 2005. The effect of customer participation on service providers' job stress. *The Service Industries Journal* 25 (7): 891–905.

Hsieh, J.K., H.C. Chiu, C.P. Wei, H.R. Yen, and Y.C. Cheng. 2013. A practical perspective on the classification of service innovations. *Journal of Services Marketing* 27(5): 371–384.

Hurmelinna-Laukkanen, P., and P. Ritala. 2010. Protection for profiting from collaborative service innovation. *Journal of Service Management* 21 (1): 6–24. https://doi.org/10.1108/09564231011025092.

Jaakkola, E., and M. Alexander. 2014. The role of customer engagement behavior in value co-creation: A service system perspective. *Journal of Service Research* 17 (3): 247–261.

Kleinschmidt, Stefan, Christoph Peters, and Jan Marco Leimeister. 2019. How to scale up contact-intensive services: ICT-enabled service innovation. *Journal of Service Management* ahead of print (ahead of print).

Lafley, A.G., and R. Charan. 2008. P&G's innovation culture. *Strategy and Business* 52 (Autumn): 41–49.

Leung, Xi Yu. 2019. Technology-enabled service evolution in tourism: A perspective article. *Tourism Review*—ahead-of-print (- ahead-of-print). https://doi.org/10.1108/tr-06-2019-0229.

Magnusson, Peter R., Jonas Matthing, and Per Kristensson. 2003. Managing user involvement in service innovation: Experiments with innovating end users. *Journal of Service Research* 6 (2): 111–124.

Morosan, C. 2018. An empirical analysis of intentions to co-create value in hotels using mobile devices. *Journal of Hospitality and Tourism Research* 42 (4): 528–562. https://doi.org/10.1177/1096348015597034.

Mustak, Mekhail, Elina Jaakkola, and Aino Halinen. 2013. Customer participation and value creation: A systematic review and research implications. *Managing Service Quality: an International Journal* 23 (4): 341–359.

Mustak, Mekhail, Elina Jaakkola, Aino Halinen, and Valtteri Kaartemo. 2016. Customer participation management: Developing a comprehensive framework and a research agenda. *Journal of Service Management* 27 (3): 250–275.

Neghina, C., J. Bloemer, M. van Birgelen, and M.C. Caniëls. 2017. Consumer motives and willingness to co-create in professional and generic services. *Journal of Service Management* 28 (1): 157–181. https://doi.org/10.1108/JOSM-12-2015-0404.

Nenonen, S.P., and Goran Lindahl. 2017. Nordic workplace concept development from office as a city to city as an office. *Journal of Facilities Management* 15 (3): 302–316. https://doi.org/10.1108/JFM-10-2016-0043.

Oertzen, Anna-Sophie, Gaby Odekerken-Schröder, Saara A. Brax, and Birgit Mager. 2018. Co-creating services—Conceptual clarification, forms and outcomes. *Journal of Service Management* 29 (4): 641–679.

Pine, Joseph, and James H. Gilmore. 1999. *The experience economy: Work is theatre & every business a stage*. Boston: Harvard Business Press.

Pinho, N., G. Beirão, L. Patrício, and R.P. Fisk. 2014. Understanding value co-creation in complex services with many actors. *Journal of Service Management* 25 (4): 470–493.

Rihova, Ivana, Dimitrios Buhalis, Mary Beth Gouthro, and Miguel Moital. 2018. Customer-to-customer co-creation practices in tourism: Lessons from customer-dominant logic. *Tourism Management* 67: 362–375. https://doi.org/10.1016/j.tourman.2018.02.010.

Roberts, D.L., and W. Darler. 2017. Consumer co-creation: An opportunity to humanise the new product development process. *International Journal of Market Research* 59 (1): 13–33. https://doi.org/10.2501/IJMR-2017-003.

Rosso, Brent D., Kathryn H. Dekas, and Amy Wrzesniewski. 2010. On the meaning of work: A theoretical integration and review. *Research in Organizational Behavior* 30: 91–127.

Sandström, Sara, Bo Edvardsson, Per Kristensson, and Peter Magnusson. 2008. Value in use through service experience. *Managing Service Quality: an International Journal* 18 (2): 112–126. https://doi.org/10.1108/09604520810859184.

Shaw, Gareth, Adrian Bailey, and Allan Williams. 2011. Aspects of service-dominant logic and its implications for tourism management: Examples from the hotel industry. *Tourism Management* 32 (2): 207–214. https://doi.org/10.1016/j.tourman.2010.05.020.

Smith, P.B., J.A. Andersen, B. Ekelund, G. Graversen, and A. Ropo. 2003. In search of Nordic management styles. *Scandinavian Journal of Management* 19: 491–507.

STAMI. 2018. *Faktabok om arbeidsmiljø og helse. Fact book on working environments and health 2018*. STAMI report, year 19, nr. 3. Oslo: Statens arbeidsmiljøinstitutt [The national institute of occupational health, Norway].

Staub-Bauer, A., ed. 2015. *Danish Leadership Style in a Global Context*. Confederation of Danish Industry.

Sthapit, Erose, and Dafnis N. Coudounaris. 2018. Memorable tourism experiences: Antecedents and outcomes. *Scandinavian Journal of Hospitality and Tourism* 18 (1): 72–94. https://doi.org/10.1080/15022250.2017.1287003.

Toivonen, M., and T. Tuominen. 2009. Emergence of innovations in services. *The Service Industries Journal* 29 (7): 887–902.

Vargo, Stephen L., Paul P. Maglio, and Melissa Archpru Akaka. 2008. On value and value co-creation: A service systems and service logic perspective. *European Management Journal* 26 (3): 145–152. https://doi.org/10.1016/j.emj.2008.04.003.

Vargo, Stephen L., and R.F. Lusch. 2004. Evolving to a new dominant logic for marketing. *Journal of Marketing* 68: 1–17.

Vargo, Stephen L., and R.F. Lusch. 2008a. Service-dominant logic: Continuing the evolution. *Journal of the Academy of Marketing Science* 36(1): 1–10.

Vargo, Stephen L., and R.F. Lusch. 2008b. Why "service"? *Journal of the Academy of Marketing Science* 36(1): 25–38.

Vargo, Stephen L., and R.F. Lusch. 2016. Institutions and axioms: An extension and update of service-dominant logic. *Journal of the Academy of Marketing Science* 44 (1): 5–23. https://doi.org/10.1007/s11747-015-0456-3.

Vargo, Stephen L., and R.F. Lusch. 2017. Service-dominant logic 2025. *International Journal of Research in Marketing* 34 (1): 46–67. https://doi.org/10.1016/j.ijresmar.2016.11.001.

Verleye, Katrien. 2015. The co-creation experience from the customer perspective: Its measurement and determinants. *Journal of Service Management* 26 (2): 321–342.

12

Seasonal Workers as Innovation Triggers

Birgitta Ericsson, Kjell Overvåg and Cecilia Möller

Introduction

Many Nordic rural communities with natural and cultural attractions try to use tourism to drive local development when employment and population are decreasing because of industrial change and urbanisation (Hall et al. 2009). Even if tourism development is successfully measured in terms of guest nights, building of second homes, etc., the result is often a highly seasonal demand structure, with either one or two peaks (winter and/or summer) (Overvåg and Ericsson 2016). In the peak

B. Ericsson (✉)
Eastern Norway Research Institute, Elverum, Norway
e-mail: birgitta.ericsson@inn.no

B. Ericsson · K. Overvåg
Inland Norway University of Applied Sciences, Elverum, Norway

C. Möller
Karlstad University, Karlstad, Sweden
e-mail: cecilia.moller@kau.se

A. Walmsley et al. (eds.), *Tourism Employment in Nordic Countries*,
https://doi.org/10.1007/978-3-030-47813-1_12

season(s) there is substantial demand for seasonal workers, as the local labour markets in these rural areas are normally not large enough to satisfy this demand (Hall et al. 2009). Thus, temporary inflows of seasonal workers are necessary for developing tourism on a larger scale in such regions, and also to obtain and keep the number of full-time permanent jobs that may contribute to permanent settlement in rural and peripheral communities.

Most large ski resorts in Norway and Sweden are highly dependent on a seasonal workforce. This is because of their peripheral location, as well as seasonal variation in the number of visitors throughout the year, with a strong peak in the winter season. The demand for labour in the winter season is also affected by the labour-intensive nature of alpine skiing activities, whereas most visitors in summer engage in less labour-intensive activities. These ski resorts operate in a highly competitive market, both domestically and internationally, and, therefore, need to innovate continuously to remain competitive.

In general, industries in peripheral and rural areas are struggling with weak networks and few knowledge flows and a labour market characterised by low turnover, high reliability and a lack of specialised knowledge, which contribute to a less beneficial environment for innovation (Virkkala 2007). This may result in labour market lock-in situations because of a lack of labour mobility. This is problematic as employees are, by virtue of tacit and codified knowledge, in general seen as one of the main sources and drivers of innovation within the service industries, including tourism (Hu et al. 2009; Williams 2007).

However, some scholars have questioned whether employees can be sources and drivers of innovation in the tourism industry. Several reasons have been given for such scepticism: employment in this industry is characterised by a lack of professional and language skills, low levels of education, low stability and high turnover, employees who often suffer unfavourable working conditions (low income, unfavourable working times, limited career opportunities, low work satisfaction), and many employees who are more focused on realising a certain lifestyle rather than a professional career (Hjalager 2002; Pegg et al. 2012; Vukasović 2018; Zampoukos and Ioannides 2011). These characteristics are argued to reduce the significance of employees' contributions to innovation in the tourism and hospitality industry (Williams 2007; Hjalager 2002; Vadell and Orfila-Sintes 2007). Furthermore, Weidenfeld (2013) questions the innovation ability in tourism enterprises. He recognises that service innovations are often dependent on tacit knowledge and human mobility. However, many of the attributes connected to the tourism industry, such as a dominance of small- or medium-size enterprises, a "lack of resources, time, money and/or know-how" and a significant number of lifestyle entrepreneurs, are attributes connected with low innovation activity, which may counter the advantages of mobility in the tourism industry (Weidenfeld 2013, p. 195).

In this chapter we suggest the situation may be quite the opposite, at least in ski resorts in Norway and Sweden. Many employees, and in particular seasonal workers, precisely because of their mobility and lifestyle-driven work motives, potentially possess knowledge that makes them particularly valuable and interesting for innovation in ski resorts. Their experiences at different ski resorts and from their own leisure activities are likely to give them a unique perspective in this context.

Our argument is mainly developed through a theoretical discussion, where we utilise theories from outside of tourism research on communities of practice and on leisure, in addition to theories on seasonal workers and innovation. Our discussion is supported by the results from a study of seasonal workers in six ski destinations in Sweden and Norway: Sälen, Tandådalen and Branäs in mid-Sweden and Skei, Kvitfjell and Trysil in south-east Norway (Ericsson et al. 2010).

We start the chapter by showing how seasonal workers are characterised by lifestyle motives (leisure and sport activities) and mobility. Then we discuss how seasonal workers, as front-line employees, might have a central role in innovation in tourism. In the next section, we use theories on communities of practice and on "specialised play" to deduce how seasonal workers, through both their work and leisure activities in different locations, can contribute to flows and diffusion of knowledge. Next, some results from a study of seasonal workers are presented to support and illustrate the theoretical discussions. We conclude with returning to our main question: May seasonal workers be innovation triggers in ski resorts?

Tourism Employment, Seasonality and Lifestyle

A significant number of seasonal workers in Nordic ski resorts move to the resorts in peak season driven by leisure motives. These groups of employees are the ones in focus in this chapter, and the definition of seasonal workers provided by Ooi et al. (2016) is applied:

> (…) individuals who migrate to tourism-dependent communities for the recreational and lifestyle opportunities on offer on a nonpermanent basis, whether for one or more seasons, many of whom adopt a cyclical pattern of mobility. (Ooi et al. 2016, p. 247)

Seasonal workers are described as being "in-between" tourists and workers (Ooi et al. 2016; Bianchi 2000). Their work motivation is considered different to other groups of tourism and hospitality workers because they are driven by leisure and hedonistic motives, rather than striving for long-term employment and income stability or a professional career. Several tourism researchers acknowledge how seasonal jobs are lifestyle motivated, as a way of combining leisure and tourism activities with temporary employment in established tourism destinations around the world (Adler and Adler 1999; Boon 2006; Janta et al. 2011; Lee-Ross 1999; Pegg et al. 2012). Ooi et al. (2016) claim that there is a lack of research on seasonal work, including the diversity of workers in terms of their experiences, motivations and mobilities. More recent studies have however included patterns of mobility, both nationally and transnationally, as well as the cyclical relocations involved in seasonal work. Some studies have also explored the transition from temporary mobility to permanency, including the potential for seasonal workers to become permanent residents in tourism resorts (Ooi et al. 2016; Möller et al. 2014). Tuulentie and Heimtun (2014) observe how place attachment and attractiveness affect both the willingness of seasonal workers to become permanent residents and the chances of them returning for seasonal work (Tuulentie and Heimtun 2014).

Other studies apply a more critical approach to seasonal work. Underthun and Jordhus-Lier (2018) use the concept of "temporary liminal spaces" to describe how workers in hospitality and tourism have different, fluid and transitional positions, where more precarious working conditions become evident. Seasonal work may be described as representing "individual liminality", constituting a temporal phase in an individual's life, associated with mobility and displacement, where the worker exists "in-between leisure lifestyle and labour market thresholds" (Underthun and Jordhus-Lier 2018, p. 14). Thus, these liminal spaces may represent "exceptional spaces where regular work life norms do not apply" (Underthun and Jordhus-Lier 2018, p. 15). However, compared with other groups of migrant workers, seasonal workers constitute "liminality of choice", where the motivation for work is leisure-oriented and represents alternative ways of working and living. Consequently, ideals of consumption, leisure and individuality remain stronger than the collective norms of worker rights, organised labour and unionisation, despite low wages, short-term contracts and inconvenient working hours (Underthun and Jordhus-Lier 2018).

In a previous study (Möller et al. 2014), we explored the heterogeneous mix of motives that serve as driving forces for seasonal work in ski resorts, by distinguishing four groups or "typologies" of seasonal workers, which also relate to the previous categorisations made by Bianchi (2000) and Adler and Adler (1999). These include (1) *Migrant tourist workers*, for whom lifestyle-related motives are prominent for taking on seasonal work, where cyclical mobility in itself remains a crucial driving force by seeking temporary stays in multiple and attractive places worldwide. (2) *Practiced seasonal workers*, who combine seasonal work during the winter with an "ordinary" job during the summer months, preferably in their places of residence, which may or may not include another established (domestic) tourism destination. (3) *Temporary seasonal workers*, short-term-employed seasonal workers who describe their work as "arbitrary", while having more pronounced economic and work-related motives for their employment. (4) *Place-attached seasonal workers*, who are recruited from the local community, combine work on the mountain with other seasonal jobs during the rest of the year, including forestry, retail or other sectors of the tourism industry. In what follows, we focus on the first two groups of seasonal workers for whom lifestyle-oriented motives remain important driving forces for seeking and combining seasonal employment in various tourist destinations. They also tend to be returning seasonal workers to the ski resorts, creating more cyclical patterns of labour mobility.

Front-Line Employees and Innovation

Within the early concept of "industrial districts", the importance of social networks and geographical proximity for dissemination of innovations has been acknowledged (Marshall 1920). Since then, many scholars have acknowledged the obvious fact that not all knowledge flows between enterprises can be separated from people's movements/involvement (e.g. Williams 2007; Weidenfeld 2013). As hardly any enterprise today can innovate based solely on their own intrafirm resources, the dissemination of knowledge is dependent on what Kacker (1988) describes as "vehicles for knowledge flows". For example, in cluster theory, knowledge transfer is focused on professional knowledge, such as technological solutions, economic systems, marketing, etc., mainly used by personnel in administrative and leadership positions, which diffuse through labour market spillovers. Transfer of such codified knowledge may occur through formal education and books as well, but the tacit knowledge and skills that develop and are required in the front-line meeting with customers in tourist enterprises have to be transferred by peoples' mobility. Tacit knowledge and skills are, above all, developed and refined in actually performing work activities (Finne and Hulbak 2005).

Williams (2007, p. 29) recognises the "limited research on the role of international migration in the transfer of tacit knowledge". He distinguishes between intra-, inter- and extra-firm migration, which he argues makes a critical difference, because different labour migrants are also potentially involved in different types of knowledge transactions. Furthermore, he finds that "migrants' knowledge transaction experiences are highly job specific", but that they anyway are potentially significant knowledge brokers (Williams 2007, p. 33).

There is some research on the role of tourism front-line employees in innovation. Jensen and Sørensen (2017) discuss how front-line personnel may elevate tourist enterprises' deliverables "from service production to experience creation" (Jensen and Sørensen 2017, p. 448). Their main message is that innovation occurs when an employee adds value to the tourists' experiences by co-creating, that is by creating emotions rather than just a service delivery. This innovation occurs in the actual encounter between tourist and employee, and does not include tangible/explicit knowledge transfers to the firm per se.

Engen and Magnusson (2015) discuss front-line employees' role as innovators in light of their boundary spanning position between firm and tourist. Their starting point is the recognition of the necessary requirements for front-line personnel to develop and apply creative ideas that benefit the firm as well as the customer. Thus, at least two types of knowledge are required to succeed in innovation, namely "supply-side knowledge" and "demand-side knowledge". Engen and Magnusson (2015) claim front-line employees are in a unique position to be able to synergise these insights from both sides of production. We would like to append, that one additional dimension of "demand-side knowledge" is the role as experienced practitioner and customer also held by migrant tourist workers.

Communities of Practice and Advantages for Society from Leisure

How can seasonal workers in front-line positions develop and diffuse knowledge that is relevant for innovation? As seasonal workers are motivated by the possibility of combining leisure activities with work, we can draw on theories related to both work and leisure. We will consider theories on "communities of practice" from work-related studies and theories connected to "specialised play" from leisure studies. Both show, but from a different perspective, how people sharing a concern or passion for something may, through participation in groups and networks, develop tacit knowledge and skills that are important for innovation.

The concept of "communities of practice" is based on research on group-based learning foremost connected to different work situations (Lave and Wenger 1991; Wenger 1998; Brown and Duguid 1991).

A “community of practice” differs from “a team” as it is kept together by the shared learning and interest of its members, and it is defined by knowledge rather than by task (Wenger 1998). The concept is useful for understanding transfer and diffusion of tacit knowledge and skills, contrary to codified and professional knowledge, because the members are practitioners with the same interests. This shows that if people who share a concern or passion for something interact regularly, they develop and share knowledge by doing and interacting (Wenger 1998). Such communities are characterised by three elements: (1) a shared identity and competence connected to a common interest (“the domain”); (2) they initiate relations that enable them to exchange information and knowledge connected to their common interest (“the community”) and (3) members’ interests are not enough to define a “community of practice”, as it is the *practice* of the interests that defines the community—they develop a shared practice by sharing experiences, techniques, tools, problem-solving methods, etc. (“the practice”).

Communities of practice are thought to play important roles in diffusion of knowledge and skills (Wenger 1998; Brown and Duguid 1991). Communities of practice are in interaction with each other, and members participating in several communities may be able to act as brokers or boundary spanners between communities (Wenger 1998; Williams 2007).

> The key individuals in this respect are those particularly well connected and able individuals who are able to act as ‘boundary spanners’ between different knowledge communities, and they are potentially important in both larger and smaller firms. (Shaw and Williams 2009, p. 330)

As tacit knowledge is “displayed or exemplified, not transmitted, sharing usually requires face-to-face-interaction” (Duguid 2008, p. 78), and, thus, the transfer usually requires geographical proximity. However, other scholars argue that cognitive closeness is as important and to some extent, or in some cases, may outweigh the necessity of geographical proximity (Amin and Cohendet 2005; Gertler 2008).

While communities of practice are based on work situations, Kjølsrød (2019) deduces how people who are seriously engaged in a leisure activity, which she labels "specialised play", are likely to generate surprising resources that can also be an advantage for society in general, and not only related to the specific leisure activity. By using these resources, play may be a source for knowledge, social resilience and public commitment for the participants. According to Kjølsrød (2019), many leisure activities are knowledge-based and require good skills, for example climbing, collecting, birdwatching, etc. From our perspective, many types of ski- and snow-related activities could also be included in this context, including skiing experiences and expertise. Investment of time, learning, determination and money may be substantial. A part of this investment is that these "players" are often engaged in sizeable international networks where they can relate to some kind of collective of like-minded people, with both offline and online communications, as they have much to learn from each other. Through these networks, they build different types of knowledge, both contextual (exchange of information, observations) and practical skills (Kjølsrød 2019).

Learning by doing through a self-directed and inquiry-based approach also characterises many leisure activities. To solve case-based problems as they arise is, thus, a competence that can be developed (Kjølsrød 2019), and that may be highly relevant for seasonal workers, for example in their direct meetings with customers on the frontline. Apart from these activities for the players being fun and an end-in-themselves as it provides gratification, identity and belonging, they also grow a crop of socially valuable resources as discussed above, which is not only valuable for the players themselves but also may be an advantage for society; however, the value for society has never been seriously assessed.

These two theories have a lot in common, as they both emphasise how knowledge and skills may be developed within groups of people with common interests. Concerning seasonal workers, we will claim that many of them are participants in such groups, connected to both work (as, for example, ski-instructors) and leisure (as dedicated skiers, snowboarders, etc.). In both instances, the discussion above shows that there is a potential for developing knowledge and skills that are relevant for innovation. In addition, they can be part of such groups or communities in several places, as they circulate between different destinations, sometimes in different countries, as both workers and players. Brown and Duguid (1991) and Wenger (1998) emphasise that these communities play a role in the diffusion of knowledge as they often are engaged with other communities of practice. They may be connected to each other through people that are members of several communities, just like our migrant seasonal workers. Such individuals may play especially important roles such as "brokers" or "boundary spanners" between different knowledge communities and communities of practice, and, thus, be potentially important for companies' ability to develop and innovate (Shaw and Williams 2009; Sørensen 2007; Wenger 1998). Most research on the diffusion of knowledge has been on mobile professionals in leading positions/higher education, and, thus, diffusion of primarily codified knowledge. Enz and Siguaw (2003) stress that it should be more focused on the front-line employees' role in this respect, as codified versus tacit knowledge and skills will diffuse in different ways.

Study of Seasonal Workers

Methodology

Survey

An electronic survey was conducted among seasonal employees in six ski resorts in Norway and Sweden: Trysil, Kvitfjell, Skei, Lindvallen, Tandådalen og Branäs (Ericsson and Hagen 2012). The survey was complimented with in-depth group interviews in Trysil and Sälen (Möller et al. 2014). The main subject for the survey was how place attachment and place attractiveness were assessed by seasonal workers, and if there was potential for eventual inward migration (involving residence acquisition). In addition, the questionnaire included a section on innovation and knowledge transfer. As this section was supplementary to the main survey, its findings are more explorative than conclusive. The data may seem a bit old, but we consider it still relevant. This approach to assess consequences of heavy dependence on a seasonal workforce as in tourism is up to today an understudied research subject.

In the analysis, the results from the two resorts Skei and Kvitfjell are merged because of the limited number of respondents. Email addresses were provided by the enterprises, and the reliability of these lists varied substantially. The sample population was the previous seasons' seasonal workers. There were a substantial number of non-deliverable addresses, and there were indications of deliverable but sleeping addresses as well, but we are not able to estimate the scope of this.

A total of 282 seasonal workers responded to the survey, representing a response rate of 30%. As mentioned above, this is a fairly low response rate but in line with response rates to electronic surveys. The relatively low response rate is partly offset by the fact that this dataset is a more appropriate sample with which to investigate whether there is any innovation potential among seasonal workers, but not to estimate the magnitude of this potential. The survey data consist primarily of dedicated seasonal workers, which is mostly consistent with the typology groups "migrant seasonal workers" and "practiced seasonal workers" mentioned above (Möller et al. 2014, p. 398).

Group Interviews

Group interviews focused on elaborating views and activating a discussion among the seasonal workers and their work experiences at the two ski resorts Trysil (Norway) and Sälen (Sweden). These group interviews, thus, provided transpersonal and multiple perspectives of the issues raised and provided an opportunity to raise issues not considered previously (Bedford and Burgess 2001).

The selection of interviewees was made in collaboration with employers in Trysil and Sälen, resulting in interviews with seven seasonal workers in Trysil and six in Sälen, conducted during the winter season 2010/2011. The interviewees included more experienced seasonal workers from restaurants, hotels, ski patrols, maintenance and ski schools. The interview themes involved motives for seasonal work as well as experiences in the ski resorts as both temporary and permanent places to live. The aim was to follow seasonal workers during the season to obtain an overview of their changing expectations, work assessments and plans. In Sälen, interviews were conducted on three different dates and in Trysil on two different dates. As several members of the groups knew each other and met for interviews several times, the interviews were held in an informal atmosphere, which means they can be labelled in-depth group interviews (Kneale 2001).

Findings

Work Experience

Motives for taking on a seasonal position in tourism are not homogeneous, but rather consist of diverse motives and driving forces. Some make seasonal work a lifestyle in itself, while others see it as a nice break while developing future life plans (Möller et al. 2014; Ericsson and Hagen 2012).

Among all seasonal workers in the survey, 28% have previous seasonal work experience, with most being in tourism (68%). Their experience is mainly in Scandinavian countries, but 20% also have worked in other countries. These include Australia and New Zealand, as well as Canada and the Alps, where the potential for knowledge transfer on snow business operations is obviously present. Even if our sample is small, it consists of seasonal workers with a broad work experience and from a broad spectre of tourist resorts.

Inclusion in Innovation Activities on Site

More than half (55%) of all seasonal workers state that they have involved in development or innovation activities in their present employment. Not all of the workers' proposals have been adopted by management, but as many as 38% states their suggestions have been implemented, meaning that the proposals of the other 17% didn't come through.

The main areas of innovation involvement are workplace organisation, 42% (organisational innovation); service measures and customer care, 33% (process innovation); and, measures to improve products and services, 32% (product innovation). In addition, there were minor suggestions regarding customer segmentation, marketing, branding and distribution. Most suggestions are incremental in nature (90%).

As front-line personnel, they can introduce innovations in line with Jensen and Sørensens' (2017) suggestion to create emotion rather than service delivery, thus, adding value to tourists' experience by co-creation. Such initiatives are described as more commonly occurring when employees have former work experience and when their leaders allow delegation of responsibility. When our front-line personnel were asked to detail their involvement in initiatives, their responses included the following:

> I have been involved in development activities by initiating several actions at my workplace.
>
> Improvement of work routines.
>
> Mostly minor issues, but they were intended to improve our work situation.

And from the interviews:

> Otherwise one meets people in the piste, also guests and tourists, and chats with them. Then you get feedback on if they are satisfied or not. This is fun. You get to hear a lot if you are together on the ski lift. This may contribute to better development at the destination. (Group interview, Trysil)

In the group interviews, the seasonal workers described how they are given greater work responsibilities and training when returning for multiple seasons. This includes more formal training in communication, leadership, language and management skills, but also more "tacit" knowledge with insights into how a restaurant or hotel is run (business thinking, budgeting), while building a network of contacts with other tourism businesses at ski resorts, nationally and internationally. On the one hand, the respondents describe in positive terms how they are encouraged to provide feedback to the management team to improve work routines or handle arising staff and organisational issues. On the other hand, the interviewees provided examples of how the inquiries and ideas for improvement were not always taken seriously or had little effect. Some comments include that ideas should be "worth making money from" and that new suggestions of work routines might be questioned because of more conservative attitudes towards "things that should remain as they have always been". This emphasises the necessity of "up-take capacity" (Sørensen et al. 2010) in destination and business management. If the management do not have the capacity to include and assess—or take up—new ideas, innovation activities may not be able to counteract the phases of stagnation, decline or rejuvenation in tourist destinations, as described in Richard Butlers (1980) model of The Tourism Area Life Cycle (TALC).

Vehicles of Knowledge—Experienced Seasonal Workers' Involvement in Innovation Activities

In this section, the analysis only involves seasonal workers with former seasonal work experience, which are corresponding to the typology "migrant seasonal workers" as detected in Möller et al. (2014). They migrate between tourist resorts in different destinations and often have a lifestyle motivation for seasonal jobs as well as the migration. This makes them familiar in several roles, which may be important in an innovation perspective. They do not only have experience from seasonal tourist work, but are able to combine these skills with expert knowledge on what a "good skiing product" is as they themselves are experienced skiers. Furthermore, they are well accustomed to how things are driven and done in other destinations, and thus are able to bring and apply new ideas between destinations. In other words, they have skills and knowledge stemming from their roles as experienced practitioners, demanding customers (in their own leisure time), and they have face-to-face contact with guests which enables them to capture complaints and suggestions as well as try out ideas. These characteristics do not apply to other groups of seasonal tourism workers. The number of respondents is smaller ($N=69$), but nevertheless provides interesting results which should be a topic for further investigation. The results from the group interviews support our survey findings.

While the innovation implementation rate is 38% among all seasonal workers, it is 54% among the experienced ones. This rate is particularly high if their former experience is from another ski resort (60%) and even higher if their experience is from an international resort (outside Scandinavia) at 71%. This indicates the importance of experience from similar resorts (ski resorts), but also that experience from any developed tourist resort—summer or winter—may be at least as important for transferring and diffusing knowledge between destinations, and that migrant seasonal workers may play a significant role in such transfer.

During the group interviews, the interviewees emphasised the benefits of combining seasonal work in different destinations, such as guiding in summer destinations and ski instruction in winter destinations. Recurring seasonal workers made it clear, that their leisure interests in various forms of skiing and the combination of two or more seasonal jobs is a driving force of their seasonal mobility between tourist destinations. The respondents highlighted that learning and knowledge transfer through seasonal work is a two-way process and that work in ski resorts benefits their work in the summer season and reverse.

> Being a tour guide and a ski instructor has mutual benefits. When I'm trained as a ski instructor I get to attend courses and listen to lectures about pedagogy, leadership and communication, etc., which I naturally have great use for as a tour guide. Such as, how to handle different situations. Things come up, both as an instructor and as a guide, and it's about how you communicate with people, so the professions are quite similar. (Stefan, group interview, Trysil)

These seasonal workers also put more emphasis on lifestyle motives for their seasonal jobs, and they value their personal development, including the ability to resist group pressure and to develop more personal independence, which is necessary to cope with life as a seasonal worker. This is considered important when living in tourist destinations, which can involve partying most days, which may be too tempting for many young people away from home for the first time.

Their competence in skiing and other winter-related issues, such as knowledge of avalanches and behaviour on alpine slopes to avoid injury, etc., is also transferrable, as many ski destinations abroad struggle with poor or dangerous snow conditions and "piste congestion" of a more serious nature. This competence is claimed to have proved beneficial for the ski resorts studied by improving routines for preventing incidents/accidents in connection with avalanches and ski patrolling, as local experience with such conditions has been limited.

The seasonal workers themselves indicate experience from other destinations as the most important determiner of their involvement in innovation (see Fig. 12.1). The most prominent factor is "industry experience in other enterprises" and this is rated higher than being either a worker or customer at similar destinations. Former work experience in the tourism industry is mentioned as being more important than specific product experience from ski resorts. Codified knowledge, indicated by education, does not seem very important. Even so, vocational

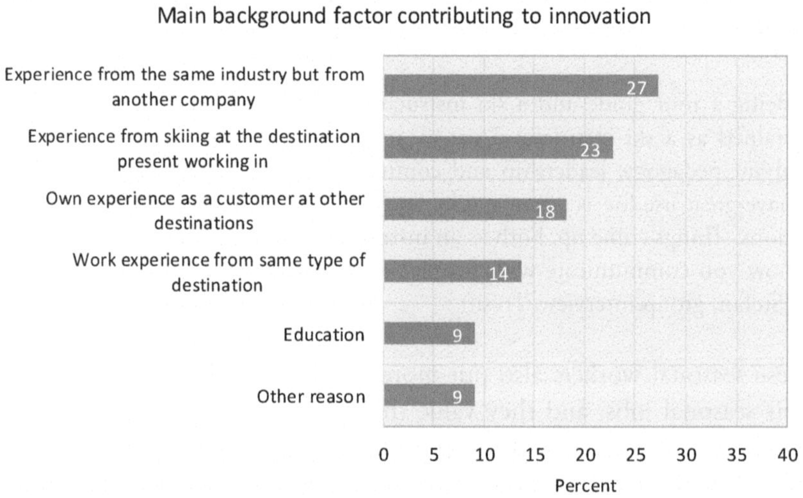

Fig. 12.1 Seasonal workers' assessment of the most decisive factor for involvement in innovation activities. Implemented actions. Percent ($N = 67$)

education related to tourism businesses seems to increase the innovation implementation rate more than any other education, even if education per se does not contribute much compared with tacit knowledge or skills.

Conclusion—Seasonal Workers as Innovation Triggers?

Employment in the tourism industry has some characteristics that are likely to limit the contribution of such workers to innovation in tourism enterprises and destinations. We have, however, in this chapter discussed whether some of these negative characteristics, namely workers' low stability and high turnover, but at the same time high mobility between destinations together with strong focus on lifestyle/leisure, may provide these workers with unique knowledge and competencies that might be highly valuable and induce innovation. Thus, for rural and peripheral destinations, some seasonal workers might be valuable innovation triggers, based on their experiences in other destinations/industries and from their strong commitment to leisure and "play". In their continuous mobility between tourist resorts as workers, demanding practitioners and customers, they may function as vehicles for knowledge flows that are otherwise unavailable (Ericsson and Hagen 2012).

In such places, a high influx of seasonal workers, therefore, may counter lock-in effects that could otherwise occur, and where the exchange of new ideas and knowledge between local enterprises becomes difficult. This implies that such enterprises have to seek new ideas in other destinations, preferably internationally, to retain their competitive advantages.

Mobile seasonal workers are members of several communities of practice related to work as well as to leisure, and at different locations because of the significant geographical distances between destinations, even if they, in many ways, belong to a common production, leisure and knowledge sphere. Nevertheless, despite the geographical distances, these communities of practice and leisure networks are close to each other cognitively, communicatively and culturally. The results from our studies also indicate that seasonal workers possess experiences, competencies and knowledge that are useful for enterprises and destinations in which they are working temporarily. While the main motives for taking on seasonal jobs are lifestyle-oriented and closely connected to developing and practising their leisure activities, it also puts them in a unique position as boundary spanners locally and vehicles for innovative knowledge transfer nationally or internationally.

Locally, migrant seasonal workers may act as boundary spanners between tourists and enterprises because they have first-hand knowledge of products as well as of customers. Their ability to promote innovations has its basis in their combined role as competent skier and expert knowledge of skiing, as well as the fact they can act as demanding customers during their own leisure activities and have in-depth knowledge of how other destinations produce "good skiing". This, in combination with the same in-depth knowledge of what other destinations do to co-create, enables them to suggest potential actions to increase value added for the tourists.

The empirical results indicate that seasonal workers with tourist industry experience may make a significant contribution to innovation and implementation within ski destinations, especially if their experience is from other ski destinations. This does not support assumptions that tourism industries lack innovation potential because of dependence on seasonal workers. On the contrary, we suggest that the presence of dedicated migrant seasonal workers may contribute to and increase innovation potential in special cases. As far as we have discussed, this conclusion depends on two important characteristics of our research. First, we have identified a group of dedicated "migrant seasonal workers" in the tourist industry that has adopted seasonal work as a lifestyle. Second, their experience, skills and knowledge enable them to play a role as vehicles to convey knowledge between different destinations. Tourism managers should be aware of this potential and give these seasonal workers sufficient attention, resources and responsibility to release as much potential for innovation as possible. As mentioned, but not elaborated on, firms' and managers' up-take capacity will decide to what extent this potential is realised and contribute to prolonging destinations' life.

References

Adler, Patricia A., and Peter Adler. 1999. Transience and the postmodern self: The geographic mobility of resort workers. *The Sociological Quarterly* 40 (1): 31.

Amin, Ash, and Patrick Cohendet. 2005. Geographies of knowledge formation in firms. *Industry and Innovation* 12 (4): 465–486. https://doi.org/10.1080/13662710500381658.

Bedford, Tracey, and Jaquelin Burgess. 2001. The focus-group experience. In *Qualitative Methodologies for Geographers*, ed. Melanie Limb and Claire Dwyer, 121–135. London: Arnold.

Bianchi, Raoul V. 2000. Migrant tourist-workers: Exploring the 'contact zones' of post-industrial tourism. *Current Issues in Tourism* 3 (2): 107–137. https://doi.org/10.1080/13683500008667869.

Boon, Bronwyn. 2006. When leisure and work are allies: The case of skiers and tourist resort hotels. *Career Development International* 11 (7): 594–608. https://doi.org/10.1108/13620430610713463.

Brown, John Seely, and Paul Duguid. 1991. Organizational learning and communities-of-practice: Toward a unified view of working, learning, and innovation. *Organization Science* 2 (1): 40–57.

Butler, Richard. 1980. The concept of a tourist area cycle of evolution: Implication for management resources. *The Canadian Geographer* 24 (1): 5–12.

Duguid, Paul. 2008. 'The art of knowing': Social and tacit dimensions of knowledge and the limits of community of practice. In *Community, Economic Creativity, and Organisation*, ed. Ash Amin and Joanne Roberts, 69–89. Oxford: Oxford University Press.

Engen, Marit, and Peter Magnusson. 2015. Exploring the role of frontline employees as innovators. *The Service Industries Journal* 35 (6): 303–324. https://doi.org/10.1080/02642069.2015.1003370.

Enz, Cathy A., and Judy A. Siguaw. 2003. Innovations in hotel practice. *Cornell Hotels and Restaurant Administration Quarterly* 44 (4/5): 115–123.

Ericsson, Birgitta, Kjell Overvåg, and Cecilia Möller. 2010. Sesongarbeidere på vintersportsteder. En undersøkelse i indre Skandinavia. ØF-rapport 15/2010. Østlandsforskning: Lillehammer.

Ericsson, Birgitta, and Svein Erik Hagen. 2012. De mobile sesongarbeidernes rolle i reiselivsbedriftnes innovasjonsarbeid. In *Innovasjon og næringsutvikling i en reiselivskontekst*, ed. Martin Rønningen and Terje Slåtten. Fagbokforlaget: Bergen. ISBN: 978-82-450-1178-4.

Finne, Håkon, and Maren Hulbak. 2005. Samarbeid mellom forskning og bedrift – mot nye praksisfellesskap? SINTEF Teknologi og samfunn, IFIM. Paper presented at HSS-konferansen 2005, Tønsberg, June 1–3. http://www.promsys.no/publish/archive/104/5887/Nr.%2036_Finne,%20Hubak_absnr.%2035.pdf. July 11, 2011.

Gertler, Meric S. 2008. Buzz without being there? Communities of practice in context. In *Community, Economic Creativity, and Organization*, ed. Ash Amin and Joanne Roberts. Oxford Scholarship Online Monographs 25: 203–227.

Hall, Colin Michael, Dieter K. Müller, and Jarko Saarinen. 2009. *Nordic Tourism: Issues and Cases*. Bristol: Channel View Publications.

Hjalager, Anne-Mette. 2002. Repairing innovation defectiveness in tourism. *Tourism Management* 23 (5): 465–474.

Hu, Mei-Lei Monica, Jeou-Shyan Horng, and Yu-Hua Christine Sun. 2009. Hospitality teams: Knowledge sharing and service innovation performance. *Tourism Management* 30 (1): 41–50.

Janta, Hania, Adele Ladkin, Lorraine Brown, and Peter Lugosi. 2011. Employment experiences of polish migrant workers in the UK hospitality sector. *Tourism Management* 32 (5): 1006–1019. https://doi.org/10.1016/j.tourman.2010.08.013.

Jensen, Jens Friis, and Flemming Sørensen. 2017. Experience innovation in tourism: The role of front-line employees. In *Proceedings Book: 5th Interdisciplinary Tourism Research Conference*, ed. Metin Kozak, Andreas Artal-tur, and Nazmi Kozak, 448–453. June 6–11, Cartagena, Spain.

Kacker, Madhav. 1988. International flow of retailing know-how: Bridging the technology gap in distribution. *Journal of Retailing* 64 (1): 41–67.

Kjølsrød, Lise. 2019. *Leisure as Source of Knowledge, Social Resilience and Public Commitment: Specialized Play*. London: Palgrave Macmillan.

Kneale, James. 2001. Working with groups. In *Qualitative Methodologies for Geographers*, ed. Melanie Limb and Claire Dwyer, 136–150. London: Arnold.

Lave, Jean, and Etienne Wenger. 1991. *Situated Learning: Legitimate Peripheral Participation*. Cambridge: Cambridge University Press.

Lee-Ross, Darren. 1999. Seasonal hotel jobs: An occupation and a way of life. *International Journal of Tourism Research* 1 (4): 239–253. https://doi.org/10.1002/(SICI)1522-1970(199907/08)1:4<239:AIDJTR169>3.0.CO;2-9.

Marshall, Alfred. 1920. *Principles of Economics*, 8th ed. London: Macmillan.

Möller, Cecilia, Birgitta Ericsson, and Kjell Overvåg. 2014. Seasonal workers in Swedish and Norwegian ski resorts—Potential in-migrants? *Scandinavian Journal of Hospitality and Tourism* 14 (4): 385–402.

Ooi, Natalie, Judith Mair, and Jennifer Laing. 2016. The transition from seasonal worker to permanent resident: Social barriers faced within a mountain resort community. *Journal of Travel Research* 55: 246–260.

Overvåg, Kjell, and Birgitta Ericsson. 2016. Fritidsboligdestinasjoner og regional utvikling i fjellområdene. In *Turisme – Destinasjonsutvikling*, ed. Arvid Viken, 226–245. Oslo: Gyldendal Norsk Forlag.

Pegg, Shane, Ian Patterson, and Pablo Vila Gariddo. 2012. The impact of seasonality on tourism and hospitality operations in the alpine region of New South Wales, Australia. *International Journal of Hospitality Management* 31 (3): 659–666. https://doi.org/10.1016/j.ijhm.2011.09.001.

Shaw, Gareth, and Allan Williams. 2009. Knowledge transfer and management in tourism organisations: An emerging research agenda. *Progress in Tourism Management* 30: 325–335.

Sørensen, Flemming. 2007. The geographies of social networks and innovation in tourism. *Tourism Geographies* 9 (1): 22–48. https://doi.org/10.1080/14616680601092857. Accessed June 27, 2019.

Sørensen, Flemming, Jan Mattsson, and Jon Sundbo. 2010. Innovation Uptake: The Embryo of Service-Encounter Based Innovation. Roskilde: Department of Communication, Business and Information Technologies, Roskilde University.

Tuulentie, Seija, and Bente Heimtun. 2014. New rural residents or working tourists? Place attachment of mobile tourism workers in Finnish Lapland and Northern Norway. *Scandinavian Journal of Hospitality and Tourism* 14 (4): 367–384.

Underthun, Anders, and David Christoffer Jordhus-Lier. 2018. Liminality at work in Norwegian hotels. *Tourism Geographies* 20 (1): 11.

Vadell, Joan B. Garau and Francina Orfila-Sintes, F. 2007. Internet innovation for external relations in the Balearic hotel industry. *Journal of Business and Industrial Marketing* 23 (1): 70–80. https://doi.org/10.1108/08858620810841506.

Virkkala, Seija. 2007. Innovation and networking in peripheral areas—A case study of emergence and change in rural manufacturing. *European Planning Studies* 15 (4): 511–529.

Vukasović, Ante. 2018. Can't wait for the winter: Self-reported effects of seasonal employment on hospitality workers' wellbeing and job satisfaction. https://login.bibproxy.kau.se:8443/login?url=https://search.ebscohost.com/login.aspx?direct=true&db=edsbas&AN=edsbas.D800FCF1&lang=sv&site=eds-live.

Weidenfeld, Adi. 2013. Tourism and cross border regional innovation systems. *Annals of Tourism Research* 42: 191–213.

Wenger, Etienne. 1998. Communities of practice: Learning as a social system. Systems Thinker. June https://thesystemsthinker.com/communities-of-practice-learning-as-a-social-system/. Accessed on September 23, 2019.

Williams, Allan M. 2007. International labour migration and tacit knowledge transactions: A multi-level perspective. Global Networks, Wiley Online Library. https://onlinelibrary-wiley-com.ezproxy.inn.no/doi/abs/10.1111/j.1471-0374.2006.00155.x. Accessed on September 23, 2019.

Zampoukos, Kristina, and Dimitri Ioannides. 2011. The tourism labour conundrum: Agenda for new research in the geography of hospitality workers. https://www.researchgate.net/publication/233579514_The_tourism_labour_conundrum_agenda_for_new_research_in_the_geography_of_hospitality_workers, https://doi.org/10.1386/hosp.1.1.25_1. Accessed on September 23, 2019.

13

Gateway, Fast Lane, or Early Exit? Tourism and Hospitality as a First Employer of Norwegian Youth

Åse Helene Bakkevig Dagsland, Richard N. S. Robinson and Matthew L. Brenner

Introduction

This chapter addresses the Nordic nexus between 'youth employment' and the 'image and attractiveness of tourism employment' focusing on a Norwegian perspective. Employment in the hospitality and tourism industry is a common gateway into working life for young people globally (Baum et al. 2016; Walmsley 2015). In fact, some have argued that there is a symbiotic relationship between the tourism industry and the young worker, to the extent that the tourism industry relies on this consistent labour pool and that young people depend on the opportunities

Å. H. B. Dagsland (✉)
University of Stavanger, Stavanger, Norway
e-mail: aase-helene.b.dagsland@uis.no

R. N. S. Robinson · M. L. Brenner
University of Queensland, St Lucia, QLD, Australia
e-mail: richard.robinson@business.uq.edu.au

M. L. Brenner
e-mail: m.brenner@business.uq.edu.au

A. Walmsley et al. (eds.), *Tourism Employment in Nordic Countries*,
https://doi.org/10.1007/978-3-030-47813-1_13

the industry provides them as a low barrier of entry into the world of work (Robinson et al. 2019). In the Nordic countries, as elsewhere, young people are 'baptised' into the labour market (Halvorsen & Hvinden 2014), *inter alia* through summer jobs, weekend engagements, and part-time work experiences during their studies. Nevertheless, international evidence is mounting that the tourism work experience of youth is increasingly fast, furious, and short, often contributing to their premature exit from the industry (e.g. International Youth Foundation 2013; Mooney 2016). This thus leads us to ask question whether the tourism work context of Norway is perpetuating the same results that are seen internationally?

In Norway, 32% of the accommodation and food service industry workforce is below the age of 24, this figure far exceeds the 13% reported of this age cohort employed in other industries (SSB 2019). These figures are also fairly consistent with those reported internationally (WTTC 2015). Trends suggest however, that young Norwegians do not stay. Turnover rates are high (DAMVAD 2014; Mykletun & Furunes 2012) and international evidence regarding attrition suggests tourism fares far worse than the all-industries average, with 45% of tourism employees having less than two years' organisational tenure, compared to 25% for all other sectors (Stacey 2015). The industry, globally, is facing difficulties in recruiting and retaining a competent workforce (Baum et al. 2019; Kusluvan et al. 2010; Tracey 2014), these difficulties are reflected in Norway with 61% of Norwegian companies reporting significant difficulties in recruiting and retaining a skilled workforce (Rørstad et al. 2018). Given the skills shortage and the pervasively high attrition rates, attention to the early perceptions of youth as they enter into tourism and hospitality industry is warranted.

Young Norwegians' prior perceptions of, and beliefs about, work in the tourism and hospitality industry are positive (Dagsland et al. 2011, 2017). They believe this industry offers interesting and exciting workplaces, and provides opportunities to develop relevant and transferable work skills. At the same time, this young Norwegian cohort is realistic, anticipating the work to be hectic and stressful. New data from 2018, collected by Dagsland with the same survey and in the same age group as described in Dagsland et al. (2017), show the same positive perceptions and beliefs about the industry as the earlier surveys. Despite the realistic as well as positive perceptions of the industry there is a steady and precipitous decline in applications for the relevant vocational foundation classes in secondary school and for the certificate of apprenticeship leading to work in the tourism and hospitality industry (Dagsland et al. 2017). The greatest decline is in the group applying for vocational foundation classes within 'restaurant and food' sector, with a decline of almost 50%, from 2002 to 2011. The numbers seem to stabilise, however, from 2011 to 2018 with only a smaller reduction. Nonetheless, in 2018, only 1.5% of Norway's secondary students chose 'restaurant and food' foundation classes as their first priority. Applications to foundation classes within 'service and transport' are relatively more stable over the same time frame, but still low with 2.2% choosing this as their first priority in 2018. This data suggests that throughput, or the supply of youth into vocational tourism and hospitality industry occupations is also low, and declining. To understand the context more fully, it is necessary to first describe the Norwegian educational system which supports the pathway to the tourism and hospitality industry.

The Norwegian Educational System

All education in Norway, from compulsory school through secondary school and university to the top-level (Ph.D.), is free. There are two main educational pathways to work within the tourism and hospitality business/industry of Norway. The first being through university-level schooling in which students complete coursework focused in hospitality and tourism preparing these individuals for entry-level managerial positions within the industry. In contrast to this purely educational path, the second pathway is more trade-based and vocationally structured. The vocational route consists of two years of foundation classes in secondary school followed by two years of industry apprenticeship. This path culminates in a certificate of apprenticeship qualifying the candidate for a specific profession within their industry. Foundational classes within 'restaurant and food' and 'service and transport' lead to certificates of apprenticeship relevant for the hospitality and tourism industry, for instance, chef, waiter, receptionist, and travel agent. The foundation classes ensure/attend to the necessary and relevant theoretical aspects of education within the field, while the apprenticeship period represents and ensures the introduction, insertion, and knowledge of practice within the chosen specialisation leading to the specific certificate of apprenticeship, the craft certificate.

As an apprentice in hospitality and tourism, one usually stays in the same enterprise, or organisation, throughout the whole apprenticeship period. For the duration, they complete a workbook describing areas and tasks they are expected to learn and practise during their apprenticeship. How this is practised and the tasks and challenges an apprentice is given are very much dependent on the enterprise or organisation they are completing their apprenticeship in. Apprentices in enterprises that are members of the 'Regional training office', are supposed to be monitored by this office with two meetings every half year. They are invited to enrol in further courses but these are voluntary. Formal apprenticeship systems therefore represent a possibility for the industry to attract and secure a stable and competent workforce, as well as the opportunity to train these occupational entrants to fit their business needs. At the same time, the apprentice's situation is very vulnerable in this system, dependent on whether the employer values and facilitates the apprenticeship experience or not (for more details see Schools in Norway 2019).

A new model of vocational education, including the apprenticeship is under consideration, combining theory and practice within the industry and interchanging between these across the four-year duration. Some pilot studies on this model are in progress but not within the tourism and hospitality field. The new model has some similarities with earlier apprenticeship models in Norway, as well as with other international contemporary models in Germany and Great Britain. This model will do a better job in incorporating theory and practice, ensuring that the theoretical and practical components are properly integrated into the apprenticeship, an issue noted by a couple of employers during the data collection of Dagsland et al. study 2015 as missing in the current apprenticeship programme. The new model further grants apprentices the opportunity to change apprenticeship employer during their apprenticeship, which aligns with perspectives voiced by many of the apprentices interviewed in the Dagsland et al. (2015). In particular, many of the apprentices in this study expressed a desire to be able to switch their apprenticeship site/employer in order to gain a variety of experiences in different professional kitchens during their training.

First Encounters—Smooth or Brutal?

Although tourism and youth are inextricably linked *vis-à-vis* employment (Robinson et al. 2019), we do not know much in general about youth's first encounters and experiences with working life, either generically (Price et al. 2011), or in the hospitality and tourism industry (Golubovskaya et al. 2019). Indeed, Yeung and Rauscher (2014) argue, that after decades of research into the impacts of early work experiences on youth, results still provide, at best, mostly contradictory findings. We do know whether young people (15–16-year old) in Norway have positive perceptions about the tourism and hospitality industry, and whether these perceptions change during their experience as an apprentice. Research indicates that Norwegian apprentices' positive perceptions of the industry are changing for the worse based on their work experiences as apprentices (Dagsland et al. 2011, 2015). Before examining the findings of these two studies (Dagsland et al. 2011, 2015), it is valuable to first examine how young people are socialised to the world of work, as well as the different aspects of the workplace, which impact this transition.

The First Encounter—The Socialisation Process into Work

The first encounters with working life play an important role in a young person's development and may have a cardinal influence on their understanding and interpretation of working life and organisations. Again, although there is a rich literature on newcomer socialisation (cf. Smith et al. 2017), there is scant research on the socialisation of the young worker (re apprenticeships see Cortini 2016) and on the transition into the tourism and hospitality labour market (Wamsley et al. 2012). First encounters are an important phase in the socialisation process of the young worker into the world of work. These encounters are characterised by various processes including the learning of an organisations' values and norms in relation to behaviour, the acquisition of new roles and tasks, and the learning of how to work and interact with co-workers within the organisational structure. Moreover, it includes comparing own expectations with reality and comparing one's personal standards to the organisations' evaluations, and may even include changes in the self-concept (Feij 1998). The socialisation process is paramount in ensuring that a new worker properly settles into a position. The process also can impact a new worker's well-being, their performance, as well as their intentions to leave an organisation. As the psychological contract literature affirms (cf. Robinson 1996), violations of trust regarding the expectation and reality nexus can have deleterious consequences for individuals and organisations alike.

Socialisation to work, *work socialisation*, can be defined as "the learning process by which young people acquire the skills, attitudes and values, and behaviour that are needed in order to be able to function as a valuable employee or as a fully integrated member of a working organization" (Feij 1998, p. 207). As such, it can include both the socialisation into the behaviours and values of a specific organisation, and occupational socialisation, which refers to "the inculcation of occupational values and skills, which might generalize across organizational settings in which the occupation may be practiced" (Fisher 1986, p. 102). Occupational socialisation into various tourism occupations, in particular cookery, has proven to be a complex and confronting process (Robinson and Beesley 2010). To compound this, research on first time workforce entrants has found that candidates that are not inherently qualified and/or prepared for their roles will socialise poorly into their organisations (Kowtha 2011). This is pertinent in the tourism context as many entrants are chosen for their value-for-money, aesthetic qualities and vibrancy (Robinson et al. 2019), but not necessarily their 'matched', 'person-fit' qualities (Kristof-Brown et al. 2005) and qualifications per se.

The apprenticeship, therefore, introduces the apprentice to the organisation with its values, behaviours, tasks and work conditions *and* at the same time to the role of the professional, the role, values, and skills of the specific profession. What happens in this period and how it is experienced, is thus important in the young person's/apprentice's development and socialisation into their chosen profession and the world of work. Even though the apprentices have some knowledge about the industry through their foundation classes and short periods of practice, what they encounter as an employee in an organisation often presents new experiences and challenges. Therefore, in spite of their training, the workplace often creates an environment, which is unfamiliar and thus can cause an apprentice to experience anxiety, uncertainty, and stress (Louis 1980; Lundberg and Young 1997; Saks and Ashforth 1996). The apprentices' acquired expectations and earlier experiences are central in how they understand and interpret, add meaning, and respond, to these situations. This will influence their further learning, their behaviour and their decision to continue or discontinue an apprenticeship and possibly leave an occupation.

The testing of expectations against reality is central in this first encounter, this first confrontation with work life, and the consequences of met or unmet expectations have been a subject of research for some time. Porter and Steers (1973) formulated a hypothesis on 'met expectations' in which a central idea was that a person's propensity to withdraw from a job would increase if the person's expectations are not substantially met, and that met expectations will contribute to job satisfaction, commitment, motivation, and reduced turnover intentions. Previous research and to a certain degree more recent research, support this hypothesis (Irving and Montes 2009). Others have questioned the hypothesis (Irving and Meyer 1994) and found stronger support for a main effect of post-entry experiences, indicating that providing positive work experiences would decrease the turnover intentions.

In this learning process, leaders and co-workers, consciously or unconsciously, are central as 'social agents' and role models (Filstad 2004; Kowtha 2011). The interaction with and support from them is probably the most significant resource in the socialisation and adjustment to the organisation (cf. Gruman et al. 2006), as they are providing information and feedback on attitudes, values, appropriate behaviours, and skills (Saks and Gruman 2011). More recent research uses the concepts 'social support' and 'perceived organizational support' to describe this relation and its importance (Allen and Shanock 2013; Chen and Eldridge 2011; Filstad 2011, Perrot et al. 2014), and a study by Kammermeyer-Mueller and Wanberg (2003) underlines the leader influence as the only socialisation factor that predicted reduced turnover.

Apprentices' Expectations and Experiences in Encounters with the (Hospitality and Tourism) Industry

As noted, two Norwegian studies (Dagsland et al. 2011, 2015) have focused on apprentices' expectations and experiences in the apprenticeship. The findings from the two studies, as we will discuss, were in accordance to a great degree.

The first (Dagsland et al. 2011), a quantitative study of 96 apprentices who encompassed a survey which was completed by apprentices within the first 2–6 months of commencing their apprenticeships. The survey asked apprentices questions in regard to their beliefs about working in the industry, their intentions and expectations of working in the industry, and about their early working experiences as an apprentice. The apprentices' expectations were measured using 17 items created for the study based partly on Hertzberg's two-factor theory (Herzberg 1966) and Hackman and Oldham's 'The job diagnostic survey' (a scale for measuring job satisfaction, Hackman and Oldham 1980). The items on expectations were formulated as statements with answers given on agreement on a Likert type of scale from one to five (1 = not at all, 5 = to a high degree) The experienced fulfilment of the expectations were measured on a similar scale with the same items now formulated as statements of fulfilment.

The findings of the first study indicated that the apprentices had high and positive expectations regarding their work and their apprenticeship. The strongest expectations were connected to learning and respect: experiencing personal and vocational development, and on being respected as a person. Following these are expectations of being included in the group as one of the others—a belongingness—given appropriate training and information, being respected for the competence they have, given varied and challenging tasks, and provided individual supervision. They further expressed expectations on working hours and rights being met, particularly with respect to long working hours and hard work.

When measuring the perceived fulfilment of their expectations, the findings showed that some of the strongest and most important expectations were perceived to not be fulfilled. These were expectations in relation to the learning aspect: personal and vocational development, receiving training and challenging tasks, and individual supervision, and on appreciation and feeling a sense of respect. These needs and expectations, via the apprentices' experiences, were largely unmet. The expectations regarding working hours and rights being met were similarly reported as not being fulfilled. In contrast to the unfulfilled expectations several expectations were determined to be fulfilled, these included their perceived inclusion into workplace team and the assistance they received during the workplace acclimation process.

Expectations concerning the characteristics of the work they expected before commencing their workplace apprenticeship including such things as; hard work, long working days, working independently, and having to do the rough/undesirable work showed no significant differences between what they had expected and their experiences. This finding demonstrated that apprentices had fairly good understanding of the work environment they were entering as apprentices.

In addition, to exploring apprentices' expectations of work and their apprenticeships, the study also examined how either these fulfilled or unfulfilled expectations influenced apprentices' intentions to continue working in the industry after their apprenticeship was completed. The findings indicated that apprentices who perceived being 'cared for' during their apprenticeships demonstrated a desire to continue working in the industry. This finding highlights the importance of inclusion, individual care and follow-up, social support, and social relations in fostering an apprentice's desire to continue their career within the tourism and hospitality industry. Similarly, responses on open-ended questions in the survey regarding positive and negative experiences or incidents, further demonstrated that negative experiences within the workplace in this early period was a reason for quitting the apprenticeship and for intentions of leaving the industry after the apprenticeship period. It is important to note, that questions on why expectations were not experienced as met or fulfilled were not included in the survey in this study. Responses on the open-ended questions on incidents that have had a great impact give, however, some indications, listing lack of respect, disparagement, and denunciation, and perceived arrogance from co-workers and leaders.

The second study (Dagsland et al. 2015) a qualitative study encompassed twelve semi-structured interviews with apprentices in both the early and late (20–22 months since commencing) stages of their apprenticeships. Data collected from the interviews was analysed using thematic analysis, deductive theoretical analysis at a semantic level and summarising content analyses (Braun and Clarke 2006; Flick 2002). The findings in this study to a great degree accorded with the findings of Dagsland et al. (2011), though the study provided greater support for the value of social relations during one's apprenticeship.

The specific findings were as follows. In relation to apprentices' expectations about their apprenticeship and the workplace, the apprentices were vague, indicating that they were just curious and excited and had no specific expectations—'none I can remember'. However, Dagsland et al. (2015) note that apprentices' expectations were revealed, as being both met and unmet, through the thick descriptions of situations. Dagsland et al. (2015) indicate that apprentices became aware of their expectations especially when describing what they missed, what was lacking, what they were displeased or dissatisfied with. The authors also reported that apprentices' expectations were related to the learning process, personal/professional development, and their interactions with their supervisor, particularly in relation to how the leader treated them and fostered their inclusion in the group. Though the experiences of the apprentices varied, in general, the majority of interviewees reported their expectations not being met.

One would expect the first reception at a new workplace and the meeting with colleagues to be welcoming and pleasant but as the study participants indicated this was not always the case. In their study, Dagsland et al. (2015) found that most apprentices felt welcomed in their workplace, but some reported feeling unwelcomed. The interviewees who reported feeling unwelcomed described co-workers and supervisors as hardly speaking to them, they recalled being bossed around from the moment they arrived in their workplace, they also described being left alone and asked to complete certain task without proper instruction. In spite of these negative experiences, most apprentices reported feeling a strong sense of inclusion in their workplace group. These findings are very similar to what was reported by Dagsland et al. (2011). In the second interview conducted near the completion of an apprentice's apprenticeship, Dagsland et al. (2015) reported apprentices feeling frustrated in their work situation, describing how though many felt a sense of belonging to their profession, they were often constantly reminded of their lower status as apprentices. Interviewees described such things as being excluded from discussions on menus and the distribution of work tasks. Dagsland et al. (2015) noted how these experiences often influenced an apprentice's perception of social inclusion/exclusion—from an initial perception of being included in the workgroup.

Dagsland et al. (2015) also noted how the lack of respect for the apprentice as an individual, and as a professional-to-be, was an important topic in the second round of apprentice interviews. Dagsland et al. (2015) described how this lack of respect was exemplified in incidents of scolding and harassment, incidents of being bossed around, and incidents in which apprentices felt they were continually expected to do the boring and menial tasks with an understanding that their opinions were not valued or wanted. Dagsland et al. (2015) noted how apprentices often felt they were expected to have the knowledge to do the varied aspects of their work, while at the same time they were primarily given menial tasks and routine work which did not provide them the opportunity to test and expand their occupational skillset. In the study, Dagsland et al. (2015) describe how these experiences impacted an apprentices' feelings of social inclusion. The authors described how many apprentices felt there was little interest in, or a low evaluation of, their knowledge and competence, as they were generally not asked their opinions or asked to complete tasks which would highlight their skills and demonstrate what they have learned. They further described how apprentices frequently felt isolated from the work team. The authors noted this isolation often stemmed perceptions of not being included in the work process of the team, and from admonishments from their fellow team members about a perceived lack of occupational knowledge or skills. Resulting from this perceived isolation, Dagsland et al. (2015) described how apprentices frequently felt they had few opportunities to take initiative, develop, and demonstrate their occupational knowledge and skills, which ultimately contributed to suboptimal learning environment which was not particularly accommodating and receptive in nature.

Dagsland et al. (2015) described how apprentices had expectations of the supervision and follow-up of the learning process they received during their apprenticeship, and how many apprentices felt these expectations were not being met. The authors noted how many apprentices felt that they were often left alone and that they felt their supervisor showed little or no involvement and concern in their development. Dagsland et al. (2015) also reported that only some of the apprentices in their study had a supervisor dedicated to them, and some of those who had one did not know who they were. To the extent of there being follow-up and supervision, this seemed to gradually fade away during the period. This resulted in an alliance between the apprentices themselves, where the second-year apprentices became the ones 'teaching' the first-year apprentices, giving them necessary information and showing them the tasks. It is important to note that Dagsland et al. (2015) also highlighted that most apprentices felt that they themselves had to take on the responsibility of guiding their learning through both the use of their training book and also by seeking out new opportunities for occupational growth at their workplace.

The findings of Dagsland et al. (2015) further show a lack of positive, and substantial feedback given to apprentices throughout their learning process. The authors reported that direct and positive feedback was rare. While indirect feedback was more prevalent, and generally came in comments that the apprentices' work should have been better or quicker. This dominant form of feedback ultimately contributed to apprentices' feelings of uncertainty and discomfort. The findings further indicate that negative feedback became more pronounced throughout the apprenticeship but not in a substantial or constructive way, rather as screaming, nagging, and verbal aggression, with negative and impolite wording. In spite of unmet expectations and negative experiences, Dagsland et al. (2015) reported that apprentices still had a desire to continue working in the industry, though contingent on future employment being outside of the enterprise/organisation they completed their apprenticeship within. Dagsland et al. (2015) found that most apprentices had a strong inner motivation, which helped them overcome their negative experiences. Moreover, the authors indicated that generally the apprentices who reported negative experiences during their apprenticeships were more inclined to attribute the negative experiences to a specific persons or situations rather than the industry as a whole.

Discussion and Conclusions

Our presentation of the data from Norwegian apprentices (Dagsland et al. 2011, 2015) and of the educational choices of Norwegian youth (Dagsland et al. 2017 and 2018 unpublished data) broadly accords with overarching narratives of hospitality and tourism not fully valuing, appreciating, and investing in its workforce. The educational system in Norway offers possibilities for the industry to employ well-qualified personnel, both with vocational education and with higher education backgrounds. There is emerging evidence that the hospitality and tourism industries are reluctant to hire educated personnel. There are two factors at play here, first wage efficiency, or unskilled labour may be cheaper (Thomas and Long 2001) and second, organisations recruiting service-orientated personnel focus primarily on attitude, aesthetics, and personality (smiling, cheerful, etc.,) for the required emotional labour (Warhurst and Nickson 2009). These factors do not necessarily accord well with investing in apprentices. Although arguably apprentices may work on subsidised wages, the organisation must factor in additional supervisory support and other resources that alter the value proposition. Yet, industry seems to consider apprenticeships as a burden. Although organisations receive economic support when training apprentices—granted not enough to cover all the expenses as the apprentices are salaried and this increases annually across the two years on the principle that the value the work of the apprentice rises as they become more experienced and proficient—but supposedly enough to cover additional expenses. This has led some scholars whether we are witnessing what has been termed the 'crisis of apprenticeships' (Schalk et al. 2017).

While the apprenticeship has worked historically, increasingly apprentices are abandoning their training when they see the opportunity of full wages. There is some evidence of this during the mining boom in Australia in the late 2000s (cf. Hajkowicz et al. 2011). On the other hand, the draconian and exploitative practices of employers make apprenticeships an unattractive option for youth (Bishop 2017). Previous research suggests during economic downturns employers respond by de-prioritising apprenticeships (Schumann 2017)—so this presents a macro-level explanation. Others opine structural failings exist in the apprenticeship system, and this was apparent in our data. From the employer's perspective there were suggestions that apprenticeships were not ideally structured to bridge the theory–practice nexus—and this diluted the value of apprentices to organisations (Dagsland et al. 2015). From the supply-side perspective, apprentices are traditionally indentured individuals to one employer for the duration of their apprenticeship, inhibiting the professional development opportunities that mobility between different organisations and industrial sectors can provide. Either way there seem to be a number of drivers for organisations, and individuals, not seeing value in apprenticeships in the same enthusiastic vein as historically.

Turning now to the broader implications of the data specifically speaking to how apprentices related their experiences, we first acknowledge it is difficult to generalise. Contexts differ and many apprentices have rewarding experiences with invested employers and supervisors, no doubt. But overwhelmingly the apprentices report a disconnect between their broadly enthusiastic expectations and the reality of their experiences after some time 'on-the-job'. It seems in the workplace they feel as though they are cheap labour as opposed to professionals-to-be—indeed the future of the industry (cf. Bishop 2017). Interestingly, apprentices in the studies reported seeking out intra-group support—that is formalised connections between apprentices, even across organisations, and this is consistent with the social support literature as tested in hospitality contexts (Fong et al. 2018).

If we now turn our attention to the socialisation literature there are some implications for employers. Our data suggest that the socialisation process for newcomer apprentices is poor, experiencing *inter alia* social exclusion, denied (or being fearful of accessing) learning experiences, not feeling respected, relying on each other rather than supervisors, experiencing abusive behaviour, delegated menial work in a hierarchical structure, and so on. This flies in the face of literature which finds that perceived organisational support, for apprentices, positively predicts important socialisation outcomes including learning (the raison d'etre of apprenticeships), and acculturation to workplace and group norms (Perrot et al. 2014). Indeed, the importance of perceived organisational support has also been found to have direct and indirect mediating effects on intention to quit (Madden et al. 2015), which is a perennial hospitality and tourism issue. Although a reasonably logical finding Tews et al. (2018), found in a study of restaurant employees, that abusive treatment of newcomers predicted turnover intent significantly more than more established employees. While this body of research suggests that the treatment of apprentices as reported in our study is discordant with positive newcomer socialisation practices, it is important to recognise that apprentices came to their work and training with expectations. If we consider the psychological contract a proxy for these expectations, then again, the most recent research finds that their breach significantly disrupts the socialisation process (Woodrow and Guest 2019)—with the obverse also holding true.

Finally, we return to the global youth in tourism and hospitality narrative, and the implications of this study on that broader conversation. Recent contributions to the literature, which suggest the industry in leveraging on the benefits of youth employment (cf. Baker et al. 2014; Warhurst and Nickson 2007), has a developmental and duty of care responsibility (Jennings et al. 2013), and even a symbiotic relationship

(Golubovskaya et al. 2017; Robinson et al. 2019) with youth. Implicit in our study and this critical discussion is something of a paradox. The tourism and hospitality industries should feel privileged that youth are attracted to work in the myriad of jobs they offer. Youth, by and large, present a good value-proposition because their labour is affordable and cheerful—meeting the prerequisites of aesthetic and emotional labour (Warhurst and Nickson 2009). Yet the industry responds with neglect verging on exploitation. In the case of apprenticeships, this is somewhat astounding. While many youth will move through tourism en route to other careers, while they study or 'find their own path' (Robinson et al. 2019), apprentices have made a conscious choice and investment into critical occupations vital to the industry. This is a perplexing paradox! On the evidence presented in this paper Norwegian employment practices *vis-à-vis* apprenticeships are contributing to the myriad challenges faced by the tourism and hospitality industry in relation to the workforce.

While space does not permit an extended discussion there are some obvious takeaways from this study. The industry needs to take greater pride, and custodianship, in providing youth with their first working opportunities and take affirmative action to ensure they are socialised in a positive manner. Leadership is a key factor here in driving attitudinal and cultural change and the adherence to more humanistic management styles (cf. Pirson 2017). These maxims will flow through to supervisors and co-workers within the apprentices' orbit. Internal educational and organisational structures, such as social support mechanisms, can be implemented to reduce the risk of psychological contract breaches, regarding apprentices' expectations (cf. Woodrow and Guest 2019), and lacking perceived organisational support, which has been linked to numerous deficit outcomes for newcomers (Madden et al. 2015; Tews et al. 2018). To conclude however, apprentices also should take responsibility for their own learning, and they should be invited to, and provided with the empowering tools to take charge of their learning, call out unacceptable behaviour, and support their peers. They are, after all, the future of the hospitality and tourism industry and the workforce at large. There is a collective response required to ensure that hospitality and tourism remain a gateway to the world of work for youth, and their journey is not marked as fast and furious leading to premature exits.

References

Allen, D.G., and L.R. Shanock. 2013. Perceived organizational support and embeddedness as key mechanisms connecting socialization tactics to commitment and turnover among new employees. *Journal of Organizational Behavior* 34: 350–369. https://doi.org/10.1002/job.1805.

Baker, T., A. Rapp, T. Meyer, and R. Mullins. 2014. The role of brand communications on front line service employee beliefs, behaviors, and performance. *Journal of the Academy of Marketing Science* 42 (6): 642–657.

Baum, T., A. Kralj, R.N.S. Robinson, and D. Solnet. 2016. Tourism workforce research: Review, taxonomy and agenda. *Annals of Tourism Research* 60: 1–22.

Baum, T., S. Mooney, R.N.S Robinson, and D. Solnet. 2019. Tourism employment challenges and paradox: A perspective article. *Tourism Review*, accepted July 4.

Bishop, D. 2017. Affordance, agency and apprenticeship learning: a comparative study of small and large engineering firms. *Research in Post-Compulsory Education* 22 (1): 68–86.

Braun, V., and V. Clarke. 2006. Using thematic analysis in psychology. *Qualitative Research in Psychology* 3 (2): 77–101. https://doi.org/10.1191/1478088706qp063oa.

Chen, J., and D. Eldridge. 2011. The missing link in newcomer adjustment: The role of perceived organizational support and leader-member exchange. *International Journal of Organizational Analysis* 19 (1): 71–88. https://doi.org/10.1108/19348831111121312.

Cortini, M. 2016. Workplace identity as a mediator in the relationship between learning climate and job satisfaction during apprenticeship: Suggestions for HR practitioners. *Journal of Workplace Learning* 28 (2): 54–65.

Dagsland, Å.H.B., R.J. Mykletun, and S. Einarsen. 2011. Apprentices' expectations and experiences in the socialisation process in their meeting with the hospitality industry. *Scandinavian Journal of Hospitality and Tourism* 11 (4): 395–415.

Dagsland, Å.H.B., R.J. Mykletun, and S. Einarsen. 2015. "We're not slaves—We are actually the future!" A follow-up study of apprentices' experiences in the Norwegian hospitality industry. *Journal of Vocational Education & Training* 67 (4): 460–481.

Dagsland, Å.H.B., R.J. Mykletun, and S.V. Einarsen. 2017. Antecedents of Norwegian adolescents' choice of educational pathway in hospitality and tourism. *Nordic Journal of Working Life Studies* 7 (4): 51–71.

DAMVAD. 2014. *Utredning om norsk servicenæring. For Nærings- og fiskeridepartementet* (The Norwegian Restaurant Business: White Paper for the Ministry of Business and Fishery) Oslo: DAMVAD.

Feij, J.A. 1998. Work socialization of young people. In *Handbook of Work and Organizational Psychology*, vol. 3, 2nd ed., ed. P.J.D. Drenth, H. Thierry, and C.J. de Wolff, 207–256. East Sussex: Psychology Press.

Filstad, C. 2004. How newcomers use role models in organizational socialization. *The Journal of Workplace Learning* 16 (7): 396–409. https://doi.org/10.1108/13665620410558297.

Filstad, C. 2011. Organizational commitment through organizational socialization tactics. *Journal of Workplace Learning* 23 (6): 376–390. https://doi.org/10.1108/13665621111154395.

Fisher, C.D. 1986. Organizational socialization: An integrative review. In *Research in Personnel and Human Resources Management*, vol. 4, ed. K.M. Rowland and G.R. Ferris, 101–145. Greenwich, CT: JAI Press.

Flick, U. 2002. *An Introduction to Qualitative Research*, 2nd ed. London: Sage.

Fong, L.H.N., P.M.W. Chui, I.S.C. Cheong, and D.K.C. Fong. 2018. Moderating effects of social support on job stress and turnover intentions. *Journal of Hospitality Marketing & Management* 27 (7): 795–810.

Golubovskaya, M., D. Solnet, and R.N.S. Robinson. 2017. *Service organisations as the caretakers for the future workforce: A CSR perspective on youth employment and frontline service roles.* 8th International Research Symposium in Service Management, 5–8 August, 1–12, Seoul.

Golubovskaya, M., D. Solnet, and R.N.S. Robinson. 2019. Recalibrating talent management for hospitality: A youth development perspective. *International Journal of Contemporary Hospitality Management.* https://doi.org/10.1108/IJCHM-11-2018-0911.

Gruman, J.A., A.M. Saks, and D.I. Zweig. 2006. Organizational socialization tactics and newcomer proactive behaviors: An integrative study. *Journal of Vocational Behavior* 69: 90–104.

Hackman J.R., and G.R. Oldham. 1980. *Work Redesign*. Reading, MA: Addison-Wesley.

Hajkowicz, S.A., S. Heyenga, and K. Moffat. 2011. The relationship between mining and socio-economic well being in Australia's regions. *Resources Policy* 36 (1): 30–38.

Halvorsen, R., and H. Hvinden. 2014. *New Policies to Promote Youth Inclusion: Accommodation of Diversity in the Nordic Welfare States*. Copenhagen, Denmark: Nordic Council of Ministers.

Herzberg, F. 1966. *Work and the Nature of Man*. Cleveland: The World Publishing Company.

International Youth Foundation. 2013. Creating opportunities for youth in hospitality. https://www.iyfnet.org/sites/default/files/library/Creating-Opportunities-for-Youth-in-Hospitality.pdf. Accessed on June 4, 2020.

Irving, P.G., and J.P. Meyer. 1994. Re-examination of the met-expectations hypothesis: A longitudinal analysis. *Journal of Applied Psychology* 79 (6): 937–949. https://doi.org/10.1037/0021-9010.79.6.937.

Irving, P.G., and S.D. Montes. 2009. Met expectations: The effects of expected and delivered inducements on employee satisfaction. *Journal of Occupational and Organizational Psychology* 82: 431–451. https://doi.org/10.1348/096317908X312650.

Jennings, J., R. Breitkreuz, and A. James. 2013. When family members are also business owners: Is entrepreneurship good for families? *Family Relations* 62: 472–489.

Kammermeyer-Mueller, J.D., and C.R. Wanberg. 2003. Unwrapping the organizational entry process: Disentangling multiple antecedents and their pathways to adjustment. *Journal of Applied Psychology* 88 (5): 779–794. https://doi.org/10.1037/0021-9010.88.5.779.

Kowtha, N.R. 2011. School-to-work transition and newcomer socialisation: The role of job-related education. *Journal of Management & Organization* 17 (6): 747–763.

Kristof-Brown, A.L., R.D. Zimmerman, and E.C. Johnson. 2005. Consequences of individuals' fit at work: A meta-analysis of person–job, person–organization, person–group, and person–supervisor fit. *Personnel Psychology* 58 (2): 281–342.

Kusluvan, S., Z. Kusluvan, I. Ilhan, and L. Buyruk. 2010. The human dimension: A review of human resources management issues in the tourism and hospitality industry. *Cornell Hospitality Quarterly* 51 (2): 171–214.

Louis, M.K. 1980. Surprise and sense making: What newcomers experience in entering unfamiliar organizational settings. *Administrative Science Quarterly* 25 (2): 226–251. https://doi.org/10.2307/2392453.

Lundberg, C.C., and C.A. Young. 1997. Newcomer socialization: Critical incidents in hospitality organizations. *Journal of Hospitality and Tourism Research* 21 (2): 58–74.

Madden, L., B.D. Mathias, and T.M. Madden. 2015. In good company: the impact of perceived organizational support and positive relationships at work on turnover intentions. *Management Research Review* 38 (3): 242–263.

Mooney, S. 2016. Wasted youth in the hospitality industry: older workers' perceptions and misperceptions about younger workers. *Hospitality and Society* 6 (1): 9–30.

Mykletun, R.J., and T. Furunes. 2012. On hospitality industries and their constant need for Workforce. In *Current Research in Hospitality and Tourism*, ed. Trude Furunes, Reidar J. Mykletun, and Einar Marnburg, 27–58. Oslo: Fagbokforlaget.

Perrot, S., T.N. Bauer, D. Abonneau, E. Campoy, B. Erdogan, and R.C. Liden. 2014. Organizational socialization tactics and newcomer adjustment: The moderating role of perceived organizational support. *Group and Organization Management* 39 (3): 247–273.

Pirson, M. 2017. *Humanistic Management: Protecting Dignity and Promoting Well-Being*. Cambridge: Cambridge University Press.

Porter, L.W., and R.M. Steers. 1973. Organizational, work, and personal factors in employee turnover and absenteeism. *Psychological Bulletin* 80 (2): 151.

Price, R.A., P.K. McDonald, J. Bailey, and B. Pini. 2011. A majority experience: Young people's encounters with the labour market young people and work. In *Young People and Work*, ed. R.A. Price, P.K. McDonald, J. Bailey, and B. Pini, 1–17. Surrey, GB: Ashgate.

Robinson, S.L. 1996. Trust and breach of the psychological contract. *Administrative Science Quarterly* 41 (4): 574–599.

Robinson, R.N.S., and L.G. Beesley. 2010. Linkages between creativity and intention to quit: An occupational study of chefs. *Tourism Management* 31 (6): 765–776.

Robinson, R.N.S., T. Baum, M. Golubovskaya, D. Solnet, and V. Callan. 2019. Applying endosymbiosis theory: Tourism and its young workers. *Annals of Tourism Research*, accepted 5th July.

Rørstad, K., P. Børing, E. Solberg, and T. Carlsten. 2018. *NHOs Kompetansebarometer 2018. Resultater fra en undersøkelse om kompetansebehov blant NHOs medlemsbedrifter i 2018* [NHO's competence inventory. Results from a survey on competence needs among NHO's federated companies in 2018], NIFU Report 2018:23.

Saks, A.M., and B.E. Ashforth. 1996. Proactive socialization and behavioral self-management. *Journal of Vocational Behavior* 48 (3): 301–323.

Saks, A.M., and J.A. Gruman. 2011. Organizational socialization and positive organizational behaviour: implications for theory, research, and practice. *Canadian Journal of Administrative Sciences/Revue Canadienne Des Sciences de L'Administration* 28 (1): 14–26. https://doi.org/10.1002/cjas.169.

Schools in Norway: The Norwegian Education System. 2019. https://www.justlanded.com/english/Norway/Norway-Guide/Education/Schools-in-Norway.

Schumann, M. 2017. The effects of minimum wages on firm-financed apprenticeship training. *Labour Economics* 47: 163–181.

Schalk, R., P. Wallis, C. Crowston, and C. Lemercier. 2017. Failure or flexibility? Apprenticeship training in premodern Europe. *Journal of Interdisciplinary History* 48 (2): 131–158.

Smith, L.G., N. Gillespie, V.J. Callan, T.W. Fitzsimmons, and N. Paulsen. 2017. Injunctive and descriptive logics during newcomer socialization: The impact on organizational identification, trustworthiness, and self-efficacy. *Journal of Organizational Behavior* 38 (4): 487–511.

SSB Statistic Norway. 2019. https://www.ssb.no/statbank/table/07972/.

Stacey, J. 2015. *Supporting Quality Jobs in Tourism*. OECD Tourism Papers. Paris: OECD.

Tews, M.J., J.W. Michel, and K. Stafford. 2018. Abusive Coworker Treatment, Coworker Support, and Employee Turnover. *Journal of Leadership & Organizational Studies*. 1548051818781812.

Thomas, R., and J. Long. 2001. Tourism and economic regeneration: the role of skills development. *International Journal of Tourism Research* 3 (3): 229–240.

Tracey, J.B. 2014. A review of human resources management research: The past 10 years and implications for moving forward. *International Journal of Contemporary Hospitality Management* 26 (5): 679–705.

Walmsley, A. 2015. *Youth Employment in Tourism and Hospitality: A Critical Review*. Oxford: Goodfellow Publisher.

Walmsley, A., R. Thomas, and S. Jameson. 2012. Internships in SMEs and career intentions. *Journal of Education and Work* 25 (2): 185–204.

Warhurst, C., and D. Nickson. 2007. Employee experience of aestehtic labour in retail and hospitality. *Work, Employment & Society* 21 (1): 103–120. https://doi.org/10.1177/0950017007073622.

Warhurst, C., and D. Nickson. 2009. 'Who's got the look?' Emotional, aesthetic and sexualized labour in interactive services. *Gender, Work & Organization* 16 (3): 385–404.

Woodrow, Chris, and David E. Guest. 2019. Pathways through organizational socialization: A longitudinal qualitative study based on the psychological contract. *Journal of Occupational and Organizational Psychology*. https://doi.org/10.1111/joop.12285.

WTTC. 2015. Travel and tourism: Economic impact 2015. Europe: The World Travel and Tourism Council. Available at www.wttc.org//media/files/reports/economic%20impact%20research/regional%202015/europe2015.pdf.

Yeung, W.J.J., and E. Rauscher. 2014. Youth early employment and behavior problems: Human capital and social network pathways to adulthood. *Sociological Perspectives* 57 (3): 382–403.

14

Labour Mobility in the Tourism and Hospitality Sector in Sweden

Mats Lundmark

Introduction

Research on labour market dynamics, workforce and work–life issues has recently been considered to be a relatively underrepresented theme in studies on hospitality and tourism (Baum et al. 2016; Zampoukos and Ioannides 2011). Labour or workforce studies in tourism and hospitality are surprisingly few and to some extent disorganized, scattered and misrepresented, according to Baum et al. (2016, pp. 4–6). The study of employee turnover and labour mobility, including the hospitality subsectors hotels and restaurants, is one of the research themes that deserves more attention, not least considering the exceptionally high levels of mobility that characterizes the sector (Baum et al. 2016, p. 12).

Labour mobility is generally an important prerequisite for the adjustment and restructuring of industries. Mobility in this sense acts as a lubricant that facilitates the transfer of workers from declining sectors,

M. Lundmark (✉)
Örebro University, Örebro, Sweden
e-mail: mats.lundmark@oru.se

A. Walmsley et al. (eds.), *Tourism Employment in Nordic Countries*,
https://doi.org/10.1007/978-3-030-47813-1_14

to industries that are currently growing. Mobility thus contributes to facilitate the expansion of viable and successful companies and businesses, and gives the workers made redundant an opportunity to find a new job (Andersson and Tegsjö 2006; Martin and Morrison 2003). A certain degree of labour market mobility is also important to contribute to a favourable diffusion of knowledge and best practice between firms and industries (Breschi and Lissoni 2001; Power and Lundmark 2004; Lundmark and Power 2008; Bienkowska et al. 2011; Östbring and Lindgren 2013).

An excessively high labour mobility is however often to the detriment of the individual business or enterprise. There are a number of costs associated with (too high) turnover—costs for search processes and recruitment, for training or education. The costs also include the risk of skill losses, and sometimes an unwanted leakage of business secrets or a loss of important contact networks (Tomlinson and Miles 1999; Tomlinson 2002; Madsen et al. 2003). Shifting to a new employer or workplace also means a certain degree of sacrifice to the individual, especially if moving to a new job also results in a longer commuting distance or the need to move to a new place of residence. Shifting to a new job normally also means investing in new work tasks and establishing new social contacts. On the other hand, changing workplace and/or employer is often (but not always) associated with a career move, including new challenges and a better income.

In research on labour issues in the tourism and hospitality sector, the negative aspects of employee turnover and labour mobility tend to dominate. The hospitality and tourism labour market show a number of distinct features. The industry is characterized by seasonal variations and other short-term fluctuations in demand. This makes the demand for labour volatile, and calls for a high degree of flexibility. This need for flexibility means that the types of jobs available in the tourism and hospitality industry often have a low entry threshold and it is also relatively easy to change jobs within the industry, or to switch to other industries (Szivas et al. 2003). Parts of the industry are also characterized by flexible employment contracts, creating important differences between those employed on long-term contracts and temporary staff. The stereotyped worker in the sector is a young, often female, person without special training, being low paid and doing part-time work during holidays and weekends. These are some of the reasons that staff turnover and mobility is higher in this industry than in most other sectors—the percentage who left his or her job from one year to the next is (with some variations over time) about 35% in Sweden. The corresponding share for the Swedish labour market as a whole is around 10–15%, and the average labour mobility in general in the service sector is around 25–30% (Statistics Sweden 2001). The industry has also long been associated with problems of black or gray labour, a situation which, however, has improved in Sweden since the mid-00s with the introduction of compulsory staff lists that has to be handed in to tax authorities (Tax Procedure Act 2011:1244). At the same time, it is evident that the tourism industry in particular destinations offer attractive workplaces for young and adventurous people, representing good opportunities for meetings and social contacts (see e.g. Ladkin 2011 for a review). It should also be recognized that the sector in many ways provides a good opportunity for younger persons and immigrants to get their first work experience in life (Walmsley 2015).

High levels of staff turnover are thus considered to be a problem that needs to be tackled by the industry or the management of the specific firm or organization. Taken together, the negative sides of high labour turnover suggest that the productivity of the firm or workplace is affected in a negative way. There are, however, claims in the opposite direction. The basic idea is that persons that leave a workplace and an organization tend to be the ones least suited for the job, and thus are not the most productive workers. It can also be claimed that, although recruiting processes may be costly, hiring new personnel is an opportunity for management to make necessary adjustments to the competence profile of the workforce.

Following the above, the aim of this chapter is twofold. First, a number of both individual and workplace-related characteristics of the hospitality and tourism workforce in four counties in central Sweden are analyzed and discussed. Second, labour mobility between workplaces in the hospitality and tourism industry is analyzed from the viewpoint of the individual worker, and in comparison with other sectors of the labour market. The high level of labour mobility that characterizes the tourism and hospitality sector is often, not least by the business itself, believed to be reducible to structural factors like a young workforce, small firms and workplaces, low paid work, part-time work and so forth. The research design used in this chapter makes it possible to evaluate such propositions.

This chapter contributes to mobility research in the hospitality and tourism sector in a number of ways. First, it complements traditionally survey-based research by using a comprehensive micro database, obtained from administrative registers held by Statistics Sweden. Second, it offers an analysis of the hospitality and tourism sector in comparison to other sectors of the labour market. Third, it tests a few independent variables not often used in mobility research (the effect of commuting distance, enrolment in studies, labour market size and tenure).

Labour Mobility Research

A number of review articles or books that summarize and evaluate labour turnover and mobility research can be found in the literature (see Rubenstein et al. 2017; Saridakis and Cooper 2016; Steel and Lounsbury 2009; Griffeth et al. 2000; Hom and Griffeth 1995; Muchinsky and Tuttle 1979; Porter and Steers 1973; see also Orkan 1972 for an early Swedish study). This meta-analytic research reveals that turnover research is firmly grounded in, and strongly dominated by, research fields such as psychology, organizational behaviour, human resource management, labour market economics and to some extent business and work–life studies. Summarizing this literature is not easy. Three core mechanisms stand out as persistent results from turnover research; attitudes towards the job and the organization where you work (e.g. job satisfaction, organizational commitment, job embeddedness), job-searching mechanisms (both objective and perceived job opportunities in the labour market) and turnover intentions, i.e. intentions to quit or stay at the job (Steel and Lounsbury 2009). Job dissatisfaction as an explanation of quitting behaviour is perhaps the most pervasive and recurrent of the three mechanisms in turnover research. Behind job (dis-)satisfaction, we find specific elements like satisfaction with pay, work content, work autonomy, work supervision, promotion and career opportunities, expectations of the present job and social interaction at the workplace (Muchinsky and Tuttle 1979).

Apart from the core explanations discussed above, we find a large number of "secondary" mechanisms that supposedly influence turnover behaviour. Some of them are linked to the individual, like personality and values, personal history, age, education, skills and training and family situation, and some of them relate to the workplace or the organization. The latter include features like work unit size, type of employment contract and working hours per month and type of industrial sector or type of technology applied at the workplace (Eriksson 2016, pp. 133–135; Steel and Lounsbury 2009, pp. 275–278; Griffeth et al. 2000; Muchinsky and Tuttle 1979; Orkan 1972, pp. 32–71). In addition, in cross-country studies on labour mobility institutional differences—like employment protection legislation—become important (Nordic Council of Ministers 2010; Dale-Olsen 2016). Although many of the variables listed above show significant relationships with employee turnover, the overall predictability of the applied statistical models is relatively low in most cases (see Griffeth et al. 2000, who reports a modest average correlation between job satisfaction and turnover). This is perhaps not surprising, given the complexity surrounding the decision to change workplace.

From an individual perspective, career paths in the hospitality and tourism industry are often vague and include frequent mobility between tasks and workplaces. The employer often places a lot of emphasis on employees' ability to be flexible, socially competent and willing to move on and learn again. This is a result of increased pressure from the outside on the hospitality industry, as well as a polarization in the hospitality industry, indicating major differences between high and low-status jobs in the industry (Baum 2007). At the same time, studies of mobility of individuals in the hospitality and tourism industry indicate that the motives for changing jobs and moving between subsectors are related to job satisfaction and work content, and linked more to lifestyle rather than rational considerations of formal qualifications and salary (Szivas et al. 2003). The tourism industry seems, according to Hjalager and

Andersen (2001), to attract ultra-mobile, cross-border and social people, but they also note that further studies are needed to study the motives and incentives to work in the industry.

Work in the hospitality and tourism industry includes a degree of uncertainty and insecurity in relation to employment. In many contexts, this is described as negative for the industry and unattractive for the individual (Szivas et al. 2003). For certain groups of young workers, however, these conditions may be seen as attractive, and high mobility is driven not only by the uncertainty in demand and employers' reluctance to hire permanent staff but also by workers' unwillingness to settle down and sign long-term employment contracts (Walmsley 2015). This mobility is not necessarily seen as disloyalty to the employer or as a source of uncertainty for the individual (Baum 2007). High mobility between jobs and workplaces, including geographical mobility, fits well into a life where work, study, travel and leisure are mixed and blurred during different periods, such as seasonal work and gap year. This phenomenon has been conceptualized and discussed recently in tourism research in terms of lifestyle mobility (Duncan et al. 2013; Thulemark et al. 2014).

Existing empirical studies differ from the investigation presented in this chapter, in the sense that they are either using survey techniques to ask employees about their views on present working conditions, intentions to leave and job search activities (Iverson and Deery 1997; Szivas et al. 2003: Sims 2007; Dipietro and Condly 2007), or they use more in-depth interview with management and/or employees (Rowley and Purcell 2001; Heldt Cassel et al. 2017). In some cases, a combination of both surveys and interviews is used (Walmsley 2004; Davidson and Wang 2011).

A standard result from studies based on questionnaires in the tourism and hospitality is that employee dissatisfaction with work conditions (such as long and unsocial working hours, low pay, stress, poor work content, etc.) is positively associated with intentions to leave the present employer, and also the business as such. The impact of job satisfaction is mediated by factors like organizational commitment, motivation, the satisfaction with supervisory staff and co-workers, interaction with customers, perceived career opportunities in the sector and job opportunities elsewhere. These are all results that more or less conforms to turnover studies of the labour market in general. It is, however, difficult to find studies that systematically compare turnover in the hospitality and tourism industry with other industries.

A few specifically tourism-related factors are also discussed in the literature. In a study using microdata from official registers, Hjalager and Andersen (2001) analyze the retention of persons with a dedicated tourism education in the Danish tourism and hospitality sector in the 1980s and 1990s. They conclude that trained cooks and waiters more often stay in the sector than the average employee, but also that surprisingly few remain in the industry; after five years only a third are still working in hotels and restaurants. Furthermore, the retention rate was declining between 1981 and 1990 (Hjalager and Andersen 2001, p. 123).

The idea that the industry is characterized by a *turnover culture* is tested in a study by Iverson and Deery (1997). Such a culture is defined as an acceptance of turnover as part of the workforce group norm, i.e. "...a normative belief held by employees that turnover behavior is quite appropriate" (Iverson and Deery 1997, p. 71). Based on a survey distributed to six five-star hotels in Melbourne, Australia, this hypothesis is confirmed. A closely related idea—a *contagion effect*—is tested in a relatively recent study by Hinkin et al. (2012). Here the simple idea is that individual workers "...are paying attention to and are influenced by the attitudes and behaviours of their co-workers" (Hinkin et al. 2012, p. 11). By calculating the variance of job satisfaction among employees in 175 business units in a US recreation and hospitality corporation, they were able to conclude that when "...cohesive unit satisfaction increases, turnover diminishes, and the reverse is also the case" (Hinkin et al. 2012, p. 11).

Data and Research Design

The analysis in this chapter is based on longitudinal geo-referenced micro data on individuals that covers the period 1993–2011. The database (Bergslagsdata/BeDa) is compiled from a number of administrative registers held by Statistics Sweden, and consists of longitudinal information, for every third year, for *all* individuals (16 years of age and older) working and/or living in one of the four counties (Värmland, Örebro, Västmanland and Dalarna) in Central Sweden. The four counties represent approximately 11% of the population of the country as a whole in 2011 (Statistics Sweden). Thus, with the exception of the large metropolitan areas in Sweden, the data is highly representative of the Swedish labour market as a whole. The database has recently been updated and the characteristics of the tourism and hospitality workforce presented in Tables 14.1 and 14.2 are thus based on the situation in 2014. An important advantage over survey-based data is that there are effectively no problems related to sampling and low response rates.

The database contains a wide range of information on *individual characteristics* (sex, age, country of birth, family composition, educational attainment, attachment to the labour market, place of residence, etc.). The database also includes *work–life-related information* such as connection to workplace and firm, industrial classification of the workplace, size of workplace, work-related income and the location of the workplace that the individual is attached to in a certain year. A disadvantage is that the database does not include information on working hours, nor does it have any information on type of work contract (although we know from the Swedish Labour Force Survey that the share holding a temporary contract is significantly higher in the hospitality sectors). Both variables are central factors influencing workplace mobility; a Nordic study using national Labour Force Survey data show that working less than 20 hours per week and being employed on a temporary contract significantly increase workplace mobility in all the Nordic countries (Nordic Council of Ministers 2010). It should be mentioned, however, that the income variable (yearly earnings from being employed) could be interpreted as a proxy for part-time working.

The longitudinal feature of the database allows us to trace different kinds of actions taken by individuals over time. The key event analyzed here is whether an individual has shifted workplace from one year to another. In operational terms, and since we only have information every third year, a workplace-to-workplace mobility is defined as a change in workplace from 2008 to 2011. The individual may thus have shifted workplace more than once during that time period.

Table 14.1 Selected subsectors of the hospitality and tourism industry according to SNI 2007. Total employment in 2008 and 2014 in the four counties

SNI2007	Economic activity	No. of employees in 2008	No. of employees in 2014
	Hospitality activities		
55101	Hotels with restaurant	4201	3883
55102	Conference hotels	403	408
55103	Hotels without restaurant	367	384
55201	Youth hostels	96	154
55202	Cabin facilities	406	436
55300	Camping sites, including caravan sites	292	356
55900	Other short-stay lodging facilities	5	–
56100	Restaurants	12,082	14,060
	Tourism and travel activities		
79110	Travel agencies	444	270
79120	Tour operators	143	131
79900	Tourism and booking services	269	269
91020	Museums	522	534
91030	Cultural heritage, preservation of historical sites and buildings	105	134
91040	Botanical and zoological gardens, nature reserves activities	13	22
93111	Operation of skiing facilities	1199	1448
93210	Amusement parks and theme parks	237	230
93290	Other leisure and entertainment activities	1411	1711
	Total	22,195	24,430

Source MIS 2007:2. Swedish Standard Industrial Classification 2007

Table 14.2 Individual and work-related characteristics in 2014 in the hospitality and tourism industry and in the labour market as a whole

Personal and work-related characteristics	Hotels	Restaurants	Tourism & travel	Total labour market
Gender				
Share (%) women	69.2	45.3	53.4	48.3
Age				
21–30 years	36.1	43.6	32.4	19.6
31–40 years	21.7	24.7	19.5	20.9
41–50 years	22.7	19.7	23.1	26.5
51–65 years	19.5	12.0	25.0	33.0
Highest education				
Primary	34.7	48.8	31.2	36.1
Secondary	56.7	46.4	55.0	42.4
Tertiary	8.5	4.9	13.8	21.5
Dedicated tourism education				
Share (%) with a tourism education	23.7	14.9	7.7	2.6
Country of birth				
Nordic	83.0	54.6	90.8	91.0
European	8.5	16.5	2.7	3.8
Non-European	8.5	28.9	6.4	5.2
Family situation				
Share (%) being single	37.4	40.2	33.5	31.8
Occupation				
Leading/professional	11.2	10.3	23.6	29.5
Income				
1st Quartile	54.7	62.7	47.2	25.0
2nd Quartile	27.2	27.1	26.0	25.0
3rd Quartile	11.4	7.2	16.2	25.0
4th Quartile	6.6	3.1	10.5	25.0
Workplace size				
1–2 empl.	2.8	6.4	3.7	2.8
3–9 empl.	13.4	42.5	20.3	13.8
10–49 empl.	54.0	50.5	28.7	33.8
50+ empl.	29.9	0.6	47.3	49.6

(continued)

Table 14.2 (continued)

Personal and work-related characteristics	Hotels	Restaurants	Tourism & travel	Total labour market
Studying[a]				
Share (%) studying and working	44.8	46.1	48.0	36.1
Labour turnover 2013–2014[b]				
Share (%) leaving the workplace	34.0	37.8	32.1	19.2

[a]Persons 30 years or younger enrolled in an education program during autumn 2014
[b]Only workplaces being active in both years
Source Statistics Sweden/BeDa. *Note* Population limited to persons 21–65 years in 2014

There are two principal categories or types of separation of an employee from his or her previous workplace. An *involuntary* job separation may be caused by downsizing or a total close-down of the workplace, or by temporary jobs not being prolonged. This type of job separation is initiated by the employer. A *voluntary* job separation is, on the other hand, initiated by the employee, and may be caused by the employee's search for another (better) job, or for ill health or retirement. For the purpose of this paper, it would be preferable to focus solely on voluntary moves, but the database used does not provide this information. However, four steps have been taken when designing the population for the analysis, which will considerably reduce this problem.

First, employees at workplaces that have closed down between 2008 and 2011 are not included in the analysis. This is made by using a relatively new method for distinguishing between new, surviving and closed workplaces developed by Statistics Sweden (Andersson and Arvidson 2006). This means that the risk of including involuntary job moves caused by closures is brought to a minimum. A remaining problem is, however, that workplaces that have reduced the workforce between 2008 and 2011 are included in the analysis. Some of those leaving a workplace might of course have been forced by the employer to quit because of redundancy or for some other reason. Thus, the exact extent of forced termination of the employment cannot be determined in the data. Second, the population is restricted to persons that were active in the labour market both in 2008 and 2011. This effectively eliminates the problem of including persons retiring or leaving the labour market for ill health during the period of study. Third, information in the database allows us to single out persons active on the labour market in November 2008 and 2011. This will exclude persons only working during the summer and winter months, thus reducing the problem of including seasonal workers in typical winter or summer destinations. Fourth, the population is also reduced to persons of the ages of 21–65 years. This will further decrease the risk of including involuntary job moves that involves temporary jobs. These four steps do not completely handle the problem of mixing voluntary and involuntary job moves, but they come a reasonably long way in narrowing down the population to those that are either staying at the workplace or voluntarily have shifted to another workplace between 2008 and 2011.

It should also be noted that the population does not include self-employed persons. People moving between employment and self-employment is an interesting research topic in itself, but it is not considered in the present study.

In the analysis, both descriptive statistics and logistic regression analysis are used. The logistic regressions use a binary dependent variable, stating whether the person has shifted to a different workplace or not between 2008 and 2011.

To identify tourism-related industries, the study uses the Swedish industrial classification system (*Svensk näringsgrensklassificering*, SNI 2007), which is equivalent to the European four-digit NACE nomenclature. A distinction is made between the hospitality sector (hotels and restaurants) and the tourism and travel sector (e.g. travel agencies, museums and leisure and entertainment activities). The selected industries are shown in Table 14.1.

The largest subsector, by far, is Restaurants, followed by Hotels with restaurants. We should keep in mind that restaurants do not only serve tourists, but also the local population, and thus that their share of tourism-related employment is exaggerated. The hospitality sector represents 80% of total employment in the selected industries in 2014. In total, the number of employees increased by 10% over the period. This increase was largely related to the restaurant sector. Considering that the financial crisis hit employment levels severely during this period, these figures indicate that the sector is an important job creator, and a gateway into the Swedish labour market.

Characterizing the Hospitality and Tourism Workforce

In Table 14.2 a number of characteristics are shown for the hospitality and tourism industry. For comparison, figures for the labour market as a whole are also shown. Most of the variables listed in Table 14.2 will later be used in the regressions in Table 14.3.

The personal characteristics of the workforce include gender, age, education, country of birth and family situation. The hospitality and tourism industry employ more women than men, in particular in the hotel sector, but the imbalance is not extreme in the case of restaurants, and tourism and travel activities. As already mentioned, workers in the tourism and hospitality sectors are younger than in most other industries, especially in the restaurant sector. The tourism and travel sector, however, show a much more balanced age structure. In the hospitality sectors, the workers are concentrated to the age group 21–30 years.

Workers in the tourism industry normally have an education from the secondary level of the Swedish educational system. Very few have a complete university degree. The exception is again the tourism and travel sector, where we find workers almost on par with the average for the labour market as a whole. Interestingly, but perhaps not surprisingly, the share of tourism workers with a dedicated tourism education is low. About a quarter of the workforce in the hospitality sectors have an educational background from a tourism study programme, and the share in the tourism and travel sector is even lower. The latter is not surprising, given that the tourism and travel sector consist of a wide range of activities, like museums, travel agencies, amusement parks, downhill skiing facilities, etc., that rarely are connected to tourism educational programmes. Perhaps more surprisingly, persons with a dedicated tourism

Table 14.3 Logistic regressions on persons changing workplace between 2008 and 2011

Variables	Model 1	Model 2	Model 3
	Exp(B)	Exp(B)	Exp(B)
Sector (ref = other)			
Hotels	1.649***	1.359***	1.276***
Restaurants	2.489***	1.935***	1.459***
Tourism & Travel activities	1.378***	1.254***	1.078
Manufacturing & Construction	0.608***	0.605***	0.814***
Wholesale & Retail	0.838***	0.742***	0.747***
Business services	1.228***	1.122***	0.947**
Health care	0.680***	0.701***	0.917***
Public administration	0.851***	0.895***	1.012
Gender (ref = Female)		1.055***	1.083***
Age (ref = 51–65 years)			
21–30 years		2.956***	1.690***
31–40 years		1.79***	1.365***
41–50 years		1.394***	1.245***
Education (ref = tertiary)			
Primary or unknown		0.789***	0.875***
Secondary		0.874***	0.956***
Studying (ref = Not studying)		2.022***	1.789***
Country of Birth (ref = non-European)			
Nordic		0.896***	1.052**
European		0.918**	0.981
Family (ref = other)			
Married or co-hab, no small children		0.895***	0.915***
Married or co-hab, small children		0.898***	0.900***
Distance to work (ref = more than 50 km)			
0–5 km			0.424***
5–10 km			0.437***
10–25 km			0.486***
25–50 km			0.592***
Leading/professional (ref = other)			1.193***
Income (ref = more than 300,000 SEK)			
0–100,000 SEK			1.949***
100,000–200,000 SEK			1.350***
200,000–300,000 SEK			1.015
Workplace size (ref = 50 empl. or more)			
1–2 empl.			2.011***
3–9 empl.			1.701***
10–49 empl.			1.402***
Jobs within 50 km (ref = more than 100,000)			
0–50,000			0.851***

(continued)

Table 14.3 (continued)

Variables	Model 1	Model 2	Model 3
	Exp(B)	Exp(B)	Exp(B)
50,000–100,000			0.917***
Tenure (ref = 12 years or more)			
Less than 3 years			3.864***
3 years			2.224***
6 years			1.608***
9 years			1.434***
Constant	0.475***	0.408***	0.249***
N	326,747	326,209	320,636
Nagelkerke (R^2)	0.02	0.08	0.173

Significance levels: ***$p<0.001$, **p<0.05, *p<0.1

education are in absolute numbers (not shown in Table 14.2) higher in manufacturing and construction, wholesale and retail, than in any of the tourism sectors; out of 7200 with a tourism education, only 1590 are found in one of three the tourism sectors (Statistics Sweden/BeDa, see also Åberg 2017).

The share of workers born in a non-European country is significantly higher in the restaurant sector (18.5%) than in any of the other sectors, whereas the share in the hotels and tourism and travel sectors is more in line with the average for the labour market as a whole. Finally, the family situation in the hospitality and tourism sectors differ in the sense that a substantially larger share of the workers are not married or co-habitants (with or without children in dependent ages). This is of course directly related to the fact that workers in the sector have a younger age profile.

The table confirms many of the standard features that are discussed in the literature on the workforce in the hospitality and tourism sector. Compared with the labour market in general, the share of persons with a leading position (or having a professional occupation) is significantly lower in hotels and restaurants, whereas in tourism and travel activities the share is closer to average. This is also reflected in the income profile in the hospitality sectors, with an overrepresentation in the lowest income group, and the opposite in the highest income group. Incomes are clearly higher in the tourism and travel sector, but still lower than average.

Tourism workplaces (and firms) are often considered to be smaller than in the rest of the economy. This is at least partly confirmed in Table 14.2. Compared to the average for the labour market as a whole, workplaces in the hospitality sectors are clearly overrepresented in the size group employing 10–49 persons, and workplaces in restaurants and the tourism and travel sector are more common than average in the 3–9 employees segment. However, hospitality and tourism workplaces are not particularly overrepresented in the smallest category.

As expected, a large share of the workforce in the hospitality and tourism industry is enrolled in some sort of educational programme at the same time as they are registered as workers in the sector.

Finally, the degree of labour mobility in workplaces is significantly higher in the hospitality and tourism sectors. Measured as the share of workers leaving the workplace from one year to another (in this case between 2013 and 2014), the average is clearly higher, in particular for the restaurant sector.

Determinants of Labour Mobility

The purpose of the analysis in this section is to see if there are still differences between industries in the probability to change workplace, when control variables are added to the models. Two sets of variables are added successively, *personal characteristics* (Model 2) and *work-related characteristics* (Model 3). Secondly, the purpose is to analyze the effects of the variables on workplace mobility of the full model (Model 3), see also Table 14.3.

As expected, in the first model we find substantial differences in the propensity to change workplace solely depending on the sector the person was attached to in 2008. The odds are highest for workers in restaurants, followed by hotels, tourism and travel activities and business services. Lower odds compared to the category "Other" are found in manufacturing and health care. Adding personal characteristics slightly increases the explanatory power of regression in Model 2, but the basic pattern of differences between sectors remains intact. However, the magnitude of differences between sectors is slightly reduced. When the second set of control variables, related to work–life, are added in Model 3 the R^2 value jumps considerably. Differences between sectors are again levelled out, but the initial pattern is to a large degree still present, i.e. the two hospitality sectors show higher odds for workplace mobility (but the effect for the tourism and travel sector has disappeared), whereas manufacturing and construction, wholesale and retail, and health care show lower odds compared with the reference category "Other". Thus, we can conclude that the hotel and restaurant industry is characterized by higher levels of mobility than other sectors, even when a number of workforce characteristics and structural features of the industry are controlled for.

The second type of observations to be made from Table 14.3 is that most of the selected control variables are significant determinants of workplace mobility. Starting with *personal characteristics*, males show a slightly increased propensity to change workplace, compared with females. There is a consistent and strong age effect, in the sense that younger persons are more likely to be mobile in the labour market. Higher education has a positive effect on workplace mobility, and being enrolled in a study programme of some sort in 2008, while working at same time (normally temporary and/or part-time work), increases the odds of changing workplace between 2008 and 2011. In model 2, persons born in a non-European country, are more likely to change workplace. Finally, family situation also seems to matter, in the sense that being married or co-habitant (with or without children of dependent age in the family) lowers the propensity to shift to a new workplace.

Most of these findings have fairly obvious rationales behind them. Age effects on labour mobility is a recurrent observation of in the literature (Nordic Council of Ministers 2010; Eriksson 2016), and the basic interpretation is that mobility follows a life-cycle pattern. The initial phase of becoming established on the labour market is characterized by a search/trial process in the early ages, followed by a phase of stabilization, and finally a situation in the later ages where the individual has more to lose than to gain from mobility. The life-cycle pattern is also reflected in the family situation, where a period of life as single for many is replaced by an intense period of family life with small children, followed by a stage in life living without dependent children in the household. Living in a relationship, especially with small children, is likely to hamper mobility, since often both persons need to find a suitable job and a move to another workplace may involve migration. Having a part-time or temporary job while studying is quite common in Sweden and it seems reasonable to assume that in many cases the job is not related to the particular education programme that the person is enrolled in. Thus, they would be expected to shift workplace when education is finished. Non-European immigrants have a disadvantaged position on the Swedish labour market in general, and their higher propensity to move between workplaces is likely to be an effect of the difficulties they face trying to find a permanent job.

The *work–life-related* variables also show some strong regularities. Long distances between work and home increase the likelihood that the individual will change to another workplace. Workers in hotels and restaurants to a larger extent live close to their workplaces, whereas workers in the tourism and travel sector show a pattern more similar to the average. This probably reflects the more urban character of the hospitality sector in general, and restaurants in particular.

Having a leading position in your workplace or being an employee with a professional education slightly increases the odds of moving to another workplace between 2008 and 2011. Persons with a low annual income are more likely to change workplace, and being employed in a small workplace increase the odds for mobility. The number of jobs found in the surrounding area within a 50 km radius influences the propensity to change workplace, or, put another way, increase the chances of finding another job. Finally, tenure is a powerful predictor of workplace mobility in the data.

Understandably, being in a position where you have to commute long distances between your workplace and your home triggers a search for a job that is located in a closer range to your residence. Long commuting distances can of course also be handled by moving closer to your workplace. Having a leading or professional position seems to result in a weak tendency for increased mobility, and could be interpreted as career moves. Low-income groups' stronger inclination for workplace mobility is likely to reflect a dissatisfaction with pay and/or type of work contract, and thus an ambition to move from a part-time or temporary job to a full-time contract in another workplace. As mentioned, we do not have information on type of employment contract in the database, but the regression results in the Nordic Council of Ministers report (2010, p. 189) show substantially higher mobility odds ratios for persons with a temporary contract, and for persons working less than full-time. The effect of workplace size on the propensity to change workplace is frequently commented upon in the literature, but the hypothesis sometimes goes in opposite directions. A frequent claim is that larger workplaces offer better career opportunities within the work unit, thus reducing the need for a workplace change when career steps are looked for. A competing proposal is that small work units may offer better social cohesion, a friendlier work climate and higher degree of loyalty with the employer, thus reducing the inclination to change workplace (Eriksson 2016, p. 147). However, the regressions reported in Model 3 show that being employed in a small workplace increases mobility.

As expected, the size of the labour market within reasonable commuting distance has an effect on mobility, in the sense that persons residing within reach of a "dense" labour market are more likely to switch between workplaces. This confirms the findings in other research emphasizing the importance of job search behaviour (Steel and Lounsbury 2009). Tenure is of course strongly related to age, but it is also likely to contain an independent effect. It seems reasonable to assume that persons having been employed at the same workplace for a long time have built up a certain amount of trust and social capital over the years. Giving up this long-term investment for a new workplace, with new colleagues and normally a new employer, is likely to have its price. Added to this is the strict employment protection legislation and seniority rules on the Swedish labour market, in combination with an increasing unemployment during the years studied (2008–2011). In a difficult labour market situation, the seniority acquired in the present job tends to be highly valued, which promotes risk-averse behaviour (Nordic Council of Ministers 2010).

It should be mentioned, finally, that the effects of the independent variables on mobility are to a large degree similar when the hospitality and tourism sector is analyzed separately. The basic mechanisms behind these variables thus seem to function in the same way, irrespective of the sector. Because of the limited space available, these results are not presented here.

Conclusion

We can conclude that the hospitality and tourism industry show higher levels of labour mobility than most other sectors. When controlling for structural factors like low paid jobs, predominantly small workplace units, a young workforce and comparatively low levels of formal education the difference in mobility between sectors is reduced. However, there is still a substantial and significant difference between sectors left to explain, indicating that there is more to it than simply the structural character of the tourism sector. The explanatory power of the estimated model is after all well below 20%.

One possibility is of course that there are structural differences between sectors not fully captured by the variables used in this analysis. For instance, information on part-time jobs and temporary employment contracts are not part of the analysis presented here. However, the effect of part-time jobs is at least partly reflected in the income variable used, and the temporary and seasonal character of jobs in the sector is to some extent kept aside when designing the studied population. Yet, there might well be other structural differences left to explore.

The results presented here thus suggests that other types of explanations are needed if we want to understand why the tourism and hospitality sectors have such high levels of mobility. One possibility, often referred to in the literature, is that the hospitality sector is ridden by less acceptable human resource practices. HR management is of course highly dependent on the general working conditions of the sector, in terms of unsocial working hours, seasonality and physically demanding work, for example. Many of these conditions are, however, similar in other service sectors (like wholesale and retail, and the transport sector), indicating that differences in HR management are only likely to be a partial explanation.

Perhaps a more promising approach would be to pay more attention to cultural differences between sectors. The idea of a "culture of mobility" that is particularly strong, especially in the hospitality sector, is worth testing on a larger scale, and in a comparative perspective. Such a culture is of course likely to be linked to other (structural) features of the sector such as the young age profile and the opportunities to work in combination with studying, a casual and part-time workforce, as well as unclear career paths in the sector. However, it is also likely that the "culture of mobility" is upheld by norms and attitudes that have been developed over time, and that these norms are highly contagious at the level of the workplace. A research design that incorporates the interplay between structural and cultural differences seems to be a worthy task for future studies on labour mobility in the hospitality and tourism sector.

References

Åberg, Kajsa. 2017. "Anyone could do that": Nordic perspectives on competence in tourism. PhD diss., Umeå University.

Andersson, Jan, and Björn Tegsjö. 2006. *En rörlig arbetsmarknad – dynamiken bland jobb, individer och företag* [A mobile labour market—The dynamics among jobs, individuals and firms]. Örebro: Statistics Sweden.

Andersson, Jan, and Göran Arvidson. 2006. *Företagens och arbetsställenas dynamik (FAD)* [Labour dynamics in firms and workplaces] Örebro: Statistics Sweden.

Baum, Tom. 2007. Human resources in tourism: Still waiting for change. *Tourism Management* 28: 379–388.

Baum, Tom, Anna Kralj, Richard Robinson, and David Solnet. 2016. Tourism workforce research: A review, taxonomy and agenda. *Annals of Tourism Research* 60: 1–22.

Bienkowska, Dzamila, Mats Lundmark, and Anders Malmberg. 2011. Brain circulation and flexible adjustment: Labour mobility as a cluster advantage. *Geografiska Annaler B* 93: 1–20.

Breschi, Stefano, and Francesco Lissoni. 2001. Knowledge spillovers and local innovation systems: A critical survey. *Industrial and Corporate Change* 10 (4): 975–1005.

Dale-Olsen, Harald. 2016. Employee turnover. In *Handbook on Employee Turnover*, ed. George Saridakis and Cary Cooper. Cheltenham: Edward Elgar.

Davidson, Michael, and Ying Wang. 2011. Sustainable labour practices? Hotel human resource managers views on turnover and skills shortage. *Journal of Human Resources in Hospitality & Tourism* 10: 235–253.

Dipietro, Robin, and Steven Condly. 2007. Employee turnover in the hospitality industry. *Journal of Human Resources in Hospitality & Tourism* 6: 1–22.

Duncan, Tara, David Scott, and Tom Baum. 2013. The mobilities of hospitality work: An exploration of issues and debates. *Annals of Tourism Research* 41: 1–19.

Eriksson, Tor. 2016. Inter- and intra-firm mobility of workers. In *Handbook on Employee turnover*, ed. Georgeand Saridakis and Cary Cooper. Cheltenham: Edward Elgar.

Griffeth, Rodger, Peter Hom, and Stefan Gaertner. 2000. A meta-analysis of antecedents and correlates of employee turnover: Update, moderator tests, and research implications for the next millennium. *Journal of Management* 26: 463–488.

Heldt Cassel, Susanna, Maria Thulemark, and Tara Duncan. 2017. Career paths and mobility in the Swedish hospitality sector. *Tourism Geographies*. https://doi.org/10.1080/14616688.2017.1402946.

Hinkin, Timothy, Brooks Holtom, and Dong Liu. 2012. The contagion effect: Understanding the impact of changes in individual and work-unit satisfaction on hospitality industry turnover [Electronic article]. *Cornell Hospitality Report* 12: 6–12.

Hjalager, Anne-Mette, and Steen Andersen. 2001. Tourism employment: Contingent work or professional career? *Employee Relations* 23: 115–129.

Hom, Peter, and Rodger Griffeth. 1995. *Employee Turnover*. Cincinnati, OH: South-Western.

Iverson, Roderick, and Margaret Deery. 1997. Turnover culture in the hospitality industry. *Human Resource Management Journal* 4: 71–82.

Ladkin, Adele. 2011. Exploring tourism labour. *Annals of Tourism Research* 38: 1135–1155.

Lundmark, Mats, and Dominic Power. 2008. Labour market dynamics and the development of the ICT cluster in the Stockholm region. In *Handbook of Research on Innovation and Clusters: Cases and Policies*, ed. Charlie Karlsson. Cheltenham: Edward Elgar.

Madsen, Tammy, Elaine Mosakowski, and Srilata Zaheer. 2003. Knowledge retention and personnel mobility: The nondisruptive effects of inflows of experience. *Organization Science* 14: 173–191.
Martin, Ron, and Philip Morrison (eds.). 2003. *Geographies of Labour Market Inequality*. Regional Development and Public Policy Series. London and New York: Routledge.
MIS 2007:2. Swedish Standard Industrial Classification 2007.
Muchinsky, Paul, and Mark Tuttle. 1979. Employee turnover: An empirical and methodological assessment. *Journal of Vocational Behavior* 14: 43–77.
Nordic Council of Ministers. 2010. Labour market mobility in Nordic Welfare States. *TemaNord*: 515. Copenhagen.
Orkan, Lennart. 1972. Regionala variationer i företagens personalomsättning [Regional variations in labour turnover]. PhD diss., Umeå University.
Östbring, Lisa, and Urban Lindgren. 2013. Labour mobility and plant performance: On the (dis)similarity between labour- and capital-intensive sectors for knowledge diffusion and productivity. *Geografiska Annaler B* 95: 287–305.
Porter, L.W., and R.M. Steers. 1973. Organizational work and personal factors in employee turnover and absenteeism. *Psychological Bullentin* 80: 151–176.
Power, Dominic, and Mats Lundmark. 2004. Working through knowledge pools: Labour market dynamics, the transference of knowledge and ideas, and industrial clusters. *Urban Studies* 41: 1025–1044.
Rowley, Gill, and Kate Purcell. 2001. As cooks go, she went: Is labour churn inevitable? *International Journal of Hospitality Management* 20: 163–185.
Rubenstein, Alex, Marion Eberly, Thomas Lee, and Terence Mitchell. 2017. Surveying the forest: A meta-analysis, moderator investigation, and future-oriented discussion of the antecedents of voluntary employee turnover. *Personnel Psychology* 1–43. https://doi.org/10.1111/peps.12226.
Saridakis, George, and Cary Cooper (eds.). 2016. *Handbook on Employee Turnover*. Cheltenham: Edward Elgar.
Sims, Wiley. 2007. Antecedents of labor turnover in Australian Alpine Resorts. *Journal of Human Resources in Hospitality & Tourism* 6: 1–26.
Statistics Sweden. 2001. Personalomsättning i olika sektorer [Labour turnover in different sectors]. Stockholm.
Steel, Robert, and John Lounsbury. 2009. Turnover process models: Review and synthesis of a conceptual literature. *Human Resource Management Review* 19: 271–282.

Szivas, Edith, Michael Riley, and David Airey. 2003. Labour mobility into tourism: Attraction and satisfaction. *Annals of Tourism Research* 30: 64–76.
Tax Procedure Act. 2011 [Skatteförfarandelagen 2011:1244].
Thulemark, Maria, Mats Lundmark, and Susanna Heldt Cassel. 2014. Tourism employment and creative in-migrants. *Scandinavian Journal of Hospitality and Tourism* 14: 403–421.
Tomlinson, Mark. 2002. Measuring competence and knowledge using employee surveys: Evidence using the British skills survey of 1997. Discussion Paper No. 50. Manchester, Centre for Research on Innovation and Competition (CRIC), University of Manchester.
Tomlinson, Mark, and Ian Miles. 1999. The career trajectories of knowledge workers in mobilising human resources for innovation. Proceedings from the OECD Workshop on Science and Technology Markets. Paris, OECD.
Walmsley, Andreas. 2004. Assessing staff turnover: A view from the English Riviera. *International Journal of Tourism Research* 6: 275–287.
Walmsley, Andreas. 2015. *Youth Employment in Tourism and Hospitality—A Critical Review*. Oxford: Goodfellow Publishers Ltd.
Zampoukos, Kristina, and Dimitri Ioannides. 2011. The tourism labour conundrum: Agenda for new research in the geography of hospitality workers. *Hospitality & Society* 1: 25–45.

15

Downshifting Dutch Rural Tourism Entrepreneurs in Sweden: Challenges, Opportunities and Implications for the Swedish Welfare State

Marco Eimermann, Charlotta Hedberg and Urban Lindgren

Introduction

Buckingham sees downshifting as "the will to give up the compulsive purchase of material things that end up owning their owners and reject the idea to sacrifice non-working activities like reading a book at night to a child in order to get a job promotion" (noted in *Shift down to gear up*, cited in Juniu 2000, p. 71).

Although downshifting is not new, it has gained renewed attention in recent decades (Hampton 2008). Journalists (Saltzman 1991), economists (Schor 1998) and psychologists (Tan 2000) have analysed its novel manifestations in capitalist societies of Australia, Great Britain and

M. Eimermann (✉) · C. Hedberg · U. Lindgren
Department of Geography, Umeå University, Umeå, Sweden
e-mail: marco.eimermann@umu.se

C. Hedberg
e-mail: charlotta.hedberg@umu.se

U. Lindgren
e-mail: urban.lindgren@umu.se

A. Walmsley et al. (eds.), *Tourism Employment in Nordic Countries*,
https://doi.org/10.1007/978-3-030-47813-1_15

the USA. Even though people with various income levels can engage in downshifting, studies emphasise professionals, executives and managers who voluntarily opt out of financially rewarding career paths to earn their living in an alternative way, even if this means lowered income and status (Hamilton and Mail 2003; Tan 2000). This renewed attention for downshifting also includes personal meanings of leisure time and the work-leisure balance (Hampton 2008). This chapter adds to this renewed attention, and studies how Dutch rural tourism entrepreneurs in Sweden reason around their motivations for migrating and their post-migration daily practices. It also contributes with insights into gendered practices regarding work in the interviewees' tourism firm.

Statistics Sweden (2019b) indicates that 12,095 people born in the Netherlands lived in Sweden in 2018, which was more than twice the volume in 2000. Many of them were tourism entrepreneurs in remote rural areas and isolated villages where they could perform their outdoor hobbies and nature-based activities in pristine environments (Carson et al. 2018; Eimermann 2015a). These migrants can be termed lifestyle migrants (Benson and O'Reilly 2009; Carson and Carson 2018; Eimermann 2013), but it is unknown whether they can also be viewed as downshifters. To address this issue, our chapter revisits in-depth interviews with Dutch migrant entrepreneurs (Eimermann 2016; Eimermann and Kordel 2018) focusing on the following questions:

1. If at all, how are the entrepreneurs' motivations for migrating and downshifting interrelated?
2. What (gendered) practices characterise their processes of downshifting, and what do these imply for their work–leisure balance?

The first question flows from the fields of lifestyle migration and downshifting. We consider whether (some of) the interviewed Dutch lifestyle migrants in rural Sweden can be downshifters before trying to characterise them according to Saltzman's (1991) typology of downshifters specified below. The second question is this chapter's core question, where "practice" encompasses the migrants' everyday agency and interactions with local and national structures (related with principles of equality in the Swedish welfare state). Rather than the formal rules and regulations that govern their practices, this question regards the interviewees' general perceptions of possibilities and challenges, with a focus on downshifting-related gender issues. This exploration renders relevant issues for further research.

After our analysis of possible interconnections between downshifting and lifestyle migration, we lift the results to a structural level as we aim to identify implications for the Swedish welfare state of the interviewees' downshifting-related practices. We relate the research questions to the Swedish model to improve our understanding of equality in general and gender equality in particular. This includes gendered differences in male and female accounts of balancing their micro firms with other income sources, leisure time and caregiving. Thus, the next section describes and problematises principles of full employment and gender equality in the Swedish welfare state. The following section presents methods, followed by the empirical section addressing the research questions. The conclusion then returns to the aim, connects our findings with broader concerns such as climate change, and suggests avenues for future research.

Lifestyle Migration, Downshifting and the Swedish Welfare State

The fields of lifestyle migration and downshifting form this chapter's conceptual basis for studying the Dutch migrants. Although downshifting research has considered how lifestyle changes may influence motivations for relocation (e.g. Hamilton and Mail 2003), lifestyle migration studies more prominently consider migration as part of achieving a better quality of life. Drawing on anthropological material and social theories, lifestyle migration studies often use a definition based on the migrants' accounts rather than the researchers' interpretations:

> Lifestyle migration is the spatial mobility of relatively affluent individuals of all ages, moving either part-time or full-time to places that are meaningful because, for various reasons, they offer the potential of a better quality of life. [...] It is about escape *from* somewhere / something,

> *to* self-fulfilment and a new life. A recreation, restoration or rediscovery of oneself, of personal potential or of one's 'true' desires. (Benson and O'Reilly 2009, pp. 2–3, original italics)

This definition indicates that social reasons rather than economic, political or other reasons are pivotal in the voluntary decision to migrate. The field of lifestyle migration has recently become included in general migration and population studies (Barcus and Halfacree 2018; Benson and O'Reilly 2016). A reason is that although social motivations may be at the core of lifestyle migration, there are e.g. relevant intersections with labour migration and refugee migration since economic, environmental and political factors also play a role in lifestyle migrants' ambitions and practices. Studying Brits in France, Benson (2010) suggests three groups of lifestyle migrants based on life course and other matters:

- young families, whose decision to migrate and choice of location is influenced by their desire for a better, healthier life where their children can enjoy more freedom,
- mid-life migrants: young, childless, ex-professionals who have left their jobs in Britain to move to France where they set up their own firms, and
- retirement migrants, who perceive that elderly in France are treated with more respect than elderly in Britain, and who live close to facilities such as general practitioners since they anticipate a future where they may not be as mobile anymore.

This chapter's interviewees match the first two groups, offering drinks, food, outdoor activities and accommodation as part of their tourism firms. They often live outside Swedish villages in search of space, tranquillity and opportunities to engage in their nature-based special interests (Carson and Carson 2018; Carson et al. 2018). They associate their migration with less traffic congestion, less stress, perceived lower crime rates and more time for family activities (Benson and O'Reilly 2009; Eimermann 2013). One of their goals is a perceived more socially sustainable and healthier way of life, which is also a main motivation found in the downshifting literature.

The field of downshifting describes and analyses processes of voluntarily making a career change in the hope that this results in having more control over one's time (at the cost of a reduced income, see Hampton 2008). Definitions of downshifting identify a long-term strategy to increase one's quality of life, which involves much less income and/or consumption than one's potential level (Larsson 2015). Studies in Western societies where many people can renegotiate the balance between work and leisure time by regulating the level of consumption to less income, underline that downshifting does not include planned retirement (Hamilton and Mail 2003; Schor 1998, 2005). While downshifting may financially be more feasible for people with high-status jobs (Tan 2000), people with other jobs can downshift as well (Hamilton and Mail 2003; Breakspear and Hamilton 2004). Downshifting is not only about working and spending less, it also means "doing things differently in a leisurely manner" (Juniu 2000, p. 71).

A further reason for downshifting encompassed within the notions of a more balanced and fulfilling life is the desire to solve clashes between personal and workplace values (Breakspear and Hamilton 2004). Downshifting can manifest in many ways, e.g. selling expensive property and moving to a cheaper house, refusing career promotions, changing jobs, working less hours in the same job or giving up work completely (Hampton 2008). Though Saltzman suggested her typology of downshifters as long ago as in 1991, it can still serve as a relevant starting point for this chapter's analysis:

- Plateauers, who avoid promotions or greater responsibilities by voluntarily remaining at a particular level or position within an organisation,
- Backtrackers, who choose to take a few steps down the organisational ladder,
- Career shifters, who use the same skills they have acquired in a new setting with less emphasis on a traditional fast track, e.g. moving from a corporate environment with its emphasis on profit and increasing market share, to a non-profit organisation,
- Self-employed, who leave organisations to work for themselves doing similar work, and

- Urban escapees, who may commute to town or set up well-considered viable businesses in rural settings, often using skills acquired in larger organisations and possibly capitalising on electronic technology to link with larger organisations or cities.

This typology relates with the downshifters' personal practices of work and leisure after migration. In Western societies, people frequently associate leisure with pleasure during recreational activities. In its full essence, leisure includes wider terms of body, mind, space, culture, society and motivations in relation to time. Rather than as downshifters' free time, leisure can be conceptualised as personal meaning of time, which often changes following career changes (Pavelka 2000, p. 25; Hampton 2008). A downshift can thus help people regain the quality of life that was lost through modern consumerism (Juniu 2000). Such reductions in consumption and attempts to live simpler lives can eventually lead to more socially and ecologically sustainable lifestyles that are in better balance with the earth's limited resources (Kennedy et al. 2013; U.N. 2015).

More specifically, studying the male and female tourism entrepreneurs' personal meanings of work and leisure time improves our knowledge of gender perspectives in downshifting. Various studies report different findings: a higher ratio of females engaged in downshifting in the USA (Schor 1998), while males were downshifting more than females in Australia (Hamilton and Mail 2003). Also, women may more often stop paid work to spend time with their family, while men change careers (ibid., pp. 22–23). Yet, we know little about how males and females perceive their practices after downshifting.

Connecting with this book's overall theme, we situate the interviewees' practices within the Swedish model's principles (Schierup et al. 2006). This model stems from the 1950s and 1960s, based on the key ideology of a large universalist welfare state. This includes equal pay for equal work in different sectors as well as an active labour market policy (ALMP) to help adapt workers to shifting industrial demands (ibid.). One goal was to keep unemployment at a minimum, as wage labour was not only encouraged for economic reasons but also for social reasons of inclusion. The welfare state could only finance benefits such as childcare subsidy, free education, subsidised health care and pension funds with income tax generated from as many employed inhabitants as possible (Shackleford and Shevlin 2001; Kolm and Tonin 2015). This led to increased labour force participation and decreased differences between the sexes: 84.7 and 80.4%, respectively of men and women aged 20–64 were employed in 2018 (Galte Schermer 2019). Unemployed men and women were often enrolled in ALMP-inspired study—or training programmes. The welfare state combats gender inequality through low thresholds for compensation eligibility and ceilings that allow maximum compensation levels for people with relatively low salaries. As men gain nothing extra from the system if they earn higher income than their female partners, their position as main breadwinners is discouraged.

One point of critique is that full gender equality encompasses more than men and women working equally and earning equal incomes (Gunnarsson 2013b; Carbin et al. 2017). This critique argues that the Swedish model's narrow interpretation of women's emancipation has contributed to a situation in which a sharp line exists between market wages for the breadwinner in the male-dominated public spheres and unpaid care work often carried out by women in the private (domestic) spheres (ibid.). Since the Swedish welfare state implies social security rights based on work performance and varying wages earned through labour market participation, many women struggle with socio-economic lags (Gunnarsson 2013a). This has led to unequal power relations where many males' social and economic positions are stronger than females'. A more comprehensive way to stimulate emancipation would be to include more issues stemming from the private sphere of society in political debates (and reforms, ibid.). Feminist critique maintains that the Swedish model views wage work as the norm that shapes men and women into active equal citizens, and the most important way to female liberation (Carbin et al. 2017). Yet, this means that reproductive work like caring for partners, children and elderly relatives, and emotional life is increasingly being avoided (ibid.). This vital work is often made invisible, and it should be more evenly distributed between men and women. Feminists contest the ideology that female liberation equals full-time work (ibid.). This connects back to our gendered perspectives on women's and men's perceived work–life and work–leisure balance after migration to Sweden.

Methods

Figure 15.1 shows Sweden and its largest cities: Stockholm, Gothenburg and Malmö. It also highlights the county of Värmland in dark grey and the county capital Karlstad (92,804 inhabitants, Statistics Sweden

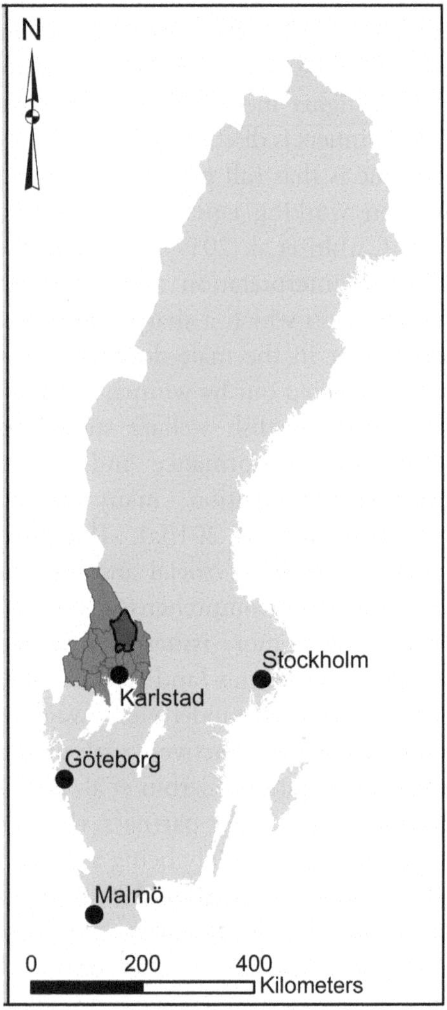

Fig. 15.1 Hagfors and Munkfors in county Värmland, Sweden. Cartography: Martin Hedlund

2019a). The municipalities with black borders are Hagfors (larger) and Munkfors (smaller), where we conducted 12 in-depth interviews with Dutch migrant tourism entrepreneurs in 2014–2015. Since early twenty-first century, a Hagfors-based real estate agent has participated in the annual Emigration Expo in the Netherlands to sell property on the Dutch market (Eimermann 2015b). Mitigating population decline through place marketing campaigns, the agency "Invest in Värmland" has also attracted new (Dutch) populations at the Expo. These are the main reasons why Hagfors (pop. 11,651) and Munkfors (pop. 3,769) are now among the Swedish municipalities with most Dutch inhabitants per capita (Eimermann 2015a; Statistics Sweden 2019a). At a one- to two-hour car/bus drive to Karlstad, they are located remotely but not in the most interior parts of Sweden. The municipalities are part of a manufacturing periphery (Hedlund 2016), indicating their less developed tourism facilities and relatively peripheral location in Sweden. Hagfors and Munkfors consist of many small villages, each with their own specific geography, history, socio-economic characteristics and social fabric.

The interview material consisted of demographic data, stories of migration decisions and reasons for starting or acquiring a lifestyle-oriented tourism firm (Ioannides and Petersen 2003). We selected the businesses within a larger pool of tourism firms since they offered food for sale, which Lundmark et al. (2014) highlight as interesting for bottom-up rural development in Sweden. We approached the entrepreneurs using the Internet, previous contacts with Dutch migrants in the area, and tourism brochures. This method is time consuming and resulted in a small sample size, which makes it hard to generalise specific findings, but the benefit of gaining in-depth insights into the interviewees' everyday post-migration life compensates for these limitations (Denscombe 2007).

The purpose of using this method was to expand previous quantitative data (e.g. Lundmark et al. 2014) with deeper understandings of motivations, emotions and lived experiences (Hennink et al. 2011). We used semi-structured life-world interviews to understand selected themes of the "lived daily world from the subjects' own perspectives" (Kvale 2007, pp. 10–11). The interview guide consisted of four parts.

With informed consent, we asked questions on the entrepreneurs' socio-economic characteristics as well as their local social and economic relations. The final part considered future plans. For example, we asked "What were the reasons for starting this firm?", "What do you find challenging?" and "Do you have any plans for the firm regarding e.g. collaborations, renovations or innovations?".

We conducted data analysis both during and after the interviews. During the interviews, we used techniques such as noting patterns and themes, and examining "plausibility" by summarising answers after each section (Miles and Huberman 1994, pp. 245–246). Afterwards, content analysis using a word-processing programme proceeded for themed interpretation (Kvale 2007). During transcription and other phases of the interview study, we increased reliability by recognising divergences between oral conversations and written texts, and we increased validity by considering the interpretative and constructive nature of transcription (ibid.). Finally, we used pseudonyms to anonymise all names.

Dutch Tourism Entrepreneurs' Downshifting Practices in Värmland

Before addressing the research questions, we first consider whether and how we can characterise the studied lifestyle migrants as downshifters. We argue that we can do so since they showed noteworthy similarities with Tan's (2000) downshifters as they accepted lowered income and status and pointed at increased personal well-being. The following examples illustrate the interviewees' reasoning: Jacob called their migration "an easy step in search of tranquil landscapes and populations". He recalled how his business venture hadn't been financially successful: "working with agricultural products will never make you rich, but reaching 50 years of age you know that that is unimportant". Shirley indicated their café was giving discounts to disabled people as they wanted to create a good name, and "we do this to be social, not to become rich". She named "the woods and health reasons" as motivations. Equally, Johanna could not believe she owned 30 hectares of land, describing her current practices as "deciding how to spend time, returning to nature and living the dream". Emma said she had "bought freedom and space". While Yvonne operated a knitting café, Robert managed a camping site during daytime and the local swimming pool in evenings and weekends. Instead of a salary, he received a percentage of the sold tickets and subscriptions. Still, he and Yvonne were content with this low income, "as long as it pays enough to buy food for us and the dogs". Also, Peter and Hermione's son had calculated in an initial business plan that they could be running with profit within four years after the start. Eight years later, this was still not the case, but they preferred their new lifestyle to their previous hectic and stressful lives. Joris and Jolene wanted to escape the rat race ("colleagues could phone me at 2am"), they based their post-migration firm on their hobbies (mountain biking and cooking) and they said that they could do so for three years thanks to the excess value of their Dutch house.

These examples around decreased income, easier living and increased personal well-being illustrate how the interviewees' reasoning resonates with motivations for lifestyle migration and downshifting as raised in our literature review. Hence, although we acknowledge that lifestyle migrants are not per definition the same as downshifters or vice versa, we do study the interviewed Dutch entrepreneurs as downshifting lifestyle migrants. Further, Table 15.1 shows the migrants' socio-economic backgrounds. Few were highly educated ("only" seven of 23 had at least three years of higher education), and very few had worked as executives or managers. Also, most interviewees were unexperienced in tourism or as entrepreneur, and all tourism firms (except Tim and Anna's hotel) were micro firms started by the migrants as part of their aspired lifestyle changes. This confirms Hampton's (2008) argument that people in all income levels can perform a downshift.

Turning more specifically to the first research question, we now consider work-, animal- and children-related motivations for migrating and downshifting. Both William & Harriet and Shirley & Paul started their bakeries together with their respective daughters, who had dreamed of this for a while in the Netherlands. They indicated institutional reasons when they explained that renting suitable space and equipment is cheaper and starting a bakery is less complicated in rural Sweden.

Table 15.1 Migrant entrepreneurs included in the study

Pseudonym, migration year	Birth-year	Level of education	Pre-migration occupations	Post-migration firm owners	Children's birth year
Tim, 2005	1960	High	ICT consultant	Hotel	–
Anna, 2005	1950	High	Dog school owner[a]		
Jacob, 2005	1954	High	ICT[a]	Dairy farm, café	'91 '96 '96 +5 in NL
Rose, 2005	1962	High	School teacher		
Peter, 2006	1954	High	School headmaster	B&B, antiques, café, camping	children in NL
Hermione, 2006	1955	Medium	Sales, voluntary work		
William, 2006	1959	Medium	Process operator	Bakery, café	'85
Harriet, 2006	1959	Medium	Computer business[a]		
Anders, 2007-10	1958	Medium	Plumber	Restaurant	Children in NL
Emma, 2007-10	1959	Medium	Human resource manager		
Jonas, 2008	1971	Medium	Electrician	Horse ranch, restaurant	–
Johanna, 2008	1972	Medium	Dance teacher[a]		
Dick, 2008	1971	Medium	Manager travel agency[a]	Guesthouse	'98 '01 '07
Max, 2010	1968	Medium	Air force officer	Restaurant, letting rooms	'98 '00 '02
Lindy, 2009	1970	Medium	Receptionist		
Paul, 2009-10	1960	Medium	ICT	Bakery, café	'89+1 child in NL
Shirley, 2009-10	1960	Low	Hospitality		
Robert, 2010	1960	Low	Military service/own firm[a]	Knitting café	–
Yvonne, 2010	1962	Low	Military service/own firm[a]		
Adam, 2011	1965	Medium	Car service	Restaurant, kiosk, pub	'92 '96
Hanna, 2011	1969	Low	Health care		
Joris, 2014	1968	High	Mountain biker	B&B, restaurant	'03 '06
Jolene, 2014	1970	High	Government services		

[a]pre-migration experience as entrepreneur

Source Fieldwork (2014–2015) (Interviewees present during the interview are underlined)

Dick had been working in the tourism industry since the 1980s, and he had lived in South Africa some years in the 1990s "to get to know the product". He motivated his move to Sweden with "better opportunities for working in tourism", which he could gain access to with relatively little effort. He also related this with a point in life (aged 37) where he wanted to get to know his new surroundings and figure out what he really wanted to do. At the time of our interview in 2015, he had owned and sold a moose park, and he was offering services to local entrepreneurs and other people such as painting, cutting grass, guiding tourist groups and managing holiday cottages. He also ran a guesthouse but tried to "get rid of it".

Related with gender issues that we address more thoroughly in the conclusion, Jacob initially saw business opportunities in Sweden offered by well-developed Swedish Broadband and ICT infrastructure. However, he gave up his ICT business ambitions to be more with animals, as he had been in his youth:

> I have worked on my father's dairy farm until I was 20. Then I worked in ICT and later started my own firm, which means high stress levels. I have always longed back to farming and animals, but I wasn't sure whether it would be a clever step. When I turned 50, I took a sabbatical year in our cottage in Sweden and after 3 to 4 months, I didn't want to return to the Netherlands. I wanted to return to my past, which is why we started our dairy farm in Värmland. (Jacob)

Similarly, Yvonne and Robert mentioned natural surroundings for their dogs. They owned 4 hectares in Sweden ("for less than a third of the price in the Netherlands"), so they could directly start their forest walk from their front door and the dogs could easily swim in the lake nearby. Max and Lindy thought about their animals and environmental issues: "Our B&B offers eggs from our own chickens, and it is part of our lifestyle to feed the sheep with left-overs from our restaurant, rather than throwing the food away". They stated that this fits well with the rural surroundings to which they migrated.

The third motivation for this type of migration is the search for good social surroundings for children's upbringing. Most interviewees originated from urban areas in the densely populated Netherlands, and they were looking for rural surroundings that fitted their aspired lifestyle changes. Not long after the interview with Shirley, their café was closed and the bakery only delivered special orders. The reason was that their daughter (whom they joined on her move to Sweden) had given birth and they wanted to focus on their grandchild growing up in tranquil surroundings rather than thinking of their firm. We (the authors) elaborate further on this choice in this chapter's conclusion.

Max and Lindy were also a household with children (under 12 years of age), who had attended primary school in both countries:

> Here classes are smaller, teaching is based on the children's needs instead of strictly following a textbook, and it is possible to take a week off during hunting season. We like the Swedish society, and our children growing up here in rural Sweden instead of near a Dutch city. (Max and Lindy)

These initial findings indicate that the interviewees were migrating from an urbanised country to a rural Swedish area to escape work pressures and to become self-employed in a way that was new to them. Their tourism firms in Sweden were often different from the sector they had previously been employed in, and this business gave them a sense of freedom, partly because of fewer rules and regulations. In Värmland, they were closer to nature and paid less for land and property, which also increased their sense of freedom and richness in non-economic terms. This shows clear connections with the lifestyle migration and downshifting literatures in a way that underlines these interviewees' perceived necessity to migrate to achieve their aspired lifestyle changes. At first sight, they appear to fit with Saltzman's (1991) categories of urban escapees and self-employed in businesses in rural settings, but few of them used skills acquired in larger organisations as they did different work than before migrating, and their firms were not always well considered and viable. Jacob and others were "capitalizing on electronic technology" (Saltzman 1991), as shown by their social media accounts through which they can reach a larger group of potential local and international customers. Broadband and ICT opportunities in Swedish rural areas are more advanced than in other European rural areas (Löfving et al. 2019), making Sweden an outstanding destination for them.

To address the second research question, we turn to the interviewees' post-migration practices, and what these imply for their work–leisure balance. The migrants' lifestyle change did not imply reduced working hours or increased leisure time. Many migrants were working for 6–7 days a week, but this was no problem, as Jolene explained: "in the beginning we were working 24 hours a day, 7 days a week, but I did that in my previous job too. The difference is that that job craved mental focus, while working in the B&B is rather physical. It is hard work, but it still feels like being on vacation". Adam indicated that he had no time for other activities as they had to work many hours in the restaurant and maintain the forest on their land. Similarly, running a bakery, William started at 3.30 am and Harriet joined him around 5.30 am before their café opened at 7 am (Mon–Sat). They worked until 6 pm and spent their leisure time at home outside the village, with no need for social contacts after a whole day's hard work and chatting with guests. Although some local clubs used the café for annual meetings, William and Harriet didn't engage with these groups.

In other villages, some interviewees did engage in local community activities during their little leisure time. For instance, Adam was a member of the local hunting team, Johanna was a member of the local tourist board, and Jolene was also asked to become a member. Joris became involved in the mountain bike club, and he wanted to contribute to the maintenance and development of local mountain bike trails. Shirley and Paul did sponsor some of the local community centre's activities, while Max and Lindy engaged similarly:

> The village consists of 15 houses, so you quickly get to know your neighbours. When we arrived here, people helped us translating texts from Swedish and filling out forms. The mind-set is that we help each other since we are living in the countryside as a small group: we engage in the local community centre's lottery offering meals at our restaurant as a price, and we participate in fund-raising activities such as the Christmas market. (Max and Lindy)

Likewise, Peter and Hermione were active in the local church: Peter was a board member and played the piano while Hermione was a member in the planning commission. They helped the community centre during summer but their engagement was also economic: "when they organise an activity, this means income for us as their guests drop by our antiques shop". Similarly, Tim and Anna collaborated with the local ice hockey gymnasium and floorball club, and offered discounts to visiting clubs that stayed in the hotel during tournaments. This shows that although lifestyle changes and downshifting are important, the interviewees still need to consider economic issues too, since they are being private individuals, but running businesses as well.

Specific examples regard bigger expenses and taxes. Interviewees indicated financial strategies since they could invest the money that they saved through lower rents and property prices in various forms of self-realisation. Considering type of business, Peter and Hermione had started as a limited company for otherwise they wouldn't be allowed to run a firm. Later they learned this was rather expensive and they became a sole proprietorship. Peter and Hermione noted that the Swedish state wanted to tax them 30% on the excess value of their Dutch house after they had sold it, and Max and Lindy had avoided this risk selling their Dutch house before they officially bought their Swedish house. This is an example of a tax-avoiding practice that might gain the individual, but with possible detrimental effects on the Swedish welfare state, as we will discuss further in the conclusions.

Other practices related with the Swedish model regard strategies to avoid employing staff due to high payroll tax. William and Harriet had previously employed temporary interns and a Swedish woman but when she injured her neck, they felt it was better she quit. Their daughter helped them when she could. It was hard to find the right balance: expanding their firm could generate more income, but this also meant increased payroll tax for hiring staff. Many female interviewees found temporary solutions through relatives and social networks. Joris and Jolene employed a friend's daughter to clean the B&B rooms during their first summer season. Jacob and Rose employed their daughter to sell ice cream. Through their trans-local links with the Netherlands, Anders and Emma offered Dutch workers food and shelter through a religious network, Johanna recruited 10–12 Dutch interns each summer for a working holiday, Max and Lindy employed Dutch interns (from a catering trade school) during summer, and their children helped in busy local weekend events. Many would like to employ a local inhabitant, but working hours were too irregular to offer a permanent contract. Peter and Hermione pointed at another issue: "the Swedish state wants to improve tourism's role for local development, but it is a catch 22 situation. Before we can invest in staff, the state needs to invest in facilities [like more public transport, longer opening hours for shops, swimming pools etc. during the final weeks of August] to extend the tourism season". Like other nearly retired interviewees looking forward to their Swedish and Dutch retirement pensions, they did not want to invest more in their firm. Yet, some female tourism entrepreneurs like Emma were more uncertain: "we don't have a social safety net. We hope to earn money from selling our restaurant, but so far we have not found a buyer". In Yvonne's words: "we could be doing this for 15 more years, but it would be nice if the next generation would take over. Our savings are also slowly running out". These examples show a lacking understanding of the Swedish model's conditions and indirect gender issues, which together constrain the interviewees' downshifting practices.

Conclusion

Although the studied lifestyle migrants may not have reduced their time spent on work, this chapter has shown that they can be termed downshifters because they chose a quiet place and a slow pace of (unprofessional) working over making a profit and a career, which often reduced their incomes. Their migration to less stressful rural settings enabled lifestyles that were closer to their personal values.

In this light, we address our aim: to identify implications for the Swedish welfare state of the tourism entrepreneurs' downshifting-related practices. One implication for the Swedish welfare state is that it appears to miss tax income as the entrepreneurs deploy strategies to avoid paying taxes. Such tax planning and tax aggressiveness are quite common for entrepreneurs, to maximise profit or wealth or to survive on a competitive market (Shackleford and Shevlin 2001; Axelsson and Hedberg 2018). Some interviewees avoided paying taxes in two ways: first when purchasing a Swedish property, but more prominently when running their firms with the help of interns and volunteers instead of hiring staff and paying payroll taxes. Like other rural microbusinesses (e.g. Ioannides and Petersen 2003), the interviewees' firms created too little economic turnover and profits to employ staff. The monetary value of the combined efforts in their firms was too low. Even after adjusting their consumption to decreased levels of income, the interviewees needed to claim bits of their savings, e.g. derived from the excess value of selling their house in the Netherlands. The interviewees' firms were important for their downshifting ambitions, but they were seldom economically sound. Some of these downshifting entrepreneurs thus experienced similar socio-economic struggles as other (tourism) entrepreneurs, as part of prioritising their individual well-being over the collective best. Further, spending-reducing practices are typical for downshifters, and if their number is rising, this can eventually imply difficulties for the Swedish model's economic sustainability in providing tax-based benefits conditional on work (e.g. childcare, education and pensions, Schierup et al. 2006; Kolm and Tonin 2015).

Additionally, this chapter confirms the subjective nature and the changing experiences of leisure for downshifters after their career change (Hampton 2008, p. 26). Following Juniu (2000), this can be a philosophical matter: people like Joris and Jolene have simplified their lives as a means to regain life quality that was lost through a modern trend of consumerism. Instead of associating leisure with fun, free time or pleasure, the interviewees related to personal meaning of time and wider terms of mind, space, culture and society (Hampton 2008). During their leisure time, interviewees engaged in their local communities in various ways. This wasn't just based on economic drivers for their businesses, it also increased social sustainability and sustainable ways of living in the communities. Despite the long working hours, many had the energy to engage in village life, as their firms were less demanding than their previous jobs. This depicts a return to more classical meanings of leisure that are not necessarily in opposition to hectic work situations in a consumption-focused society (Juniu 2000). Thus, we argue that the contributions of new residents to their neighbours, clubs, churches and community centres are important to keep the villages alive, which is in the interest of the Swedish welfare state.

More specifically, we now focus on downshifting-related gender issues that arise in connection with the tourism businesses. Such issues emerged when the males engaged in external income sources while the females operated their firms on location. Paul was not as fond of rural areas as Shirley was and he worked with IT in Stockholm (previously Oslo), Tim combined his hotel work with being an IT consultant, and Anders was working as a plumber 10 months a year. Robert managed the camping site and worked in the local swimming pool while Yvonne ran the knitting café. Peter worked some hours per week as a Dutch language teacher, while Hermione was in charge of the B&B and antiques shop since "this was more her dream from the beginning" (Peter explained). As the opportunities to complement income from their firms with a local job were limited, many interviewees needed other income sources elsewhere. Many males were working in non-local jobs (seasonally or part-time), while the females continued running the tourism firms locally. In Saltzman's (1991) terms, many interviewed men were self-employed or career shifters, while many women were urban escapees. This confirms previous critique on household situations in which the male wageworker operates in the public sphere of the labour market, while the female operates the more homebound firm in the private or semi-public sphere (Gunnarsson 2013a). As the females in this study at times indicated that they were caregivers (Shirley and Harriet helping their daughters in caring for their newly born child), this chapter also confirms and critiques unbalanced gender power relations (Carbin et al. 2017). Hence, to further our understanding of how and why male and female members of a household engage in downshifting, we should develop gender perspectives on male and female ambitions and downshifting practices. Including age, education, employment history, entrepreneurial experience, individual and household income levels, stages in the life course and gendered migration motivations and decisions will result in more nuanced knowledge in this field (Schor 1998; Hamilton and Mail 2003; Hampton 2008). This will also contribute to a wider plethora of migration studies (Benson and O'Reilly 2016).

Further, this conclusion boils down to questions regarding desirable future societies, and how they may handle expected global climate challenges. If the effects of global warming are as severe as stated in the IPCC report (2018), then it will be impossible to continue the Swedish model as we know it. In that sense, downshifters may well be part of a solution to future challenges since they consume less and spend more time on ecologically sustainable alternatives such as growing own vegetables and keeping animals like sheep and chickens as important parts in their business ideas (illustrated by Max and Lindy). Some downshifting entrepreneurs do develop their businesses with limited resources and in less favoured areas, where full-employment jobs are scarce and self-employment might be the only alternative, although it is unsure whether their firms will come to employ staff. As the Swedish model can benefit from such risk-takers who diversify otherwise homogenous rural local economies, it should support new business endeavours even when some of them turn out to be unsuccessful. Since many tourists value accessibility to B&B's and cafés in sparsely populated areas (e.g. Ioannides and Petersen 2003), downshifters running such firms as part of their lifestyles can contribute to a socially and ecologically more sustainable society.

Since this offers important opportunities for the Swedish welfare state's adaptation to future climate challenges (U.N. 2015), future research on downshifters can contribute with insights for welfare state and local officials as well as individual entrepreneurs. Overall, such research should be sensitive to the role of downshifting and sustainable lifestyles in meeting future challenges induced by climate change. From regional development perspectives, such studies could balance challenges of potential individual tax-avoiding behaviour with gains of attracting new populations in general and risk-taking creative downshifters in particular. Planning policy analysis could e.g. study what impetus downshifters bring to rural areas and how their contribution to local economies can be supported. Such studies can also question the validity of post-migration expectations triggered through place marketing events like the Emigration Expo (Eimermann 2015b). Thus, this chapter has explored challenges, opportunities and implications of downshifting for both the concerned individuals, the households and interrelated current and future concerns for changing welfare states.

References

Axelsson, Linn, and Charlotta Hedberg. 2018. Emerging topologies of transnational employment: 'Posting' Thai workers in Sweden's wild berry industry beyond regulatory reach. *Geoforum* 89: 1–10.

Barcus, Holly R., and Keith Halfacree. 2018. Lifestyle migrations. In *An Introduction to Population Geographies*, ed. Holly R. Barcus and Keith Halfacree, 195–232. Abingdon: Routledge.

Benson, Michaela C. 2010. The context and trajectory of lifestyle migration: The case of the British Residents of Southwest France. *European Societies* 12: 45–64.

Benson, Michaela C., and Karen O'Reilly (eds.). 2009. *Lifestyle Migration: Expectations, Aspirations and Experiences*. Farnham: Ashgate.

Benson, Michaela C., and Karen O'Reilly. 2016. From lifestyle migration to lifestyle *in* migration: Categories, concepts and ways of thinking. *Migration Studies* 4: 20–37.

Breakspear, Christie, and Clive Hamilton. 2004. *Getting a life—Understanding the downshifting phenomenon in Australia*. Discussion Paper Number 62. Manuka: The Australia Institute.

Carbin, Maria, Johanna Overud, and Elin Kvist. 2017. *Feminism som lönearbete*. Stockholm: Leopard förlag.

Carson, Doris A., and Dean B. Carson. 2018. International lifestyle immigrants and their contributions to rural tourism innovation: Experiences from Sweden's far North. *Journal of Rural Studies* 64: 230–240.

Carson, Doris A., Dean B. Carson, and Marco Eimermann. 2018. International winter tourism entrepreneurs in northern sweden: Understanding migration, lifestyle, and business motivations. *Scandinavian Journal of Hospitality and Tourism* 18: 183–198.

Denscombe, Martyn. 2007. *The Good Research Guide for Small-Scale Social Research Projects*. Maidenhead, UK: Open University Press.

Eimermann, Marco. 2013. *There and back again? Dutch lifestyle migrants moving to rural Sweden in the early 21st century*. PhD diss. Örebro University, Sweden.

Eimermann, Marco. 2015a. Lifestyle migration to the north: Dutch families and the decision to move to rural Sweden. *Population, Space and Place* 21: 68–85.

Eimermann, Marco. 2015b. Promoting Swedish countryside in The Netherlands: International rural place marketing to attract new residents. *European Urban and Regional Studies* 22: 398–415.
Eimermann, Marco. 2016. Two sides of the same coin: Dutch rural tourism entrepreneurs and countryside capital in Sweden. *Rural Society* 25: 55–73.
Eimermann, Marco, and Stefan Kordel. 2018. International lifestyle migrant entrepreneurs in two New Immigration Destinations: Understanding their evolving mix of embeddedness. *Journal of Rural Studies* 64: 241–252.
Eimermann, Marco, and Svante Karlsson. 2018. Globalising Swedish countrysides? A relational approach to rural immigrant restaurateurs with refugee backgrounds. *Norsk Geografisk Tidsskrift* 72: 82–96.
Galte Schermer, Isabelle. 2019. *Kvinnor på arbetsmarknaden*. www.ekonomifakta.se. Accessed on August 23, 2019.
Gunnarsson, Åsa. 2013a. Introduction. In *Tracing the Women-Friendly Welfare State*, ed. Åsa Gunnarsson, 9–13. Gothenburg: Makadam.
Gunnarsson, Åsa (ed.). 2013b. *Tracing the Women-friendly Welfare State*. Gothenburg: Makadam.
Hamilton, Clive, and Elizabeth Mail. 2003. *Downshifting in Australia—A sea-change in the pursuit of happiness*. Discussion Paper Number 50. Manuka: The Australia Institute.
Hampton, Roy S. 2008. *Downshifting, leisure meanings and transformation in Leisure*. PhD diss. Pennsylvania State University.
Hedlund, Martin. 2016. Mapping the socioeconomic landscape of rural Sweden: Towards a typology of rural areas. *Regional Studies* 50: 460–473.
Hennink, Monique, Inge Hutter, and Ajay Bailey. 2011. *Qualitative Research Methods*. London: Sage.
Ioannides, Dimitri, and Tage Petersen. 2003. Tourism 'non-entrepreneurship' in peripheral destinations: A case study of small and medium tourism enterprises on Bornholm, Denmark. *Tourism Geographies* 5: 408–435.
IPCC. 2018. *Global Warming of 1.5 °C*. https://www.ipcc.ch/sr15/. Accessed on October 8, 2019.
Juniu, Susana. 2000. Downshifting: Regaining the essence of Leisure. *Journal of Leisure Research* 32: 69–73.
Kennedy, Emily Huddart, Harvey Krahn, and Naomi T. Krogman. 2013. Downshifting: An exploration of motivations, quality of life, and environmental practices. *Sociological Forum* 28: 764–783.
Kolm, Ann-Sofie, and Mirco Tonin. 2015. Benefits conditional on work and the nordic model. *Journal of Public Economics* 127: 115–126.

Kvale, Steinar. 2007. *Doing Interviews*. London: Sage.
Larsson, Jörgen. 2015. Downshifting. In *The Wiley Blackwell Encyclopedia of Consumption and Consumer Studies*, ed. Daniel T. Cook and J. Michael Ryan, 248–250. Chichester: Wiley.
Löfving, Linnea, Gustaf Norlén, and Timothy Heleniak. 2019. *Case Study Report: Digital Västerbotten*. Stockholm: Nordregio.
Lundmark, Linda, Marcus Ednarsson, and Svante Karlsson. 2014. International migration, self-employment and restructuring through tourism in sparsely populated areas. *Scandinavian Journal of Hospitality and Tourism* 14: 422–440.
Miles, Matthew B., and A. Michael Huberman. 1994. *Qualitative Data Analysis*. Thousand Oaks: Sage.
Pavelka, Joseph P. 2000. *It's Not About Time! Rediscovering Leisure in a Changing World*. Carp, ON: Creative Bound Books.
Saltzman, Amy. 1991. *Downshifting: Reinventing Success on a Slower Track*. New York: HarperCollins.
Schierup, Carl-Ulrik, Peo Hansen, and Stephen Castles. 2006. *Migration, Citizenship, and the European Welfare State*. Oxford: University Press.
Schor, Juliet B. 1998. *The Overspent American*. New York: HarperCollins.
Schor, Juliet B. 2005. Sustainable consumption and worktime reduction. *Journal of Industrial Ecology* 9: 37–50.
Shackelford, Douglas A., and Terry Shevlin. 2001. Empirical tax research in accounting. *Journal of Accounting and Economics* 31: 321–387.
Statistics Sweden. 2019a. *Population*. www.scb.se. Accessed on July 5, 2019.
Statistics Sweden. 2019b. *Born abroad*. www.statistikdatabasen.scb.se. Accessed on July 5, 2019.
Tan, Philomena. 2000. *Leaving the rat race to get a life: A study of midlife career downshifting*. PhD diss. Swinburne University.
U.N. 2015. *Sustainable development goals*. https://sustainabledevelopment.un.org. Accessed on August 20, 2019.

16

Sustainable Tourism Employment, the Concept of Decent Work, and Sweden

Tara Duncan, Anna Gudmundsson Hillman and Jörgen Elbe

Introduction

It is now seemingly clear that the sustainability of tourism is intensely important to societies, economies and the environment. Defined by the UNWTO as tourism 'that takes full account of its current and future economic, social and environmental impacts, addressing the needs of visitors, the industry, the environment and host communities' (UNEP and UNWTO 2005, pp. 11–12), sustainable tourism is often seen as a (policy) tool that benefits 'communities in destinations around the world' (World Bank Group 2017 in Saarinen and Gill 2019, p. 4).

T. Duncan (✉) · A. G. Hillman · J. Elbe
Dalarna University, Falun, Sweden
e-mail: tdu@du.se

A. G. Hillman
e-mail: pag@du.se

J. Elbe
e-mail: jel@du.se

A. Walmsley et al. (eds.), *Tourism Employment in Nordic Countries*,
https://doi.org/10.1007/978-3-030-47813-1_16

From a Swedish context, the importance of sustainability is the starting point of the recent Official Report of the Swedish Government (SOU 2017:95, p. 23) on tourism and hospitality. The goal of this report was to identify ways in which the Swedish government could strengthen the tourism industry and create conditions for growth, enterprise and employment through sustainable (tourism) development (SOU 2017:95, p. 19). This further links with Sweden's aim of being a leader in the implementation of the 2030 Agenda for Sustainable Development (SOU 2017:95, p. 114) which is directly tied to growing industries such as tourism (WTTC 2018).

Within discussions on the sustainability of tourism lie questions about the sustainability—socially and economically—of tourism employment. As Baum et al. (2016a, p. 808) state, issues of stable employment, income earning opportunities and 'decent work' lie 'at the heart of the sustainability debate'. However, research into tourism employment remains a relatively neglected area (Baum 2015; Duncan et al. 2013: Solnet et al. 2014) despite continued calls for more detailed and relevant knowledge (Baum 2015; Baum et al. 2016b).

This is the point of departure for this chapter. Whilst tourism employment is seen as a route for sustainable development that can offer many positives for destinations and communities (such as increasing value to GDP and exports, SOU 2017:95; WTTC 2018), questions remain, particularly surrounding employment in tourism. In addressing these questions through considering what 'decent work' is and what sustainable tourism employment might be, this chapter will reiterate that there remains a dearth of empirically pertinent research into the sustainability of tourism employment, particularly in the Swedish context. Utilising results from a qualitative systematic review of literature, the chapter will highlight that the piecemeal and economic focus of much tourism employment research has allowed the focus to remain on managerial issues rather than focus on the sustainability of employment for tourism workers. The chapter will conclude with a discussion about whether current conceptions of tourism employment recognise wider interdependent political, socio-cultural and economic drivers necessary to sustain a 'decent' tourism workforce and where Sweden 'sits' within these discussions.

Theory and Context

As Robinson et al. (2019) state, tourism employment consists of a wide variety of jobs over a range of skill levels in an increasing number of sub-sectors. These jobs exist at multiple levels, from local, regional and national and exist in businesses of all sizes from micro entrepreneurial businesses through to large multinational conglomerates (Baum 2015; Robinson et al. 2019). However, tourism employment often remains characterised by seasonality, variable hours, low pay, limited desirability as a career choice, uncertainty and precarity (Baum 2015; Mooney 2018; Robinson et al. 2019). Tourism employment is often also seen as exploitative (Janta et al. 2011), and resistant to trade unionisation (Bergene et al. 2015), although Sweden is unusual in having fairly high, though decreasing, union membership (Fulton 2013; Gustafsson and Kullander 2015).

With these criticisms in mind, an understanding of 'decent work' seems imperative. The International Labour Organisation's (ILO) mandate for decent work 'sums up the aspirations of people in their working lives' (ILO, n.d.) to encourage productive work opportunities, security in the workplace, social protection for families, prospects for personal development and social integration, freedom to express concerns, organise and participate in decisions affecting their lives and equal opportunities for women and men (ILO, n.d.). Thus, the focus becomes the quality of the employment (Winchenbach et al. 2019) not employment as job creation. This continues in the ILO's (2019, p. ix) most recent report which reports on progress towards achieving Goal 8 Decent Work and Economic Growth of the UN's Sustainable Development Goals. In this report, the ILO's Centenary Declaration for the Future of Work has a strategy for a human-centred approach to the future of work.

Yet both the issues outlined above concerning tourism employment and the aspirational goals of the ILO and its agenda for 'decent work' are not unproblematic. Firstly, the issues with tourism employment have been challenged as Western-centric (Nickson et al. 2003; Robinson et al. 2019). Perceptions of tourism jobs as low-value or low-quality may be reliant on the wider context (e.g. McDowell et al. 2009) or individual perceptions and perspectives. Secondly, the concept of decent work and its relationship to Goal 8 offers its own set of challenges. How to 'measure' decent work is problematic. From a Swedish perspective, the 2017 follow-up report to the 2030 Agenda for Sustainable Development (Statistics Sweden 2017), simply talks of promoting or developing sustainable tourism and only two indicators, neither of which the report suggests are possible to easily measure statistically, relate to issues of employment (Statistics Sweden 2017, p. 27). Winchenbach et al. (2019, p. 1028) reiterate this point suggesting we are still 'some way off developing comparable data on objective measures… on the quality of work globally'. Alongside this is the concern that decent work has been 'paired' with economic growth in the SDGs (Baum 2018; Scheyvens et al. 2016). This concerning dichotomy is perhaps most apparent in the description of the goal. The Goal is entitled 'Decent Work and Economic Growth' but decent work is immediately subsumed in the more detailed description which reads that the Goal 'promotes sustained, inclusive and sustainable economic growth, full and productive employment and decent work for all' (United Nations 2019).

In spite of these ongoing discussions, there continues to be an 'ineffectual engagement with workforce issues' (Baum 2018, p. 3) in tourism policy and planning, within the private sector and within academic research. Tourism employment research remains a somewhat neglected area (Baum 2007, 2015; Ladkin 2011; Solnet et al. 2014) that has tended to take a piecemeal approach (Baum et al. 2016b). More recent articles by Baum et al. (2016a, b) illustrate a certain amount of consolidation of existing literature and a focus on aspects of decent work (Baum 2018; Robinson et al. 2019; Winchenbach et al. 2019). At the same time, Ladkin's (2011) critique that management and economic perspectives still dominate much research remains valid. Linked to this are Joppe's (2012) concerns about the challenges and possibilities of migrant labour in tourism and her recognition that whilst many countries may now have Tourism Satellite Accounts to measure the economic contribution of tourism, they still lack the human resource 'bit' of this equation. Finally, and perhaps most frustratingly, Baum's refrain from 2015 attests the lack of progress in dealing with human resource issues over longer periods of time; as Baum (2015) says, his earlier paper from 2007 ended on a sombre note and his update in 2015 was no more optimistic.

This lack of research and engagement with tourism employment is perplexing given the centrality of work to people, communities and destinations and the continued growth and reliance on tourism as an economic driver by many countries globally, including Sweden (Baum et al. 2016a; Duncan et al. 2013; SOU 2017:95). This may be due, in part, to the complexity and heterogeneous nature of the tourism industry and as already suggested, the somewhat piecemeal (Baum et al. 2016b) approach that has been taken towards tourism employment research. As Baum et al. (2016a, p. 2) state, much of the research on tourism work has been done at organisational or management level, has been limited to human resource management research, has been classed as hospitality rather than included in a wider tourism concept or has critiqued research through an atheoretical lens (Baum 2007, 2015; Ladkin 2011).

However, what continues to remain missing within discussions of tourism employment are ideas of the sustainability of tourism employment. In addressing this issue one could, as Kramar (2014) suggests, consider adopting aspects of sustainable human resource management. Whilst still problematic to define (Kramar 2014, p. 1075), common elements of research into sustainable human resource management revolve around the development of human and social capital within organisations and acknowledge the role of HRM practices on environmental, social and human aspects of society (Kramar 2014). Thus sustainable human resource management emphasises human and social outcomes rather than focussing on individual or fiscal rewards. Sustainable human resource development advocates a focus on the tourism employee, the host community and the tourist (Jitherndran and Baum 2000), highlighting the necessity to see tourism employment as a central part of communities, destinations, regions and nations. These discussions remind us that sustainable employment, in any industry, exists at many levels and solutions to prevailing challenges need to be integrated across these levels.

Alongside this is Baum's (2018) recent paper considering sustainable tourism employment that argues for an outward focus looking at the society within which tourism work is located as a basis for more 'meaningful searches for explanations and the formulation of policies and practices that can affect change' (Baum 2018, p. 5). What both Baum's call and sustainable human resource management research demonstrates is the need for a multi-disciplinary lens with which to consider tourism employment where a focus on social and cultural sustainability becomes increasingly significant.

Baum's (2018) suggestion of a complex ecosystem within which tourism employment is located resonates. The existing interdependencies between in- and out-migration of working age populations; job quality; local transport options; housing availability, location and affordability; access to education, both vocational and tertiary; intersections with leisure and culture, retail and social welfare provision (Baum 2018, p. 5) impact on the resilience of communities to become, or remain sustainable. The term community perhaps also needs interrogating. As van Dyk (2018) suggests, it is a term that used in a common-sense way and 'seems never to be used unfavourably' (Williams 1976 in van Dyk 2018, p. 531); it is a feel-good potpourri accentuating ideas of harmony, solidarity, identity and shared knowledge and interest (van Dyk 2018, p. 531). The romance of community ignores power relations, conflicts of interest and growing perceptions of insecurity among workers (Alberti et al. 2018). Processes are normalising the precariousness of work, specifically in industries such as tourism, through extensions of standard employment relations and the construction of precarious work (short term, temporary or zero-hour contracts for instance) as an alternative to unemployment (Rubery et al. 2018). Add to this mix increasing digitisation which, in Sweden for instance, is projected to continue to reduce the size of the workforce in the accommodation and restaurant sectors (BFUF 2018). These issues indicate that our current understandings of the paid (and unpaid) work within tourism, hospitality and events simply do not provide enough insight into this complex eco-system.

To return to the bigger picture, as Robinson et al. (2019, p. 1009) suggest, the 'all-pervasive neo-liberal paradigm that governs contemporary political, economic and social discourses, particularly in tourism, inhibits the ideals of sustainability'. Baum et al. (2016a) point out that sustainable tourism remains an economic exercise that minimises aspects of environmental and sociocultural sustainability. Political manoeuvring, policy compliance (whether mandatory or as part of a voluntary commitment) and positive marketing have become the heart of tourism's engagement with sustainability 'rather than genuinely working to achieve the benefits of a sustainable approach' (Baum et al. 2016a, p. 3). What this broader discussion indicates is the ongoing troubling discourses surrounding tourism employment. As an example, in Sweden the proportion of part-time workers in tourism is almost twice as high as in any other industry (Robinson et al. 2019, p. 1016). As Sweden's demographics continue to shift (Statistics Sweden 2015), what does that mean for the communities where tourism is growing or being encouraged? As many businesses struggle to find suitably skilled employees and potential workers with transferable skills show little or no interest in tourism and hospitality as a valid career choice, how do communities and destinations find ways to sustain tourism employment in any sensible way?

What becomes clear is that, as Hollenhorst et al. (2014, p. 306) suggest, tourism often seems to hide 'behind a mask that is all the more beguiling because it appears so sustainable'. In trying to take a sustainable lens, the complex, entangled nature of tourism employment allows recognition of the role tourism and hospitality labour play within contemporary societal debates (Duncan et al. 2013; Ladkin 2011). Gaining an understanding of sustainable tourism employment can disrupt the challenges and dichotomies (such as urban-peripheral) currently central in work on sustainable tourism. Thus, it becomes necessary to recognise the importance of social and cultural sustainability within workforce planning through acknowledging the interdependencies of wider political, social, economic and cultural drivers. As Baum (2018, p. 13) suggests, recognising these key drivers in the effort towards 'the creation and sustainability' in workforce terms, makes 'the notion of achieving a decent work culture a realistic proposition for tourism employment'.

Duncan et al. (2013) have suggested that there is a need to re-conceptualise how tourism and hospitality workers are considered by researchers and stakeholders. As the above discussions illustrate, there is a continuing imperative to understand and challenge not only the way in which tourism employment has and is being researched but also how such outcomes are being utilised by researchers, government and industry (if at all) (Lugosi et al. 2009). There are ongoing troubling discourses surrounding tourism employment. Questions remain that need answering. What does sustainable tourism employment look like? How can tourism employment be sustainable? What does 'decent work' in tourism (and in Sweden) look like? It is clear that Sweden's growing focus on tourism does consider broader issues of tourism employment; the 2017 report devotes considerable space to consider the competencies needed for a growing tourism industry (SOU 2017) and the need for increased training and education opportunities to ensure future industry growth. However, what also become clear is that there remains an economic focus to much of the anticipated development. Whilst the report (SOU 2017) does raise some concerns and issues around future tourism employment growth, there is little discussion about aspects of sustainability with this tourism employment growth, especially for those who are likely to be employed in front-line tourism jobs, namely young people and new migrants.

Therefore, in Sweden and beyond, questions around the sustainability of tourism employment and what decent work 'is' still need answering to ensure that tourism work does not become less than 'decent work'. As Robinson et al. (2019, p. 1020) suggest in their conclusions, there is a need for a 'recalibration of sustainability in tourism relative to the social dimension, and specific to the workforce'. The rest of the chapter will attempt to reflect on some of these issues, taking a Swedish context. In order to do this, results from a recent research project are incorporated into the discussions to continue to illustrate the lack of empirically grounded, relevant literature on the sustainability of tourism employment.

Method

The discussions in this chapter revolve around a piece of research funded by FORMAS, the Swedish Research Council for Sustainable Development. The research involved a qualitative systematic review of the literature that intended to interrogate academic literature to provide an understanding of the current state of tourism employment. The aim was to focus on the sustainability of tourism employment with the goal of linking to the UN's Sustainable Development Goals, specifically Goal 8: Decent Work and Economic Growth (United Nations 2019). The purpose was to be able to provide information relevant for tourism practitioners within academia and industry in Sweden.

The systematic literature review adapted review processes from Mairs et al. (2016) utilising a narrative approach (Popay et al. 2006) where texts and words are used to 'tell the story' (Popay et al. 2006, p. 5) rather than taking a more quantitative approach. What this allowed was a move away from more simplistic summaries of findings to a broader analysis that allows the findings to highlight new insights or knowledge (Mair et al. 2016, p. 4). The five-step process involved defining research terms, identifying databases, searching and extracting literature, analysing the data and ensuring the robustness of the research process.

The research utilised the Scopus database as it covers a higher proportion of Social Science journals than Web of Science (Mongeon and Paul-Hus 2016) and the initial literature search (up to February 2018) resulted in 3592 articles matching our criteria. The synthesis included research articles between 1983 and October 2018. Conference abstracts and conference reports were excluded from the beginning as they proved difficult to access, book chapters were excluded in the second and third rounds of analyses as many were not accessible through our institution. The search also limited text to the English language and it is recognised that this could be a limitation of the research. The search terms used in the search string included English and American spellings of words (e.g. labour/labor) and reflected the difficulty in trying to define sustainability within tourism employment. The search string included the terms 'human rights', 'decent work', 'sustain*[1]', 'work* conditions' and 'work* environment' in order to try and capture

[1] The asterisks indicates a word that can have multiple endings. Using an asterisk allowed the database to search for any version of the word in relation to our other criteria. For instance, sustain* could be sustainable, sustainability or sustain; using the asterisks allowed the search string to look for all variations.

all aspects of sustainability within a tourism context. The search string excluded volunteer work. To ensure continued relevance, an RSS feed alert, using our search string, identified a further 247 articles to the end of October 2018. The process of refining the inclusion criteria and reducing the number of relevant articles involved numerous rounds of discussion, negotiation and (dis)agreement by the authors. Over time, and after much discussion and close-reading of abstracts, keywords, aims, methods and conclusions, the authors identified and sourced a final 142 articles relevant for the research.

Two rounds of coding were undertaken. The first used SPSS and this was undertaken to understand some basic trends within the articles identified. Whilst this gave a more quantitative and descriptive aspect to the findings, highlighting some interesting observations, this is not the focus for this chapter (although a few relevant points will be discussed below). The second round of coding used NViVo and was based on variables designed from our understanding, and definition of, decent work and led to the identification of two focal streams around the dominant themes in the articles and how findings were actioned in implications and relevance.

Quantitative Findings

Before moving on to the thematic findings and wider discussion about the sustainability of tourism employment, it is worth noting a few of the quantitative results in light of the previous theoretical insights. Firstly, Table 16.1 shows that the majority of articles have been published since 2000 with an increasing number being seen from 2008. This is interesting to note in light of the UN Millennium Declaration, signed in September 2000 from which the Millennium Development Goals were derived (United Nations, n.d.a) and the subsequent transformation of these eight goals into the 17 Sustainable Development Goals in 2015 (United Nations, n.d.b).

Table 16.2 illustrates where the majority of the articles were published, with 69% of all articles published in tourism and hospitality-related journals. What is interesting to note is the number of

Table 16.1 Publication year

Year of publication	Frequency	%
1980–1984	2	1.4
1985–1989	0	0.0
1990–1994	2	1.4
1995–1999	3	2.1
2000–2004	7	4.9
2005–2009	25	17.6
2010–2014	44	31.0
2015–October 2018	59	41.5
Total	142	100.0

Table 16.2 Top 10 publishing journals

Rank	Journal	# of articles
1	*International Journal of Contemporary Hospitality Management*	15
2	*International Journal of Hospitality Management*	14
3	*Tourism Management*	11
4	*Journal of Hospitality and Tourism Management*	6
5	*Worldwide Hospitality and Tourism Themes*	6
6	*Asia Pacific Journal of Tourism Research*	5
7	*Journal of Human Resources in Hospitality and Tourism*	5
8	*The Service Industries Journal*	5
9	*Sustainability*	4
10	*Anatolia*	3

publications in the top five Tourism and Hospitality journals (as defined by the 2019 Australian Business Deans Council ranking) (Australian Business Dean's Council 2019). Only *Tourism Management* and the *International Journal of Hospitality Management* feature in the top ten journals. This perhaps reiterates Baum's (2007, 2015; see also Duncan et al. 2013) contention that much research into employment has been classified as 'hospitality' work rather than included within a wider 'tourism' context. The worrying conclusions that we might draw from this separation, and the number of hospitality journals listed in Table 16.2, is that both tourism researchers and the tourism industry are somehow deluded in how they perceive the issues and challenges facing both the tourism and hospitality sectors of the wider visitor industry.

Qualitative Findings

The findings from the NVivo analysis reiterate the discussions above about the fragmented and troubling nature of tourism employment research. Split into two focal streams around the dominant themes in the articles and the dominant actions (see Fig. 16.1), these categories in themselves illustrate the complexity of trying to examine tourism employment. The dominant themes identified in the literature tend to focus on aspects that might be expected to be found in the literature, although as Figs. 16.2–16.4 and the following discussions suggest, the dominant themes are far from straightforward. The dominant actions are even more problematic. This stream developed through analysis of how the articles tended to talk of findings and implications of research. The stream developed through a realisation that words like solutions, engagement, development or mitigation were being used frequently to suggest, for instance, future recommendations or managerial implications. However, what became apparent was that these actions were often

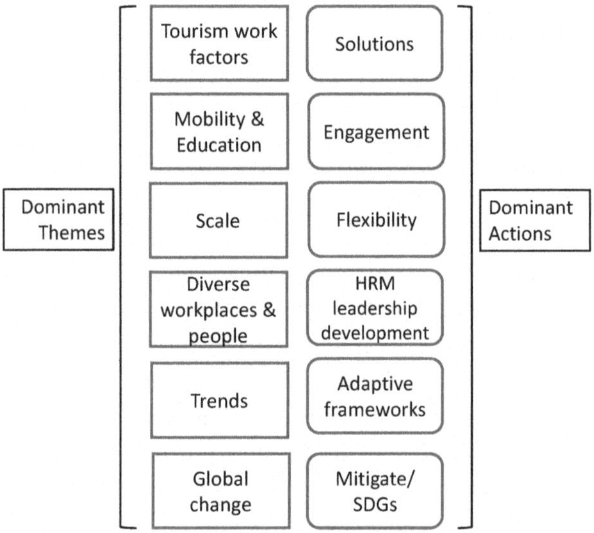

Fig. 16.1 Dominant themes and actions

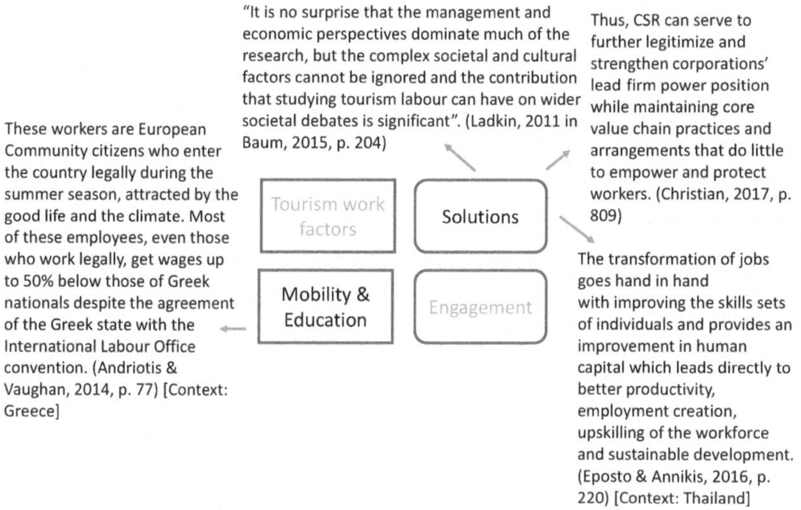

Fig. 16.2 Detailed analysis of themes I—an example of conflicting findings

more commentary and did not necessarily offer solid, workable ways for industry, or the academy, to move the findings or research forward.

What does become obvious as each theme is examined more closely is the often contradictory nature of findings. An example of this is provided in Fig. 16.2 which suggests that adopting sustainable practices through corporate social responsibility may lead to a reduction in employee rights and empowerment (Christian 2017), in other contexts, tourism has provided positive and sustainable employment opportunities (Eposto and Annakis 2016). Figure 16.2 also illustrates that common conceptions of migrant workers from less-developed economies being exploited by the tourism industry can also be disrupted. Andriotis and Vaughan's (2004) research highlights that unfair employment practices exist for many types of workers in many different situations. They take this further, as illustrated in Fig. 16.3, when they suggest that these tourism jobs are 'less worthy' (Andriotis and Vaughan 2004, p. 82)

The analysis of the 142 articles highlights the issues for human resources (HR) within the tourism industry (see Fig. 16.3). In varying contexts globally, HR practices are being stretched, either as HR

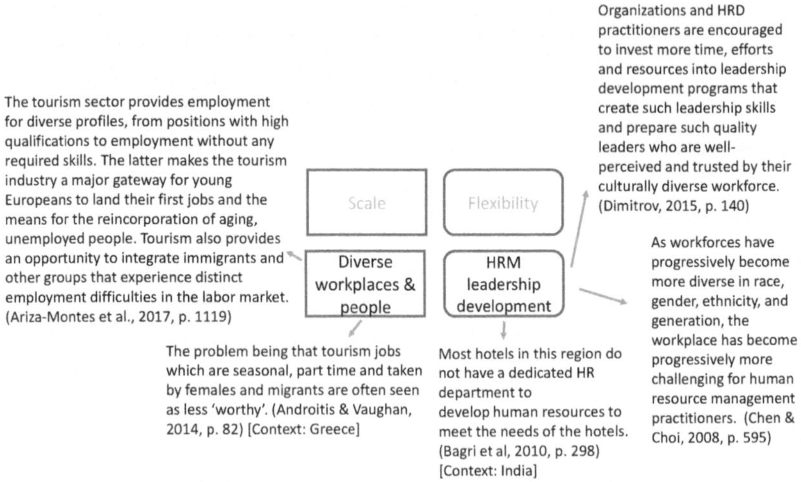

Fig. 16.3 Detailed analysis of themes II—an example of conflicting findings

functions are consolidated, delegated to supervisors and managers rather than an HR professional, or given wider remits allowing for less time to focus on, for example, an increasing diverse workforce. At the same time, as Ariza-Montes et al. (2017) suggest in Fig. 16.3, tourism can offer employment opportunities to many different groups of people within society. This is discussed in more detail below.

Figure 16.4 highlights one aspect not covered in the quantitative findings earlier. Whilst we can see a growth in literature on tourism employment, specifically in the last ten years and around aspects of sustainability and decent work, much of the literature featuring in our review stems from a relatively small number of researchers (Tom Baum's work stands out in particular in this regard), though an increasing number of other researchers are focussing on the sustainability dimension of tourism employment. Figure 16.4 also helps to illustrate that much of the research into sustainable tourism employment is conceptual and has yet to be backed up with empirical research. Whilst this might help engage industry in finding ways to start tackling the sustainability of tourism employment, it perhaps challenges any conception that decent work in tourism is actually happening 'on the ground' (for an exception to this, see De Beer et al. 2014).

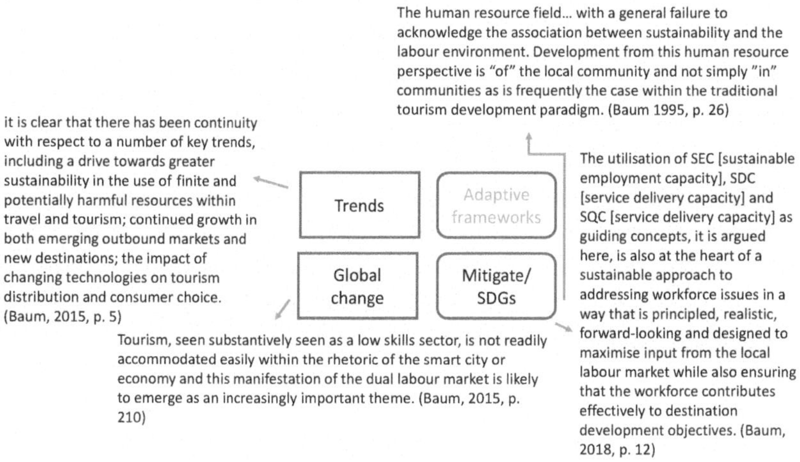

Fig. 16.4 Detailed analysis of themes III

Discussion and Conclusions

The qualitative systematic review and its findings were intended to provide relevant information for tourism practitioners within academia and industry in Sweden. Results have been presented in a number of industry forums, including at Almedalsveckan, a yearly political festival that involves all of the major political parties and large numbers of meetings organised and hosted by businesses, NGOs, academic institutions, trade unions, industry associations, charities and other organisations (see Almedalsveckan 2019). However, how the systematic review fits within the Swedish purview is perhaps most germane.

Sweden's tourism industry is the fastest growing service industry (Business Sweden 2016) and its export value has increased at a faster rate than other export goods and services: in 2015 there was a growth of 17.9% for tourism compared to a 7.7% growth for other export products and services (Swedish Agency for Economic and Regional Growth 2016). Whilst domestic tourism still dominates, the Swedish government is increasing its focus on international visitors (SOU 2017:95). This focus sees tourism as a sustainable development tool for Sweden, specifically in more rural and peripheral areas (SOU 2017: 95).

Alongside this are a number of industry initiatives that promote better employment conditions. For instance, in late October 2019 HRF, the Swedish Hotel and Restaurant Union, in conjunction with Visita, a Swedish tourism industry association, launched an industry code to provide support and guidance to (specifically small and medium-sized) businesses in tourism on dealing with human resource issues, specifically the hiring, development and dismissal of employees (HRF 2019a). HRF has also recently extended their programme 'A World of Opportunities' which provides professional Swedish language tuition to housekeeping staff (HRF 2019b).

Whilst these initiatives illustrate that decent working conditions are being sought after and encouraged in Sweden, what is less obvious are the issues behind these initiatives. For instance, Visita suggests that over 40% of those employed in hotels and restaurants were born overseas, i.e. they are migrants. This equates to over 17% of Sweden's total workforce. Add into worrying figures of labour slaves in hotels and restaurants in Europe (O'Regan 2019) and we can perhaps wonder if these initiatives hide wider problems within tourism. Linge, Furunes, Baum and Duncan's chapter in this volume highlights issues with integration of migrant housekeeping staff in Norway for instance. Whilst HRF's programme of professional Swedish language tuition is a move in the right direction, it perhaps hides a wider problem about who is employed in these roles and how they are integrated into the workplace and wider society. Secondly, as Heldt Cassel et al. (2018) suggest, career paths in Swedish hotels are problematic. Outside of larger urban areas, occupational mobility and professional development opportunities remain very limited. Thus whilst Ariza-Montes et al. (2017) illustrate that tourism can provide job creation for different groups in society, we still need to consider, as Joppe (2012) and Mooney (2018) highlight, the differential opportunities for migrants, women and other minority groups employed within the tourism industry.

Perhaps a more promising outlook can be seen through looking at the industry. Sigtunahöjden, a hotel and conference centre on the outskirts of Stockholm, aims to be a role model for sustainable work. They have chosen to adopt the Sustainable Development Goals within their sustainability vision and strategy (Sigtunahöjden 2019) and are focussing specifically on Goals 8 (Decent Work and Economic Growth), Goal 15 (Life on land) and Goal 17 (Partnerships for the Goals). In Goal 8 they state that they are inclusive workplace with a relaxed atmosphere and two of their current areas of focus include offering opportunities to people with special needs and youth employment (Sigtunahöjden 2019). Sigtunahöjden's work, showcased in the industry newletter 'Turism Nytt (2019, October 27) is illustrative of Sweden's wider engagement with sustainability (Schieler 2019). It is this wider engagement with sustainability that is perhaps also the 'problem'. The long-term engagement with sustainability in Swedish—and all of the Nordic countries—(Schieler 2019) means that government and industry are aiming to increase their already good practices, as evidenced, for instance, by Sweden's meaningful engagement with the UN's Sustainable Development Goals (Statistics Sweden 2017). However, in doing so, as suggested above, there are elements that remain difficult to 'measure' or quantify and decent work or sustainable tourism employment is one of these areas (Statistics Sweden 2017). Perhaps the question that needs to be asked is what 'decent work' in tourism looks like in Sweden? Who is responsible for ensuring that there is decent work? Companies such as Sigtunahöjden, who are engaging with the SDGs and working towards ensuring decent work for their employees is a step in the right direction. But how, in rural and peripheral areas, do communities ensure that tourism employment is 'decent' and, again, who is responsible for ensuring that the work is 'decent' in these areas, regions and destinations? If this is linked back to aspects of sustainable human resource development, then the sustainability of tourism employment is an issue that involves industry, local, regional and national government and local communities at the very least (Jitherndran and Baum 2000).

To conclude, we can perhaps take the view, as Hollenhorst et al. (2014, p. 312) do, that 'tourism in its current form poses a serious threat not only to global sustainability, but also to the sustainability of our homeplaces'. If we consider this specifically in relation to the sustainability of tourism employment, then as Robinson et al. (2019, p. 1021) suggest, the growing precarity of much tourism employment is underpinned because tourism has 'seeming grown incompatible with an authentic sustainability vision'. However, it seems pointless to simply point out how 'badly' tourism continues to do in its engagement with sustainability and, despite best efforts, with the SDGs. Rather, and specifically in the Swedish context, initiatives such as those outlined above by Visita, HRF and businesses such as Sigtunahöjden illustrate the possibilities of how we can engage significantly with the issues of decent work and the sustainability of tourism employment. Two points remain. Firstly, the neglect of research into decent work and sustainable tourism employment needs to be addressed—both globally and in the Swedish, and Nordic context. As the chapters in this volume illustrate, that neglect is being rectified. Secondly and finally, it worth reiterating Baum's (2018) point: If there is meaningful engagement at all scales on the issues surrounding the drive for employment to be sustainable—including local infrastructure, housing, social welfare and transport—then perhaps it is more than possible for decent work to be a realistic outcome for businesses and communities in Sweden and globally.

Acknowledgements The research used in this chapter was funded by FORMAS (the Swedish Research Council for Sustainable Development) under grant 2017-02038.

References

Alberti, G., I. Bessa, K. Hardy, V. Trappmann, and C. Umney. 2018. In, against and beyond precarity: Work in insecure times. *Work, Employment & Society* 32 (3): 447–457.

Almedalsveckan. 2019. Almedalen Week. Retrieved from https://www.almedalsveckan.info/english.

Andriotis, Konstantinos, and David Roger Vaughan. 2004. The tourism workforce and policy: Exploring the assumptions using crete as the case study. *Current Issues in Tourism* 7 (1): 66–87. https://doi.org/10.1080/13683500408667973.

Ariza-Montes, Antonio, Juan M. Arjona-Fuentes, Rob Law, and Hessop Han. 2017. Incidence of workplace bullying among hospitality employees. *International Journal of Contemporary Hospitality Management* 29 (4): 1116–1132. https://doi.org/10.1108/IJCHM-09-2015-0471.

Australian Business Dean's Council. 2019. ABDC Journal Quality List. Retrieved from https://abdc.edu.au/research/abdc-journal-list/.

Baum, T. 2007. Human resources in tourism: Still waiting for change. *Tourism Management* 28 (6): 1383–1399.

Baum, T. 2015. Human resources in tourism: Still waiting for change? A 2015 reprise. *Tourism Management* 50: 204–212.

Baum, T. 2018. Sustainable human resource management as a driver in tourism policy and planning: A serious sin of omission? *Journal of Sustainable Tourism* 26 (6): 873–889. https://doi.org/10.1080/09669582.2017.1423318.

Baum, T., C. Cheung, H. Kong, A. Kralj, S. Mooney, N. Thi Thanh, S. Ramachandran, and M. Ružic Siow. 2016a. Sustainability and the tourism and hospitality workforce: A thematic analysis. *Sustainability* 8: 809.

Baum, T., A. Kralj, R. Robinson, and D. Solnet. 2016b. Tourism workforce research: A review, taxonomy and agenda. *Annals of Tourism Research* 60: 1–22.

Bergene, A.C., K. Boluk, and E. Buckley. 2015. Examining the opportunities and challenges of union organisation within the hospitality industry. In *A Hospitable World? Organising Work and Workers in Hotels and Tourist Resorts*, ed. D. Jordhus-Lier and A. Underthun, 195–212. London: Routledge.

Christian, Michelle. 2017. Protecting tourism labor? Sustainable labels and private governance. *GeoJournal* 82: 805–821. https://doi.org/10.1007/s10708-016-9717-z.

BFUF. 2018. #8 Tillväxtbransch I förändring: prognoser för kometensbehov inom hotel- och restaurangbranschen till år 2030. Sammanfattning.

Business Sweden. 2016. Travel and tourism in Sweden: Sector overview. Retrieved from https://www.business-sweden.se/globalassets/invest-new/reports-and-documents/travel-and-tourism-in-sweden—sector-overview-2016.pdf.

De Beer, A., C. Rogerson, and J. Rogerson. 2014. Decent work in the South African tourism industry: Evidence from tourist guides. *Urban Forum* 25 (1): 89–103.

Duncan, T., D. Scott, and T. Baum. 2013. The mobilities of hospitality work. *Annals of Tourism Research* 41: 1–19. https://doi.org/10.1016/j.annals.2012.10.004.

Eposto, Alexis, and John Annakis. 2016. From standard jobs to 'green' jobs? A strategy for developing markets. *Corporate Ownership and Control* 14 (1): 219–229.

Fulton, L. 2013. Worker representation in Europe. Labour Research Department and ETUI. Produced with the assistance of the SEEurope Network. Retrieved from http://www.worker-participation.eu/National-Industrial-Relations.

Gustafsson, A.-K., and M. Kullander. 2015. Sweden: Fall in union density. Brussels: Eurofound. Retrieved from https://www.eurofound.europa.eu/publications/article/2015/sweden-fall-in-union-density.

Heldt Cassel, S., M. Thulemark, and T. Duncan. 2018. Career paths and mobility in the Swedish hospitality sector. *Tourism Geographies* 20: 29–48. https://doi.org/10.1080/14616688.2017.1402946.

Hollenhorst, S.J., S. Houge-Mackenzie, and D.M. Ostergren. 2014. The trouble with tourism. *Tourism Recreation Research* 39 (3): 305–319. https://doi.org/10.1080/02508281.2014.11087003.

HRF. 2019a. Fack och arbetsgivare lanserar brascnhod [Unions and employers launch industry code]. Retrieved from https://news.cision.com/se/hotell–och-restaurangfacket/r/fack-och-arbetsgivare-lanserar-branschkod,c2934141.

HRF. 2019b. Kostnadsfri yrkessvenska för hotell- och restaurangbranschen i Jönköping, Kalmar, Västervik och Växjö [Free professional Swedish for the hotel and restaurant industry in Jönköping, Kalmar, Västervik and Växjö]. Retrieved from https://news.cision.com/se/hotell–och-restaurangfacket/r/kostnadsfri-yrkessvenska-for-hotell–och-restaurangbranschen-i-jonkoping–kalmar–vastervik-och-vaxj,c2930805.

International Labour Organisation. n.d. Retreived from http://www.ilo.org/global/About_the_ILO/Mainpillars/WhatisDecentWork/lang–en/index.htm.

International Labour Organisation. 2019. Time to act for SDG 8: Integrating decent work, sustained growth and environmental integrity. Geneva: International Labour Office. Retrieved from https://www.ilo.org/wcmsp5/groups/public/—dgreports/—inst/documents/publication/wcms_712685.pdf.

Janta, H., A. Ladkin, L. Brown, and P. Lugosi. 2011. Employment experiences of Polish migrant workers in the UK hospitality sector. *Tourism Management* 32 (5): 1006–1019.

Jithendran, K., and T. Baum. 2000. Human resources development and sustainability—The case of Indian tourism. *International Journal of Tourism Research* 2: 403–421.

Joppe, M. 2012. Migrant workers: Challenges and opportunities in addressing tourism labour shortages. *Tourism Mangement* 33 (3): 662–671.

Kramar, R. 2014. Beyond strategic human resource management: Is sustainable human resource management the next approach? *International Journal of Human Resource Management* 25: 1069–1089.

Ladkin, A. 2011. Exploring tourism labour. *Annals of Tourism Research* 38 (3): 1135–1155.

Lugosi, P., P. Lynch, and A. Morrison. 2009. Critical hospitality management research. *The Services Industry Journal* 29 (10): 1465–1478.

Mair, J., B. Ritchie, and G. Walters. 2016. Towards a research agenda for post-disaster and post-crisis recovery strategies for tourist destinations: A narrative review. *Current Issues in Tourism* 19 (1): 1–26. https://doi.org/10.1080/13683500.2014.932758.

McDowell, L., A. Batnitzky, and S. Dyer. 2009. Precarious work and economic migration: Emerging immigrant divisions of labour in Greater London's service sector. *International Journal of Urban and Regional Research* 33 (1): 3–25.

Mongeon, Phileppe, and Adèle Paul-Hus. 2016. The journal coverage of Web of Science and Scopus: A comparative analysis. *Scientometrics* 106: 213–228.

Mooney, S. 2018. Jobs for the girls? Women's employment and career progression in the hospitality industry. In *Handbook of Hospitality Human Resource Management*, ed. R. Burke and J.M. Christensen Hughes, pp. 184–215. London: Edward Elgar Publishing.

Nickson, D., C. Warhurst, A.M. Cullen, and A. Watt. 2003. Bringing in the excluded? Aesthetic labour, skills and training in the 'new' economy. *Journal of Education and Work* 16 (2): 185–203.

O'Regan, Michael. 2019. Airbnb must face the facts: Human trafficking an dmodern slavery happen in rented accommodation. *The Conversation*. Retrieved from https://theconversation-com.cdn.ampproject.org/c/s/theconversation.com/amp/airbnb-must-face-the-facts-human-trafficking-and-modern-slavery-happen-in-rented-accommodation-124933.

Popay, J., Roberts, H., Sowden, A., Petticrew, M., Arai, L., Rodgers, M., and Duffy, S. 2006. Guidance on the conduct of narrative synthesis in systematic reviews. A product from the ESRC Methods Programme. Version 1.

Robinson, R.N.S., A. Martins, D. Solnet, and T. Baum. 2019. Sustaining precarity: Critically examining tourism and employment. *Journal of Sustainable Tourism* 27 (7): 1008–1025. https://doi.org/10.1080/09669582.2018.1538230.

Rubery, J., D. Grimshaw, A. Keizer, and M. Johnson. 2018. Challenges and contadictions in the 'normalising' of precarious work. *Work, Employment & Society* 32 (3): 509–527.

Saarinen, J., and A.M. Gill. 2019. Introduction: Placing resilience in the sustainability frame. In *Resilient Destinations and Tourism: Governance Strategies in the Transition Towards Sustainability in Tourism*, ed. J. Saarinen and A.M. Gill, 3–12. Abingdon: Routledge.

Scheyvens, R., G. Banks, and E. Hughes. 2016. 'The private sector and the SDGs: The need to move beyond 'business' as usual'. *Sustainable Development* 24 (6): 371–382. https://doi.org/10.1002/sd.1623.

Schieler, Max. 2019. Sweden again tops national sustainability poll. Robeco. Retrieved from https://www.robeco.com/en/insights/2019/07/sweden-again-tops-national-sustainability-poll.html.

Sigtunahöjden. 2019. Sigtuna and the sustainable development goals. Retrieved from https://sigtunahojden.se/en/sustainability/sustainable-development-goals/.

Solnet, D., D. Nickson, R. Robinson, A. Kralj, and T. Baum. 2014. Discourse about workforce development in tourism—An analysis of public policy, planning, and implementation in Australia and Scotland: Hot air or making a difference? *Tourism Analysis* 19: 609–623.

SOU (Statens Offentliga Utredningar). 2017:95. Ett Land att Besöka: En samlad politik för hållbar turims och växande besöksnäring. Stockholm: SOU.

Statistics Sweden. 2015. The future population of Sweden 2015–2060. Stockholm, Sweden. Retrieved from http://www.scb.se/Statistik/_Publikationer/BE0401_2015I60_BR_BE51BR1502.pdf.

Statistics Sweden. 2017. Statistical follow-up of the 2030 agenda for sustainable development. Stockholm, Sweden. Retrieved from http://www.scb.se/contentassets/4861167095b54ea686074be19d75a5cb/agenda_2030_eng.pdf.

Swedish Agency for Economic and Regional Growth. 2016. *Tourism in Sweden 2015*. Stockholm, Sweden. Retrieved from https://tillvaxtverket.se/download/18.59ff66f1157d260336b52d0c/1477920508755/info_0657_tillg.pdf.

UNEP and UNWTO. 2005. *Making tourism more sustainable—A guide for policy makers*. Paris, France and Madrid, Spain. Retrieved from http://www.unep.fr/shared/publications/pdf/DTIx0592xPA-TourismPolicyEN.pdf.

United Nations. n.d.a. Millenium development goals 2015 and beyond. Retrieved from https://www.un.org/millenniumgoals/.

United Nations. n.d.b. The sustainable development agenda. Retrieved from https://www.un.org/sustainabledevelopment/development-agenda/.

United Nations. 2019. Sustainable development goal 8. Retrieved from https://sustainabledevelopment.un.org/sdg8.

van Dyk, S. 2018. Post-wage politics and the rise of community capitalism. *Work, Employment & Society* 32 (3): 528–545.

Winchenbach, A., P. Hanna, and G. Miller. 2019. Rethinking decent work: The value of dignity in tourism employment. *Journal of Sustainable Tourism* 27 (7): 1026–1043. https://doi.org/10.1080/09669582.2019.1566346.

WTTC (World Travel and Tourism Council). 2018. Travel and tourism: Economic impact 2018 Sweden. Retrieved from: https://www.wttc.org/-/media/files/reports/economic-impact-research/countries-2018/sweden2018.pdf.

17

A Labour Regime Perspective on Workforce Formation in Nordic Tourism: Exploring National Tourism Policy and Strategy Documents

Dorothee Bohn and Cecilia De Bernardi

Introduction

Perhaps the most prevalent argument in support of tourism development all over the globe is the creation of employment. Indeed, tourism and hospitality (T&H) is very labour intensive and currently one out of ten jobs worldwide, equalling the occupation of 292 million people, is linked to the T&H sector (World Bank Group 2017). As a characteristically fragmented industry, tourism generates a wide spectrum of jobs. People pursue careers in a diverse range of tourism sub-sectors, such as programme services, hospitality, transportation or entertainment, which require many different professional skills. Moreover, wage labour is not just tied to formal employment in multinational corporations or

D. Bohn (✉)
Department of Geography, Umeå University, Umeå, Sweden
e-mail: dorothee.bohn@umu.se

C. De Bernardi
Dalarna University, Falun, Sweden
e-mail: cdb@du.se

A. Walmsley et al. (eds.), *Tourism Employment in Nordic Countries*,
https://doi.org/10.1007/978-3-030-47813-1_17

small- and medium-sized firms but micro-entrepreneurship and informal work are on the rise, also due to the growing platform economy (Robinson et al. 2019). Regarding the Nordic countries, the T&H labour market is structurally determined by seasonality, setting it apart from other economic sectors. This holds ramifications for employers and destination communities regarding labour availability and capacity building as well as for employees with respect to income security (Rantala et al. 2019). High staff turnover poses challenges for firms and destinations to retain a skilled workforce and to maintain competitive tourism production processes (Heldt Cassel et al. 2018). From many workers' standpoint, non-permanent employment offers an insufficient basis for livelihood sustenance. Thus, the Nordic tourism workforce is usually young and mobile without any explicit tourism education (e.g. Chen and Wang 2015) whereas those people who received tourism specific training often do not remain employed in the sector as Åberg (2017) shows in the case of Sweden.

Within tourism research, the sector's workforce is a generally neglected area. The majority of studies adopt applied or micro-level perspectives on themes such as employee behaviour or management practices while macro-level examinations that uncover organisational and institutional issues shaping T&H labour are fairly scarce (Baum et al. 2016). In addition, there exist significant practical as well as research gaps with respect to the role of workforce in sustainable tourism policy and planning (Baum 2018). Notwithstanding sustainability schemes are administratively and operationally executed by employees, Robinson et al. (2019) note that contemporary tourism research favours the economic-environmental dimensions while overlooking the social including the increasingly precarious nature of much tourism employment. In their view, a critical step forward in contemporary tourism workforce research is to elucidate the factors leading to and maintaining the common stigma of work in the T&H sector, namely poor employment conditions, meagre salaries, limited career options, insufficient union representation and loose regulation (Robinson et al. 2019).

Hence, this chapter presents a macro-outlook to tourism workforce formation by drawing upon the concept of labour regimes. We analyse how the reproduction, mobilisation, motivation, and utilisation of workforces for tourism are considered and advanced on a public policy level in Nordic countries. Empirical materials consist of Finnish, Swedish and Norwegian national tourism strategies and public policy documents. The content analysis contributes to a better understanding of tourism policy efforts that precondition the tourism labour regime. These results are then discussed with respect to the implications of tourism workforce policies for Nordic welfare states and society.

A Labour Regime Perspective on Workforce Formation

In Western nation states, the shift from Fordist to Post-Fordist production and consumption along with the effects of globalisation and the embrace of neoliberalism has altered labour markets drastically. Formal work, characterised by permanent, full-time and socially protected employment, has become increasingly replaced by flexible and non-standard (e.g. part-time, temporary work or zero-hour contracts) employment relationships (Cicca 2015). These new modes of labour have not just transformed the professional lives of people, flexible and non-standard work relations continue to affect also the livelihoods, mobility and household strategies of individuals, families and societies (Taylor and Rioux 2018). Moreover, flexible work is frequently associated with the term precariousness because unstable and contingent employment undermines socio-economic security and deprives workers of control over their working conditions (Sippola and Kall 2017). Precarious work is particularly widespread in primary production and manufacturing but also deeply engrained in the global T& H sector and on the rise in the platform or gig economy (Robinson et al. 2019).

In much of the traditional economic literature, social and political dynamics necessary to produce workforces that operate inside aforementioned flexible labour markets remain unmentioned because labour is simply considered a "factor of production" (Taylor and Rioux 2018). The concept of labour regimes offers in this regard a more holistic analytical framework. According to Andræ and Beckman (1998, p. 23), labour regimes

> summarize the complex of institutions, rules and practices that regulate the relations between labour and capital as they manifest themselves in the work place. (…) [Central to the analysis of labour regimes is how] this "mode of regulation" is influenced and conditioned at the level of society at large, by the nature of enterprises and entrepreneurial classes, local and national politics, the interventions of the state and organized interests, as well as by the way in which labour is recruited, trained and supported by family, community, and state outside the work place.

Knutsen and Hansson (2010, p. 156) highlight that the labour regime concept unites "the micro-politics of the workplace and the macro-politics of the state" and permits the exploration of how workforces are reproduced, mobilised, motivated and utilised through the interaction of social institutions, ranging from the family and the educational sector to the social security system, state policy and legislation (Cicca 2015). Taylor and Rioux (2018) outline these conceptual components of labour regimes in the following way:

- Reproduction of workforce happens in the private sphere of the household but also through the social interventions of the state for instance through welfare schemes.
- Mobilisation refers to how a specific segment of the population is drawn into a particular occupation which has distinct professional qualifications and criteria plus social status. The state regulates the conditions of each occupational branch.
- Motivation of workforce can be understood as the socio-political discourse around what constitutes legitimate work and how workers perceive their roles.
- Utilisation denotes how workforce is deployed and controlled for the production of goods and services. Central is the length and intensity of the work, the timing and the specific work task including the use of technology.

A widespread approach in respective research is to conceive labour regimes as hegemonic schemes seeking to maintain workers' consent to uneven and exploitative employment conditions particularly in transition economies (e.g. Andræ and Beckman 1998; Arnold and Campbell 2017; Smith et al. 2018). In the context of Western nations, Cicca (2015) observes that specific labour regime analyses are fewer in number because scholars address work and labour market generally under the umbrella of the welfare state assuming that the latter creates distinct employment structures. However, such take might fall short in uncovering the commodification of labour, the mobilisation of different social groups into the workforce plus the influence of socio-political and economic processes generated at higher geographical scales on local workplaces (Knutsen and Hansson 2010).

Although there exist organisational as well as normative differences between the Nordic states' welfare models, in its essence the traditional Nordic welfare state seeks "to improve the ability of society to master its problems and to enrich and equalize the living conditions of individuals and families. In social policy, the cornerstone of the model is universalism" (Erikson et al. 1987, cited in Greve 2007 p. 44). The comprehensive welfare state is a crucial component in the success and the international attractiveness, referring to the high living standards, family friendliness and free education, of the Nordic counties. This model is largely financed through taxation. Yet, the sustainability of the welfare state is not only increasingly challenged by demographic changes, globalisation and financial crises but also by a growing tendency towards neoliberal policy-making and short-term planning cycles evolving around privatisation, marketisation and deregulation (Peck 1996). Market solutions are on the rise in governing the social sphere but the gap between upper-middle class and marginalised groups is widening (Abrahamson 2010). Respective research focuses more on the resilience of the welfare state to external events whereas internal developments, for instance the effects of growing precariousness and low-quality employment, receive less attention. Tourism is in this respect an interesting case because the sector is hailed as a generator of employment while jobs are mostly located along the lines of low-income service work. Given the public–private governance structures plus the heterogeneity of T&H, labour regimes present therefore an expedient theoretical platform for the enquiry of the sector's workforce formation arising in the intersection of capitalist accumulation, state intervention and society. Before examining what kind of labour regimes are aspired in governmentally ratified tourism planning documents and how these polices support Nordic welfare states' principles, the chapter provides brief overviews of workforce and working conditions in Finland's, Norway's and Sweden's tourism industry.

Workforce in Nordic Tourism

With respect to the peripheral North, tourism is often emphasised as a sector that can substitute the loss of workplaces in primary production, public administration and extraction industries (Müller and Ulrich 2007). Indeed, tourism is currently a segment of the economy showing strong growth in Nordic countries and government policies aim at further expansion of the sector (Árnadóttir 2019). Nonetheless, T&H creates principally low-paid and seasonal jobs that are usually filled by women, youngsters or migrants (Underthun and Jordhus-Lier 2018). This Nordic tourism workforce is characterised by a high level of mobility, which is, on the one hand determined by the seasonal nature of employment (Tuulentie and Heimtun 2014) but on the other also a precondition for career advancement (Heldt Cassel et al. 2018). Lifestyle-driven motivations for seeking employment in tourism are also common in Nordic countries. Chen and Wang (2015) report that migrant seasonal tourism workers are not only attracted by the high wage levels in northern Norway, but the amenity rich Arctic landscapes and the social ties formed during previous work stays are decisive as well. Consumption-driven lifestyle entrepreneurship among in-migrants is also a common phenomenon in Nordic regions due to the low entry barriers of T&H (e.g. Carson et al. 2018). Many companies, especially safari firms, do not even rely on established tourism education, but recruit people who pursue outdoor activities in their leisure time or train staff entirely themselves (Valkonen et al. 2013). Among professionals who received a formal tourism education, the tendency of leaving the sector for employment in other industries is high as Åberg (2017) finds in the context of Sweden. A crucial factor might be that even full-time jobs in Sweden's tourism industries are generally lower paid than professions in other areas (Brandt 2018).

In addition to meagre salaries and zero-hour working contracts, there are also other examples of precarious working conditions and social dumping in Nordic tourism. For instance, OnsØyen et al. (2009) report that hotel room attendants in Norway complain about poor management practices, insufficient communication, time pressure, unfamiliarity with health, environment and safety regulations plus confusion around wages and work hours. Edelheim (2019) describes the case of a resort in Finnish Lapland where the owner avoided Finnish-born staff but hired first generation immigrants and temporary workers, posted by foreign staffing companies, in order to cut down labour expenses and to circumvent unionisation attempts. Another ethically questionable practice concerns the replacement of paid seasonal workers with volunteers during the peak winter season. Brennan (2018) illustrates how some dog-sledding firms in Finnish Lapland promise the worker-tourist unforgettable Arctic winter adventures, food and accommodation in exchange for free but fairly unprotected work. Unionised worker protection, which has a strong tradition in Nordic countries, is not only difficult to access but also impractical for a highly mobile and fragmented workforce (Underthun and Jordhus-Lier 2018).

The available literature portrays the Nordic tourism workforce as highly mobile and oftentimes driven by lifestyle-consumption desires while precarious working conditions are not uncommon. However, these descriptive studies usually do not examine the underlying structural and institutional arrangements in a specific geographical location plus the interests of capital and state at play in the formation of tourism workforces. These macro issues are also central in unfolding how the common stigma of tourism labour (Baum et al. 2016) arise and are perpetuated. Lastly, examining these matters offers insights into the political nature and the societal implications of labour.

Research Framework

In order to gain insights into how the reproduction, mobilisation, motivation and utilisation of workforces for tourism is advanced on a public policy level in Nordic countries, national sector specific strategy and policy documents were analysed according to the following research questions:

- What do Nordic strategy and policy documents tell about how workforces for tourism are reproduced, mobilised, motivated and utilised?
- What are the implications of these tourism labour regimes for Nordic welfare states?

Comparing tourism workforce formation in the Nordic countries makes sense because these nations are similar in their social welfare approach but there exist some differences regarding labour market policies (Elvander 2002). National tourism policy documents, outlined in Table 17.1, were chosen because workforce is emphasised as a crucial input to tourism production and labour planning is usually a national-level responsibility. All selected documents are either produced or ratified by Nordic governments. Due to the limits of this chapter, only Finland, Norway and Sweden were included in the analysis. For Finland, the national tourism strategy for 2019–2028 (TEM 2019) plus the most recent tourism industry report (Jänkälä 2019), both created by the Ministry of Economic Affairs and Employment (TEM), were chosen.

There are three different tourism planning documents available for Norway, namely a national destination strategy and a white paper published by Norwegian Ministry of Trade, Industry and Fisheries in addition to a roadmap to sustainable tourism jointly crafted by Norway's key public–private tourism stakeholders. Regarding Sweden, the objective is to commence the working process for a national tourism strategy in 2020 (Swedish Tourism Ltd., n.d.) but the sample for this analysis includes a tourism policy inquiry delivered to the Swedish government, the national strategy for the Swedish tourism industry produced by

Table 17.1 Selected Nordic tourism strategies and policy documents

Country	Tourism strategy/tourism public policy document
Finland *Ministry of Economic Affairs and Employment (TEM)*	Achieving more together—Sustainable growth and renewal in Finnish Tourism. Finnish tourism strategy 2019–2028 and measures 2019–2023 (2019) Industry report tourism (Jänkälä 2019)
Norway	Destination Norway
Norwegian Ministry of Trade, Industry and Fisheries	National strategy for the tourism industry (2012)
Norway's public-private tourism stakeholders	Towards sustainable travel and tourism in Norway—A roadmap (2017)
Ministry of Trade, Industry and Fisheries	White paper: Experiencing Norway—A unique adventure (2017)
Sweden *Swedish Governmental Offices*	A land to visit—An overall policy for sustainable tourism and a growing tourism industry—Inquiry delivered to the Swedish government in 2017
Swedish tourism Ltd. (Public–private stock company owned by Jämtland Härjedalen tourism, Visita and SCR Svensk Camping)	'National strategy for the Swedish tourism industry' Interim analysis 'National strategy for the Swedish tourism industry'—this is how we increase the pace before 2020
Swedish Agency for Economic and Regional Growth	Competency supply and need in the Swedish tourism industry

public–private tourism stakeholders and an analysis report of competency supply and education for the Swedish tourism industry issued by the Swedish Agency for Economic and Regional Growth. Common to all analysed Nordic policy documents is the objective to foster a sustainable and economically growing tourism sector. Concrete strategies focus primarily on digitalisation, entrepreneurship, coordination, funding, accessibility, marketing as well as product and destination development.

Qualitative content analysis was chosen to answer the research questions in order to provide replicable and valid textual inferences (Krippendorff 2004). The aim of qualitative content analysis is to identify categories that summarise the informational content of the data set (Drisko and Maschi 2016). First, we selected all paragraphs that contained the words "work", "labour", "workforce", "employment" and “job”. Secondly, the content of the relevant paragraphs was coded. These in vivo codes were then in the third step summarised into themes.

Workforce Formation from a Public Policy Viewpoint

The different strategic approaches of Finland, Norway and Sweden to tourism workforce policy and the resulting labour regimes are depicted in Table 17.2. Mobilisation as well as motivation of people to join and to stay employed within tourism are emphasised the most. Utilisation and reproduction of workforce remain less discussed as these issues might be regarded as lying outside the reach of a tourism strategy that primarily targets immediate matters affecting the economic situation and competitiveness of the industry. In Nordic countries, the utilisation of workforce is generally regulated by sector-specific agreements and labour union representation in addition to common labour laws while the reproduction of workers is determined principally by the welfare state and social institutions.

Finland

The main difference of Finland’s recently launched tourism strategy to its previous versions concerns a stronger emphasis on a sustainable T&H industry. Economic growth is still the core of the strategy but responsibility is added as a qualitative component as the mission statement illustrates. Tourism is envisioned as “a responsible and growing service business that creates wellbeing and employment throughout the year in Finland” (TEM 2019, p. 23 author’s translation). Indeed, the generation of employment along with the economic potency of tourism, including aggregate and multiplier effects, are put centre stage to justify the sector’s expansion in the country. Yet, the Finnish tourism industry, especially hospitality businesses, currently face difficulties in recruiting enough skilled employees and this lack of workforce is identified as a

Table 17.2 Tourism labour regimes created in Nordic tourism policy and strategy documents

Country	Mobilisation	Motivation	Utilisation	Reproduction	Labour regime characterisation
Finland	• Improving the T&H sector's attractiveness and image in order to secure a year-round skilled workforce • Recruitment of international workforce. Promoting tourism employment among international students in Finland • Strengthening and diversifying flexible education and vocational training on all educational levels in order to satisfy companies' skill demand (preferred is short-term, flexible and working-life oriented training)	• Imbalance in wealth motivates foreign workers to seek employment in T&H • Sustainability values held by tourism firms might attract employees with the same values • Lifestyle and leisure motivation for tourism employment	–	–	Mobile and segmented labour regime that meets the T&H industries' requirements for different types of employees
Norway *Destination Norway National strategy for the tourism industry (2012)*	• T&H as a career path for the young • Education at different levels traineeships in the tourism industry, tourism employee competence development and life-long learning	Enhancing the status of tourism work both financially and in terms of stimulating tasks	• Staff-driven innovation •Entrepreneurship	• Good wages are a precondition for making a living in a high-cost country • Stable jobs are a requirement for achieving sustainable tourism • T&H offers livelihood opportunities in rural regions	Protected, standard employment

continued

Table 17.2 continued

Country	Mobilisation	Motivation	Utilisation	Reproduction	Labour regime characterisation
Towards sustainable travel and tourism in Norway—A roadmap (2017) and White paper: Experiencing Norway —A unique adventure (2017)	Education	–	• Deregulation inside the Working Environment Act • Tourism work contributes to national economy	–	Workforce excluded from sustainability planning
Sweden	• Keeping graduates from tourism-specific education employed in the sector • State funding for professional and adult education, tourism-related projects and research • Improved coordination and simpler hiring procedures • Workforce entry jobs for immigrants	• Good jobs with adequate salaries for qualified personnel • Workforce entry jobs for immigrants	–	–	Segmentation in order to provide better jobs for qualified personnel while generating also low-skill employment for immigrants

substantial factor hampering the future growth of tourism (Jänkälä 2019). The reasons for labour shortage in T&H originate in "job seekers' skill deficiency, difficulties in finding open vacancies, the welfare trap, the sector's weak profitability and bad image as an employer, irregular and seasonal work and residency related problems" (TEM 2019, p. 52, author's translation). In 2018, the Ministry of Economic Affairs and Employment of Finland funded therefore the project "Matkailudiili" [tourism deal], which aims at optimising the link between employers and employees through image campaigns, recruitment events and match-making activities that seek to attract young people by emphasising the lifestyle attributes of seasonal T&H work (matkailutyo.fi, n.d.).

Mobilisation of workforce is also a main policy focus in the national tourism strategy. The overall goal for workforce formation stated in the national Finnish tourism strategy is to "develop research-based labour market dynamics, occupational and regional mobility and a sufficient availability of a skilled tourism workforce in the right time in the right place" (TEM 2019, p. 52 author's translation). Strategic action, jointly executed by the Ministry of Economic Affairs and Employment, the Ministry of Education and Culture, the Finnish National Agency for Education, the Finnish Immigration Service, educational institutions, the employment agencies and lobby organisations, is channelled into the improvement of the tourism sector's image, easier access to a foreign workforce and flexible, working life oriented education (see Table 17.2).

Hence, the advanced labour regime for tourism can be characterised as mobile and segmented so that the T&H industries' requirements for different types of employees are met. Skilled workers for more continuous positions ought to be retained while the strategy aspires also to make it easier for tourism firms to recruit people for seasonal, low-skilled jobs from international labour markets. This access to foreign workforces was not mentioned in the antecedent tourism strategy. Segmentation is also advanced with respect to education as teaching contents vary according to the educational level but the emphasis in the document is placed again on satisfying the tourism sector's needs for job specific vocational expertise.

Moreover, the close relationship of mobilisation and the motivation of workforces becomes evident in Finland's tourism strategy. As workers are active agents in choosing a certain profession over another (Taylor and Rioux 2018), it is seen strategically important to improve the T&H industries' image as an employer to counteract high staff turnover. Motivation is also at play in Finland's effort to invite foreign workers. Labour mobility in the service sector is usually driven by an imbalance in living standards and richer countries try to attract workforce from poorer countries who accept working conditions that citizens of the host nation refuse (Taylor and Rioux 2018). Seasonal T&H jobs fall in this category of unfavourable employment. Another dimension of drawing people into the tourism workforce is to promote seasonal T&H jobs in terms of lifestyle or serious leisure among young people (see Matkailutyo.fi, n.d.).

Norway

There is a great disparity regarding the importance given to workforce in Norway's tourism strategies and the aspired labour regimes. The national destination strategy published by the Norwegian Ministry of Trade, Industry and Fisheries in 2012 pays the most attention to workforce issues inside the strategic area of "destination and experience development". The strategy highlights not only that "the industry's employees are its most important resource" but also that a permanent tourism workforce is vital for keeping rural areas populated and preserving local knowledge (Norwegian Ministry of Trade, Industry and Fisheries 2012, p. 62). Nonetheless, retaining competent workers is challenging for the sector because of seasonality. The main emphasis is on the mobilisation of workforce through promoting tourism education among "clever young people" (ibid.).

In order to motivate people to join the tourism workforce, the strategy underlines that T&H needs to be an attractive industry that provides not only good salaries but also challenging tasks (Norwegian Ministry of Trade, Industry and Fisheries 2012). Moreover, the document presents a trial project called "Tourism Development Centre" that seeks to offer training courses or work in other professions for tourism employees during the off-season while the worker remains officially employed in a tourism company. This initiative is not just supporting workforce supply for the tourism sector and cutting down public spending on unemployment benefits but a step towards the creation of better working conditions because seasonal employment "is an unsatisfactory situation for the individual employee, since it results in an unpredictable working life and little personal development" (Norwegian Ministry of Trade, Industry and Fisheries 2012, p. 15).

Concerning the utilisation of workforce, the document mentions the aspiration of Norway to foster staff-driven innovation and entrepreneurship as means to enhance the economic value of tourism. The reproduction of workforce is touched upon in terms of Norway being a high-cost country and sufficient wages are a precondition for making a living. "Good and stable jobs" are even pointed out as a requirement for achieving sustainable tourism while T&H offers livelihood opportunities which contributes to sustaining rural regions (Norwegian Ministry of Trade, Industry and Fisheries 2012, p. 24). The bodies responsible for executing these workforce policies include various government agencies, employer and employee organisations, the education sector, local, regional and national administrations and the private tourism industry.

Although the Government's White paper published in 2017 adopts a sustainability framework for its policies, not much consideration is given to the tourism workforce. Only one programme aims at utilisation, referring to pro-business deregulation inside the Working Environment Act, and at mobilisation in terms of supporting tourism education (Norwegian Ministry of Trade, Industry and Fisheries 2017). Similarly, Norway's roadmap to sustainable travel (2017) does not embrace the well-being of the workforce in sustainable planning efforts. Work occurs thematically only one time in the context of utilisation:

> Sustainable travel and tourism require that we take care of the nation's nature and culture capital, strengthen the social values, bolster pride in local communities while developing new jobs with a focus on value creation that makes travel and tourism economically viable. (p. 8)

This almost entire neglect of workforce is in stark contrast to the approach taken in the National strategy for the tourism industry (2012) which advances protected and standard employment as part of achieving sustainability in tourism.

Sweden

In the Swedish policy documents, much emphasis is placed on the mobilisation of workforce and competence development because the tourism industry will be in need of workers in the near future (Swedish Tourism Ltd. 2016). Mobilisation is expressed in three ways that are relevant to discuss. First, graduates from tourism specific education should be attracted to enter employment and stay employed in the sector. State funding is provided for professional and adult education, tourism-related projects and research (Swedish Agency for Economic and Regional Growth 2016; Governmental Offices of Sweden 2017). Secondly, the state institutions that deal with workforce mobility should improve coordination and simplify hiring procedures (Swedish Agency for Economic and Regional Growth 2016). Thirdly, Swedish T&H is seen as a sector which offers the opportunity to employ immigrants in the otherwise highly skilled Swedish job market (Bussi and Pareliussen 2015). Indeed, people who are not born in the country are statistically higher represented in certain T&H professions. "Many occupations [in the tourism sector] have lowered the requirements of formal competencies" (Swedish Agency for Economic and Regional Growth 2016, p. 30, author's translation) and the goal is to offer "work duties that do not require extensive education" (Swedish Agency for Economic and Regional Growth 2016, author's translation). There are also different programmes to assist newcomers to integrate into the workforce, such as the initiative to help foreign chefs to find work in Sweden (Swedish Tourism Ltd. 2016). Yet, immigrants are at a disadvantage since they need to acquire first country-specific skills in order to integrate in the workforce (Duvander 2001). Fast-tracking newcomers into work does not always render positive effects in terms of future labour market prospects (Butschek and Walter 2014). The Swedish strategies for immigrant tourism workforce bear therefore the risk of excluding newcomers from the higher skilled labour market through an endless cycle of state-funded working life entry programmes or keeping them in low-paid, precarious jobs (Wiesbrock 2011; Kahn 2012). These segmentation practices, foster not only inequality inside the tourism workforce, but might lead also to societal tensions in terms of creating low-quality employment for certain social groups.

Regarding the motivation of people to join the tourism workforce, the same discrepancy can be observed. On the one hand, there is the call for education, competence development and jobs that reward skills with adequate salaries. In the Inquiry (2017), the need for "good and fair terms in the work life, good work environment and there have to be possibilities of re-employment" (Government Offices of Sweden 2017, p. 136 author's translation) are addressed. The latter refers to the Swedish employment laws which stipulates that people have to be re-employed as soon as the employer has the possibility to re-hire. On the other hand, the need to employ immigrants creates the prerequisite for low-quality jobs in which people might be trapped at least until country-specific skills are acquired. Indeed, low-qualifications jobs are what first generation immigrants usually attain (e.g. Duvander 2001; Rooth and Ekberg 2006). The necessity for high employment is rooted in the so-called "Swedish model". The government defines this model as its "strategy for inclusive growth" which rests upon "a labour market that that [sic] facilitates adjustment to change, a universal welfare policy and an economic policy that promotes openness and stability" (Government Offices of Sweden, n.d., p. 1). High employment is a prerequisite in order to finance the Swedish welfare state (Government Offices of Sweden, n.d., p. 9) and state intervention in the Swedish labour market is generally extensive (Kahn 2012).

Nevertheless, there is the danger of Swedish T&H becoming cheap labour industry, which is not seen as a positive outcome in the document itself. A report published a year before from another government agency highlights that the sector does actually not lack personnel with low qualifications (Swedish Agency for Economic and Regional Growth 2016), but T&H requires higher education and advanced qualifications (Government Offices of Sweden 2017).

Discussion and Conclusions: Implications of Tourism Labour Regimes on Nordic Welfare States

The analysis revealed that primarily mobilisation of workforce is thematised in the Nordic national tourism strategies and policy documents. This stems from the overarching objective to spur the growth of the T&H sector and sufficient labour supply is portrayed as a precondition for a competitive operational environment. On the one hand, the tourism industry has a strong influence on policy-making through the common public–private partnership structures while on the other, the key premise of neoliberal governance holds that citizen well-being can be best achieved through liberated markets and entrepreneurial freedom (Harvey 2005). Hence, tourism policy and planning are seen to be most efficient when industry-friendly strategies, catering to the private sector's needs, are pursued.

In this sense, Finland and Sweden opt both for a segmented, flexible and mobile workforce so that different people fill in the T&H industries' wide range of jobs. Strategic action in Finland aims at retaining skilled year-round workers through improving tourism's image as an employer plus easing the access to an international, mobile workforce that satisfies seasonal demand. Similarly, Sweden seeks to keep people who have received a tourism education employed in the sector through more attractive wages and employment initiatives while low-skilled T&H jobs are one of the few possibilities in Sweden's high-skill labour market to include immigrants without necessary qualifications into the workforce. However, these segmentation strategies might lead to accelerating the polarisation among workers (Peck 1996) and running counter to the Nordic welfare states' socio-economic foundation of equality. Moreover, making it easier for firms to hire foreign workers opens the door for social dumping when employers try to circumvent sectoral worker protection agreements and ancillary labour costs (Frazer and Marlier 2010). At an individual level, workers might be exposed to precarious and exploitative working conditions while at a collective level, social dumping undermines the redistribution mechanisms, such as tax payments, required for a functioning welfare state system. Likewise, large-scale state-funded labour market entry programs are a burden for the welfare state as well when misused by employers and people remain trapped in such schemes (Wiesbrock 2011; Kahn 2012). In turn, eradicating the common stigmata of tourism employment, namely low wages, absent career options and irregular working times, through innovative labour market policies targeting the reduction of in-work poverty would prove more sustainable from a welfare state point of view in terms of avoiding additional expenses arising from low quality employment.

The labour regime approach outlined in Norway's national strategy for the tourism industry (2012) aiming at protected, full-time and well-paid work supports welfare state principles. However, the country's subsequent strategies address workforce issues only scantily although sustainability is presented as a core for tourism development. Indeed, not only labour but also the relationships of employment to the welfare state's social system is commonly overlooked in sustainable tourism planning (Baum 2018; Robinson et al. 2019). The only references made to societal matters in Norway's Sustainable Tourism Roadmap (2017), beside cultural protection, are unspecified goals that the T&H sector should render social and economic value to the nation. Neoliberal policy-making, including labour deregulation and the general turn towards redistributing the responsibility for sustainability execution from the state to businesses and individuals (see *Towards Sustainable Travel and Tourism in Norway* 2017, p. 2; TEM 2019) is presented in all examined tourism strategies as a way to achieve greater economic viability of tourism that serves the well-being of all people. Yet, the aspired labour regimes do not necessarily support traditional welfare state ideas and reproduction and utilisation, except for labour deregulation, of workforce receive extremely limited consideration in the analysed documents.

A major barrier to year-round employment in Nordic tourism is seasonality. Strategic action to counteract seasonal fluctuations is mostly concentrated on product development in order to sustain operations throughout the year. Thus, the improvement of employment conditions is initiated via the market instead of state-led development initiatives. The state provides principally the legal labour framework in addition to T&H education and skill development. Solnet et al. (2014) remark that training initiatives are not powerful enough to transform the problematic workplace reality of the tourism industry and they advocate for fundamental workforce polices facilitating sustainable jobs. Nonetheless, in Nordic countries there is a strong lifestyle motivation driving people to seek flexible and seasonal tourism employment. These jobs might enable people to make a living in the present but in the long run, low-income work usually does not generate sufficient savings for future retirement and represents a clear burden for the welfare state.

With respect to the outlook on workforce and tourism labour presented in the analysed documents, there might be some truth in Solnet et al.'s (2014) observation that precarious tourism employment and the sector's stigma as an unfavourable employer are most likely here to stay. Workforce policies suggest rather a lock-in situation than innovation. Except for one Norwegian tourism strategy, there is no policy action directed towards changing tourism working conditions profoundly. On the contrary, accelerating deregulation and flexibilisation serves usually capital more than the worker. Moreover, the strategies' sustainability and responsibility initiatives are solely focused on the industry and consumer plus long-term societal consequences of a seasonal, low-paid workforce are not elaborated.

As tourism workers become increasingly mobile, a crucial step for better employee protection could derive from labour unions taking not only a proactive stakeholder role in public tourism strategy-making but also in modernising their structures. For instance, portable social entitlements, internationally valid union memberships and better access to information regarding worker's rights and work legislation of a country could assist in counteracting precarious employment.

References

Abrahamson, Peter. 2010. European welfare states beyond neoliberalism: Toward the social investment state. *Development and Society* 39 (1): 61–95. http://www.jstor.org/stable/deveandsoci.39.1.61.

Andræ, Gunilla, and Björn Beckman. 1998. *Union Power in the Nigerian Textile Industry: Labour Regime and Adjustment.* Uppsala: Nordiska Afrikainstitutet.

Arnold, Dennis, and Stephen Campbell. 2017. Labour regime transformation in Myanmar: Constitutive processes of contestation. *Development and Change* 48 (4): 801–824. https://doi.org/10.1111/dech.12315.

Åberg, Kajsa G. 2017. "Anyone could do that": Nordic perspectives on competence in tourism. PhD diss. Umeå University.

Árnadóttir, Ragnheidur Elin 2019. Nordic tourism policy analysis. Nordic Council of Ministers. https://www.norden.org/sv/node/35903. Accessed on July 5, 2019.

Baum, Tom. 2018. Sustainable human resource management as a driver in tourism policy and planning: A serious sin of omission? *Journal of Sustainable Tourism* 26 (6): 873–889. https://doi.org/10.1080/09669582.2017.1423318.

Baum, Tom, Anna Kralj, Richard N.S. Robinson, and David J. Solnet. 2016. Tourism workforce research: A review, taxonomy and agenda. *Annals of Tourism Research* 60: 1–22. https://doi.org/10.1016/j.annals.2016.04.003.

Brandt, Daniel. 2018. Wage determinants in the Swedish tourism sector. *Scandinavian Journal of Hospitality and Tourism* 18 (1): 18–38. https://doi.org/10.1080/15022250.2016.1206832.

Brennan, Christopher. 2018. Northern lights instead of workers' rights: Volunteer working tourists in Finnish Lapland. *Nordic Journal of Working Life Studies* 8 (2): 43–60. https://doi.org/10.18291/njwls.v8i2.106154.

Bussi, Margherita, and Jon Kristian Pareliussen. 2015. Skills and labour market performance in Sweden. OECD Economics Department Working Papers No. 1233. OECD Publishing, Paris, https://doi.org/10.1787/5js0cqvnzx9v-en. Accessed on September 24, 2019.

Butschek, Sebastian, and Thomas Walter. 2014. What active labour market programmes work for immigrants in Europe? A meta-analysis of the evaluation literature. *IZA Journal of Migration* 3 (1): 48.

Carson, Doris Anna, Dean Bradley Carson, and Marco Eimermann. 2018. International winter tourism entrepreneurs in Northern Sweden: Understanding migration, lifestyle, and business motivations. *Scandinavian Journal of Hospitality and Tourism* 18 (2): 183–198. https://doi.org/10.1080/15022250.2017.1339503.

Chen, Joseph S., and Wei Wang. 2015. Foreign labours in Arctic destinations: Seasonal workers' motivations and job skills. *Current Issues in Tourism* 18 (4): 350–360. https://doi.org/10.1080/13683500.2014.894499.

Ciccia, Rossella. 2015. Classifying labour regimes beyond the welfare state: A two dimensional approach. In *The Changing Worlds and Workplaces*

of Capitalism, ed. Seán Ó Riain, Felix Behling, Rossella Ciccia, and Eoin Flaherty, 58–78. London: Palgrave Macmillan.
Drisko, James W., and Tina Maschi. 2016. *Content Analysis*. Oxford: Oxford University Press.
Duvander, Ann-Zofie E. 2001. Do country-specific skills lead to improved labor market positions? An analysis of unemployment and labor market returns to education among immigrants in Sweden. *Work and Occupations* 28 (2): 210–233.
Edelheim, Johan R. 2019. Worker protection, entrepreneurship and loopholes: New challenges for Finnish trade unions. In *Justice and Ethics in Tourism*, ed. Tazim Jamal. Abingdon: Routledge.
Elvander, Nils. 2002. The labour market regimes in the Nordic countries: A comparative analysis. *Scandinavian Political Studies* 25 (2): 117–137. https://doi.org/10.1111/1467-9477.00066.
Frazer, Hugh, and Eric Marlier. 2010. In-work poverty and labour market segmentation in the EU: Key lessons. Synthesis Report. European Commission. https://ec.europa.eu/social/BlobServlet?docId=8992&langId=en. Accessed on July 5, 2019.
Government Offices of Sweden. n.d. The Swedish Model. https://www.government.se/information-material/2017/06/the-swedish-model/. Accessed on October 7, 2019.
Government Offices of Sweden. 2017. *Ett land att besöka – En samlad politik för hållbar turism och växande besöksnäring*. SOU 2017:95 [A land to visit—An overall policy for sustainable tourism and a growing tourism industry—Inquiry delivered to the Swedish government in 2017]. https://www.regeringen.se/rattsliga-dokument/statens-offentliga-utredningar/2017/12/sou-201795/. Accessed on July 11, 2019.
Greve, Bent. 2007. What characterise the Nordic welfare state model? *Journal of Social Sciences* 3 (2): 43–51. https://doi.org/10.3844/jssp.2007.43.51.
Harvey, David. 2005. *A Brief History of Neoliberalism*. Oxford: Oxford University Press.
Heldt Cassel, Susanna, Maria Thulemark, and Tara Duncan. 2018. Career paths and mobility in the Swedish hospitality sector. *Tourism Geographies* 20 (1): 29–48. https://doi.org/10.1080/14616688.2017.1402946.
Jänkälä, Susanna. 2019. Toimialaraportit. Matkailu [Industry reports. Tourism]. *Työ- ja elinkeinoministeriön julkaisuja* 2019 (3). http://urn.fi/URN:ISBN:978-952-327-390-0. Accessed on November 27, 2019.

Kahn, Lawrence M. 2012. Labor market policy: A comparative view on the costs and benefits of labor market flexibility. *Journal of Policy Analysis and Management* 31 (1): 94–110.

Knutsen, Hege M., and Eva Hansson. 2010. Theoretical approaches to changing labour regimes in transition economies. In *Missing Links in Labour Geography*, ed. Ann Cecilie Bergene, Sylvi B. Endresen, and Hege Merete Knutsen, 155–169, Surrey: Ashgate.

Krippendorff, Klaus. 2004. *Content Analysis: An Introduction to Its Methodology*. Thousand Oaks, CA: Sage.

Matkailutyo.fi. n.d. Matkailudiili. https://www.matkailutyo.fi/matkailudiili/. Accessed on December 5, 2019.

Ministry of Economic Affairs and Employment (TEM). 2019. *Yhdessä enemmän –kestävää kasvua ja uudistumista Suomen matkailuun Suomen matkailustrategia 2019–2028 ja strategiset toimenpiteet 2019–2023* [Achieving more together—Sustainable growth and renewal in Finnish tourism. Finnish Tourism Strategy 2019–2028 and measures 2019–2023]. http://urn.fi/URN:ISBN:978-952-327-472-3. Accessed on November 27, 2019.

Müller, Dieter K., and Philip Ulrich. 2007. Tourism development and the rural labour market in Sweden, 1960–1999. In *Tourism in Peripheries: Perspectives from the Far North and South*, ed. Dieter K. Müller and Bruno Jansson, 85–106. Wallingford: CABI.

Norwegian Ministry of Trade, Industry and Fisheries. 2012. Destination Norway. National strategy for the tourism industry. https://www.regjeringen.no/contentassets/1ce1d6cdcbac47739b3320a66817a2dd/lenke_til_strategien-engelsk.pdf. Accessed on July 5, 2019.

Norwegian Ministry of Trade, Industry and Fisheries. 2017. White paper: Experiencing Norway—A unique adventure. https://www.regjeringen.no/en/topics/business-and-industry/travel-industry/experiencing-norway–a-unique-adventure/id2578740/. Accessed on July 5, 2019.

OnsØyen, Lars E., Reidar J. Mykletun, and Trygve J. Steiro. 2009. Silenced and invisible: The work-experience of room-attendants in Norwegian Hotels. *Scandinavian Journal of Hospitality and Tourism* 9 (1): 81–102. https://doi.org/10.1080/15022250902761462.

Peck, Jamie. 1996. *Work Place: The Social Regulation of Labor Markets*. New York: The Guilford Press.

Rantala, Outi, Suzanne de la Barre, Grånas Brynhild, Gunnar Pár Jóhannesson, Dieter K. Müller, Jarkko Saarinen, Kaarina Tervo-Kankare,

Patrick T. Maher, and Maarina Niskala. 2019. Arctic tourism in times of change: Seasonality. Nordic Council of Ministers. https://doi.org/10.6027/TN2019-528.

Robinson, Richard N.S., Antje Martins, David Solnet, and Tom Baum. 2019. Sustaining precarity: Critically examining tourism and employment. *Journal of Sustainable Tourism*, January, 1–18. https://doi.org/10.1080/09669582.2018.1538230.

Rooth, Dan-Olof, and Jan Ekberg. 2006. Occupational mobility for immigrants in Sweden. *International Migration* 44 (2): 57–77.

Sippola, Markku, and Kairit Kall. 2017. Locked in inferiority? The positions of Estonian construction workers in the Finnish migrant labour regime. In *Labour Mobility in the Enlarged Single European Market*, ed. Jon Erik Dølvik and Line Eldring, 215–240, Bingley: Emerald.

Smith, Adrian, Mirela Barbu, Liam Campling, James Harrison, and Ben Richardson. 2018. Labor regimes, global production networks, and European Union Trade Policy: Labor standards and export production in the Moldovan clothing industry. *Economic Geography* 94 (5): 550–574. https://doi.org/10.1080/00130095.2018.1434410.

Solnet, David, Dennis Nickson, Richard N.S. Robinson, Anna Kralj, and Tom Baum. 2014. Discourse about workforce development in tourism—An analysis of public policy, planning, and implementation in Australia and Scotland: Hot air or making a difference? *Tourism Analysis* 19 (5): 609–623. https://doi.org/10.3727/108354214x14116690097936.

Swedish Agency for Economic and Regional Growth. 2016. *Kompetensförsörjning och kompetensbehov inom svensk besöksnäring* [Competency supply and need in the Swedish tourism industry]. https://tillvaxtverket.se/download/18.448d2df3160e422a573b787c/1516609938333/Kompetensförsörjning_och_kompetensbehov_inom_svensk_besöksnäring_Rapport_00197_final.pdf. Accessed on July 11, 2019.

Swedish Tourism Ltd. n.d. *Strategiarbete* [Strategy work]. https://www.svensk-turism.se/strategiarbete. Accessed on July 5, 2019.

Swedish Tourism Ltd. 2016. *Halvtidsanalys – "Nationell strategi för svensk besöksnäring" - så ökar vi tempot inför 2020* ["National strategy for the Swedish tourism industry" Interim analysis "National strategy for the Swedish tourism industry"—This is how we increase the pace before 2020].

Taylor, Marcus, and Sébastien Rioux. 2018. *Global Labour Studies*. Cambridge: Polity Press.

Towards Sustainable Travel and Tourism in Norway. A roadmap. 2017. Oslo. https://destinationcenter.org/wp-content/uploads/2017/12/Tourism-Road-Map-Norway.pdf. Accessed on July 5, 2019.

Tuulentie, Seija, and Bente Heimtun. 2014. New rural residents or working tourists? Place attachment of mobile tourism workers in Finnish Lapland and Northern Norway. *Scandinavian Journal of Hospitality and Tourism* 14 (4): 367–384. https://doi.org/10.1080/15022250.2014.967998.

Underthun, Anders, and David Christoffer Jordhus-Lier. 2018. Liminality at work in Norwegian hotels. *Tourism Geographies* 20 (1): 11–28. https://doi.org/10.1080/14616688.2017.1314546.

Valkonen, Jarno, Heikki Huilaja, and Saara Koikkalainen. 2013. Looking for the right kind of person: Recruitment in nature tourism guiding. *Scandinavian Journal of Hospitality and Tourism* 13 (3): 228–241. https://doi.org/10.1080/15022250.2013.837602.

Wiesbrock, Anja. 2011. The integration of immigrants in Sweden: A model for the European Union? *International Migration* 49 (4): 48–66.

World Bank Group. 2017. Twenty reasons sustainable tourism counts for development. *Tourism for Development Knowledge Series*. Washington, DC: World Bank. https://openknowledge.worldbank.org/handle/10986/28388. Accessed on July 5, 2019.

18

Battling the Past: Social, Economic, and Political Challenges to Indigenous Tourism Employment

Ellen A. Ahlness

Introduction

Arctic Drive, which offers tourism experiences in the Arctic, urges visitors to its website to 'Go with the most adventurous Arctic destination management company for your perfect authentic experience' (The Arctic Drive 2019). If one is to purvey the numerous websites and literature of the ever-increasing number of companies offering Arctic tourism activities, lodging, and transit, they would notice that the concept of an 'authentic' Arctic experience is both a highly desired *and* hotly contested concept for tourists and members of the tourism industry alike seeking to gain their business. What makes the concept of 'authentic' Arctic tourism so challenging are the discrepancies in images that exist between Arctic and non-Arctic peoples, as well as the many imaginaries that southern actors maintain about a region that few have visited before. A particular challenge in conversations about

E. A. Ahlness (✉)
University of Washington, Seattle, WA, USA
e-mail: eahlness@uw.edu

A. Walmsley et al. (eds.), *Tourism Employment in Nordic Countries*,
https://doi.org/10.1007/978-3-030-47813-1_18

Arctic imaginaries is over what degree Indigeneity should (or must) feature in an experience for it to be an 'authentic' Arctic experience (Penikett 2017). It is true that some non-Indigenous tourism experiences can certainly be authentic (e.g. when travelling to Iceland), so the crux of the problem emerges when people believe there is a unified, authentic 'Arctic' to be experienced, rather than diverse landscapes and rich mosaics of cultures that are so unique that one could argue they are only linked together by their latitude. When tourism companies suggest there is a singular Arctic experience that their work best encapsulates, they contribute to a legacy of southern discourses that present a flat portrait of Arctic culture ('Explore Sami, Maasai and Hadzabe' 2017; 'Scandinavian Tours & Nordic Travel' 2019; 'Arctic Scandinavia: Climbing, Hiking & Mountaineering' 2018). Contributing to this effect is the trend that more frequently than not, these companies are owned and operated by southern and non-Indigenous people.

Tourist decision-making processes are shaped by over one hundred years of region-building that has overwhelmingly drawn from southern framing and understandings of the Arctic. The way the region has been defined—its importance, character, and value has a history of being assigned and perpetuated by non-native players, to the exclusion of Indigenous and other regional residents as southern actors identified their economic interests and territorial delimitations (Huggan 2016; Keskitalo 2007; Ahlness 2019). It is nothing new to recognize and criticize the fetishization of Indigenous peoples and livelihoods that goes on in the tourism industry. What follows the criticism, then, is a rich and diverse body of literature (theoretical, empirical, and practical) that explores the many ways in which the Scandinavian tourism industry may be decolonized. While of high importance to the Indigenous peoples of Scandinavia (the Sámi and Greenlandic Inuit), the impact of decolonizing tourism practices and research have been stifled by their lack of applicability to the tourism industry at-large: companies and industries that conduct their affairs below the Arctic circle among predominantly non-Indigenous peoples and along non-Indigenous livelihoods. While the Sámi are Europe's only Indigenous people, they are largely situated in the far north, geographically distant from the southern economic and political hubs of the Scandinavian states.

Consequently, state and market understandings of the Arctic north are dominated by constructions from southern imaginations (Penikett 2017; Hough 2013).

In modernity,[1] Nordic states have viewed the north as a realm for national glory. Arctic explorers and researchers such as Amundsen, Freuchen, and Nansen largely hailed from southern Scandinavia (and indeed, contributed to the dominance of 'Nordic' Norwegian scientific and rhetorical discourses of the Arctic in social and political thought) (Friedman 2010). As technological advances were made in transportation, mapping, navigation, and visual documentation, southern visions of the north changed. No longer was it seen as a realm historically inhabited by Indigenous inhabitants, but as an unexplored and viable frontier. While early expeditions involved Indigenous peoples as guides and topographical experts, later expeditions saw the minimized role of Indigenous peoples (Lewis-Jones 2013). Today, northern Nordic tourism minimally employs and under-employs Indigenous peoples, continuing a pattern of marginalization. This trend is particularly troubling as scholarly and practical interest in Northern tourism grows (Sahlberg 2000). Moreover, what attracts tourists to the Arctic is not a single form of tourism, but different geophysical features, varying socio-economic, institutional and geopolitical preconditions, implying that Arctic tourism is shaped through its hinterlands setting (Müller 2015).

Three questions emerge from these marginalization trends: What is the impact of historical Nordic perceptions of the North on Indigenous employment within the tourism industry? What is the nature of the challenges faced by Indigenous communities developed from these perceptions? Finally, what solutions hold the most promise in correcting these social, political, and economic-rooted marginalizing legacies? I offer a three-part response to these questions that considers how historical region-building, dominated by colonial language and framing, manifests in the contemporary Nordic Arctic tourism industry.

This chapter begins by examining the exploration-hinterlands framing that persists in how the Nordic public views the Arctic. This

[1]Beginning in the 1800s, according to many historians and anthropologists (Ruggie 1993).

perception shows that while the Nordic countries are indeed *northern* countries, the way their general publics consume narratives of the Arctic is shaped by southern imaginaries. These southern-originating narratives have negative impacts for Arctic Indigenous communities. Next, drawing from a review of case study literature on Arctic tourism, I categorize the natures of the challenges faced by Indigenous communities in finding employment in the Arctic tourism industry. I present three 'challenge archetypes' and address the rhetorical content of each archetype through their historical, political, and economic components. Third, I present considerations for ongoing and future tourism development in the Far North that considers cultural values within the Nordic model.

Hinterlands in History: The Southern Narrative

Nordic perceptions of the Arctic are defined through a history of economic domination and intrigue. The pursuits of Nansen and Amundsen painted the North as a place to champion one's nation: the insertion of southerners brought value to the region[2] by ascribing desires and fulfilment to the hinterlands (McCannon 2012). Southern actors projected their own needs and desires onto the Arctic landscape. Through the twentieth century, the dominant image of the Arctic in the south was that of a pristine and untouched land. Unable to traverse the seas and ice themselves, laypeople relied on images developed by those who had travelled northward to inform their own portraits of the Arctic. With such images depicting landscapes (as was the artistic fashion of the era), this framework excluded Indigenous communities (whose presence falsified the mentality that the Arctic is an unpopulated hinterland) from the works that shaped its era's understandings of the North. Moreover, this created a generation of visual reference materials that inform today's ideas of what the Arctic 'used to be'—shaping nostalgia (McCannon 2012).

In art and expedition reports, southerners tended towards one of two conclusions that drew from a pristine frontier mentality: some viewed it

[2]Value that was designed for and consumed by other southerners.

as a serene, untouched hinterlands (Steinberg et al. 2015), while others took it to mean the North was a wild region that needed to be tamed (Lewis-Jones 2013). These reports and visuals created a high demand for other forms and images of the Arctic that could be consumed by the Scandinavian (and broader European) public, which in turn directed the shape of the public's imagination (Potter 2007). This public perception has been remarkably durable over the course of the past hundred years. In this way, the modern public imagination of the Arctic—defined largely through romanticized and pristine imagery—varies significantly from dominant political and policy perceptions among decision-makers in the very same countries (Steinberg et al. 2015; Fjellestad 2016; Ryall et al. 2010). We must acknowledge and recognize that in the context of the Arctic, political decision-making uses different lenses and framing processes than the privileged colonial public. These lenses imagine the Arctic as a realm of urgency and needed action, as a realm for economic benefit, or for peaceful cooperation, among other imaginaries (Steinberg et al. 2015; Langhelle and Mikkelsen 2008; Exner-Pirot 2013). The general public is not particularly interested in or informed about what happens in Arctic, be it politics or colonial tourism practices.[3]

The exploration imaginary supports a disregard for the contributions of Indigenous peoples in northern industries for several reasons. First, it suggests that southern activity is what inserts value to a region. It holds little recognition of the peoples of the Arctic and their contributions to southern explorations. Second, the dominant public discourse lets southerners determine *what* the value ascribed to a region is, as well as what southern actors seek to 'get out' of the region. Third, the discourse perpetuates the idea of the Arctic north as a virgin area, supporting industries that play on tourist desires to experience something 'new', 'exotic', and 'unique'. Today, northern tourism is overwhelmingly run by non-Indigenous peoples. Indigenous Nordic peoples are underemployed and underserved by the tourism industry. Indeed, tourism development often threatens Sámi land and turns culture into a marketing

[3]Moreover, there is a lot of data to support this lack of knowledge, as discussed by J. D. Ford and Pearce (2010), Penikett (2017), Hamilton (2017).

tool in a legacy of disregard for Indigenous communities. At many tourist locales in Norway and Sweden, non-Indigenous peoples dress in inaccurate replications of Sámi clothing. This has resulted in mixing gender-specific clothing items and combining regional items, to the frustration of local Native communities (Saarinen 2003). In Rovaniemi, a popular tourist destination containing the 'Santa Claus Village' theme park, Santa and his elves are dressed in Sámi-style garb (Tervo-Kankare, Hall, and Saarinen 2013). Moreover, in some Finnish and Swedish locales, a popular 'ceremony' consisting of ceremonially 'crossing' the Arctic Circle is practised, despite the ceremony having no significance or root in Sámi spirituality.

At a deeper level, this southern-led model of tourism, built upon a history of southern interjections and expedition mentalities, poses serious problems for the health and success of Northern communities. In some parts of Sápmi and the Kola Peninsula foreign travel agencies have concessions to organize and sell fishing rights and hunting opportunities, which dually infringe upon Sámi hunting and land claims while also excluding them from participating in any economic benefit from the tourism. At the same time, such tourism activities and recreational hospitality locations, such as the building of rural cabins in Northern Norway, result in the reduction of reindeer grazing lands, to the further economic detriment of Indigenous peoples engaged in traditional livelihoods. This southern-led model of tourism, built upon a history of southern interjections and expedition mentalities, poses serious problems for the health and success of Northern communities. A deeper discussion provides greater detail on the challenges Northern Indigenous communities face in the tourism industry, drawing from personal experiences and the small-but-growing case study literature focusing on the Nordic Arctic.

Contemporary Challenges, Historical Roots

Individual communities face unique and prolonged challenges in advancing Indigenous-inclusive tourism. In categorizing the nature of the challenges faced by Indigenous communities in tourism employment, I present three challenge archetypes drawn from interviews and case study research. Each archetype sits on a foundation of historical, political, and economic disparities between Indigenous and colonial populations. Inherently intertwined with these challenges is the basic problem surrounding the concept of Indigenous tourism. The label of Indigenous tourism can be applied to activities in which Indigenous peoples are involved with varying degrees of control (Pettersson 2006). This is a challenging definition because *employment* does not mean control of or benefit from tourism. Decolonizing tourism literature seeks to use critical explorations of Indigenous-*inspired* tourism to create a set of guidelines to aid the transition towards more Indigenous-*responsive* tourism practices.

To discuss the primary challenges Indigenous communities and individuals involved in the Scandinavian Arctic tourism industry face, a review of tourism employment case study literature was conducted, spanning 21 articles centred on Indigenous employment in Arctic tourism (Table 18.1). 17 articles focus on Sami experiences with the tourism industry, and 4 look at Greenlandic Inuit experiences with tourism industries. 8 look at tourism cases in Norway, 13 in Sweden, 6 in Finland, and 4 in Greenland (accounting for works that conducted transboundary comparisons). The research focuses on the period of 2002 to 2017 a 15-year period chosen because of the transformative legal and normative changes that occurred during this time impacting Indigenous legal statuses and livelihood, which informed researcher analyses (e.g. the Finnmark Act [2005], UN Declaration on the Rights of Indigenous Peoples [2007], and ongoing conversations regarding the proposed Nordic Sámi Convention forum), as well as the period's concern with contemporary events. All works but one were discussed in academic journal articles—the one book chapter included in the analysis was chosen as a critical perspective on worn materialities (clothing and appropriation) alongside appropriative practice case studies. This recent time range these publications fall within reflects the recent growth of concern that has emerged among Indigenous peoples and scholars (Indigenous and non-Indigenous) in decolonizing not only research on the tourism industry, but on the actual employment practices of the tourism industry itself (Abram 2016).

Table 18.1 Challenges to Indigenous peoples in the tourism industry, by themes in literature

Broad theme	Sub theme	Cases containing theme
Extractive tourism	Economic extraction	(Pedersen 2004) (Abram 2016) (Heberlein et al. 2002) (Saarinen 2003) (Byström and Müller 2014) (Halpern 2008)
	Cultural extraction	(Müller and Huuva 2009) (Müller and Pettersson 2006) (Kramvig and Flemmen 2016) (Zhang and Müller 2018)
Cultural norms	Miscommunication	(Pettersson 2002) (Engström and Boluk 2012)
	Creating effective agreements	(Viken and Müller 2006) (Leu and Müller 2016) (Lemelin et al. 2012) (Tommasini 2013)
Legal obstacles		(Müller 2015) (Rybråten 2013) (Pedersen and Viken 1996) (Pettersson 2006) (Fay and Karlsdóttir 2011)

The research done on colonial practices in Nordic tourism industries identifies many case-specific and industry-specific challenges, and to go into them comprehensively would be beyond the scope of a single chapter. Instead, I focus on three common narratives that emerged across the breadth of literature. Each narrative represents a challenge to Indigenous communities gaining greater economic benefit and representative power that manifests at a systemic and actor level. The following table illustrates the breakdown of the three core narratives as they are identified and discussed within the literature before moving into a discussion of these challenges, their content, and histories. Each challenge-based discussion brings in additional literature on employment and hiring in the Scandinavian Arctic and Greenland to expand on prevalent themes and present additional illustrations of problematic behaviours and responses within the Far North tourism industry.

Extractive Tourism

Extractive tourism is the continuation of colonial practices across social, economic, political, and cultural lines. This contains two sub themes: economic extraction and cultural extraction. Economic extraction has serious and long-lived historical roots in the taxation of Sámi and Sámi livelihoods across Norway, Sweden, and Finland. Taxation discrepancies between Sámi and ethnic Swedes began in the fifteenth century, with similar systemic discrepancies emerging in Norway at the dawn of the sixteenth century. Swedish settlers were encouraged to expand northward through land rights, water rights, tax exemptions, and military exemption incentives. These benefits were not offered to pre-existing northern Sámi residents. The northern expansion was an infringement on Sámi land and cultural rights. In the 1840s, colonization rhetoric was formally replaced by assimilation policies (Kvist 1993). Economic and political components interacted with one another to create comprehensive and pervasive economic extraction from an already vulnerable minority population.

Throughout the following decades, Sámi that did not assimilate lost ownership of their lands through a series of grazing acts. Early acts were established in 1886 and 1889 with later 1928 and 1971 acts also severely limited grazing rights. These new restrictions forced mixed livelihood Sámi to choose between herding and other forms of livelihood, an extension of the Swedish-Sámi distinction laws that emerged in the mid-1800s. The state's rhetoric was that Sámi livelihood was incompatible with civilized living. Many of these taxation and grazing acts persisted well into the second half of the 1900s. Currently, the legacy of colonialism and assimilation has contributed to the persistence of economic penalties and disincentives to traditional livelihoods. In 2014, these economic penalties came to a head in an internationally publicized court case. Norwegian Sami Jovsset Ante Sara, a reindeer herder, was ordered to cull 41 of his 116-reindeer herd. He brought his case to the Norwegian Supreme Court, arguing that the law meant he cannot sustain himself and his family with so few reindeer and that the government's policy infringed on Indigenous rights. While the Supreme Court ruled against Sara, stating in the ruling that the right to herd reindeer does not encompass 'a right to keep enough reindeer to enjoy a financial return', Sara has expressed his intent to take the case to the European Court of Human Rights.

While Sara's pursuit in mainstream court channels is explicitly political, global awareness of this case was made in large part to the protest art of Marat Anne Sara, an artist, activist, and Jovsset Ante's sister (Ahlness and Gauto 2019). When Jovsset Ante took his case to the Finnmark District Court in 2016, Marat Anne attended, bringing the first iteration of her culling protest art: a pile of bloodied reindeer heads with a Norwegian flag mounted at the top. Jovsset Ante's case won against the district and was eventually passed up to the Supreme Court in December 2017. Dr. Nicholas Tyler at the Arctic University of Norway elaborated: 'the Sámi herders are being squashed in physically—by roads, dams, mines, railroads, and wind farms. Then they are told they are the unsustainable ones' (Martyn-Hemphill 2017). At the same time, these economically marginalizing processes are going on, the image of the reindeer herder is romanticized and fetishized in advertisements and commercialization of Arctic tourism (Kramvig and Flemmen 2016). The herding experiences tourists are exposed to do not represent the reality of herding Sámi: because of economic disincentives, in 2004, only one in ten Sámi participated in the herding industry (and even fewer participated in herding-as-tourism) in Norway and Sweden (Pettersson 2004). Many Sami have moved out of reindeer herding because the market is now restructured and market-oriented, demanding larger areas and fewer herders. Those who want to succeed in tourism are tempted to adapt past traditions when cultural encounters are desired outcomes of tourism (Robinson 1999). Findings also show that the type of tourism professions people are engaged in is more a result of the available geographical resource than an inherent inclination among reindeer herders to work with specific fields such as nature-based attractions (Leu and Müller 2016).

These economic incentives demonstrate how economic extraction is intertwined with cultural extraction. Tourism companies hire non-Indigenous peoples to do jobs and tasks that have otherwise Indigenous origins, or portray non-Indigenous peoples as Indigenous in a fetishization with little benefit to communities (Hollenhorst et al. 2014).

This is particularly the case when it comes to selling handicrafts. Few Sámi are hired in the selling process, meaning that while they are employed, they are in non-visible ways. Consumers thus have the space to image the Sámi producers of handicrafts as holding certain characteristics: dressing in a particular way and meeting expectations of an imagined, traditional way of life (Lyngnes and Viken 1998; Proshansky et al. 1983). In Finland, many fortune hunters (largely, but not entirely, from outside the Arctic) have developed controversial ventures in tourism that have been criticized as being exploitative in employing Sámi themes (and Sámi individuals); however, this problem is not unique to Finland. A case in Sweden in the late 2000s demonstrates the challenges of overcoming these essentialist practices. Sámi went from behind the scenes to selling cultural handicrafts themselves; however, they were only allowed to wear traditional dress when selling the items. While the Indigenous employees were moved to positions of visibility, the nature of their employment retained its imposed essentialism (Kramvig and Flemmen 2016). Arctic tourism is an industry where hiring is largely shaped based on who you know and existing social networks—and the sad reality is that Indigenous community networks often lack the same cohesive networks among non-Indigenous and southern business networks (Janta et al. 2012).

Moreover, economic factors incentive the fetishization of 'authentic' experiences, turning cultural practices and products into objects to be consumed and demonstrating how economic and cultural extraction are difficult to address in vacuums. When the experiences and products in question hold significant cultural or spiritual meaning to a group (particularly a historically marginalized group), control over the commodification becomes a point of importance *and* contention. When out-group members (e.g. non-Sámi) control the commodification of these experiences—and more so, are the primary economic beneficiaries of these experiences and products—the industry extracts from local communities (Forsyth 2017). When Indigenous peoples play a greater role in the tourism industry, particularly in selling handicrafts and cultural objects, they not only control commodification in ways that are more concerned with preventing appropriation (e.g. 'we will sell traditional fabrics and dress materials, but *not* ceremonial garb to tourists'), but also become greater or primary beneficiaries of the economic activity.

Cultural Norms

Variations in cultural values and norms result in two sub-thematic challenges for Indigenous peoples in the Nordic tourism industries. Nordic and Indigenous norms dually challenge growth in Indigenous tourism employment by presenting communication and agreement-crafting difficulties. Miscommunication in tourism employment is not the result of a failure of expression or language barriers, but the result of tourism leaders and company owners misunderstanding the intentions and desires of both tourists and Arctic communities. A basic lack of knowledge about what tourists want combined with disagreements among Arctic (particularly Indigenous) communities leads to high levels of uncertainty about *what* goals should be implemented, as well as *how* they should be reached. In these cases, tourism industry players who have been transplanted from the south too often try to adapt southern strategies to northern environments, failing to recognize how the presence of Indigenous communities complicates the provision of tourist-desired activities (Veijola and Strauss-Mazzullo 2019).

In Swedish Sápmi, an increasing number of Sami are involved in small tourism companies that focus on their culture (Aikio 2015; Olsen 2019). What the tourists demand and appreciate, on the other hand, is still a point of confusion and growth for these companies, according to the findings of papers analysed through the challenges framework (notably evident in Kramvig and Flemmen 2016; Leu and Müller 2016). While there is work yet to be done among these players in filling the gap between supply and demand, they start from a point of strength. Knowing their community needs means there is one less point for miscommunication in developing regional tourism industries, thus they may focus more attention on identifying tourist desires. Meanwhile, actors moving northbound from the south (and thus bringing with them non-Indigenous, southern frameworks of operation) must divide their attention and energy between the two groups. Increased communication and network linkages between community-based companies and larger transplanted players can provide distinct advantages for both sides in reducing the costs associated with *identifying* tourist desires and *fulfilling* them in community-responsive ways.

Additionally, challenges exist in creating effective employment agreements and standards between Indigenous individuals and non-Indigenous tourism companies. There are simple differences among value systems that emerge in the hiring and employment process. Semi-structured phenomenological interviews with six Norwegian Sámi involved in the tourism industry resulted in discussions of the differences these individuals perceived between their time employed by non-Indigenous individuals (bosses or founders of companies) and Indigenous individuals (Sámi bosses or founders of companies) in the tourism industry. The interviewees were primarily women ($n=4$), with most falling between 20 and 30 years old ($n=4$), with one between 40 and 50 years old, and another between 50 and 60 years old. The interviewees were employed in transportation ($n=1$), performance ($n=3$), art and handicrafts ($n=1$), and outdoor sport ($n=1$) within the tourism industry. Over half had worked in non-Indigenous owned or operated businesses before moving into Indigenous-owned or operated workplaces ($n=4$) while the remainder had worked in Indigenous-owned or operated workplaces in the past, but were now working in non-Indigenous ('southern') owned workplaces ($n=2$).

Through conversations with more seasoned attendees of that conference, the attendee found that a common theme among interviews was that a greater satisfaction with work/life balance and a work environment that supported family and community commitments was much more possible when the employees had Indigenous bosses, or when a majority of the employees were also Sámi. Female interviewees shared that they felt more comfortable sharing their family makeup (whether or not they had a partner or children, whether or not they planned to have children in the near future, and if so, what their childcare plans were) to bosses and peers in Indigenous-operated companies. This strongly contrasted with their concerns over sharing personal details (particularly child-rearing or family planning information) when looking for jobs in non-Indigenous companies or businesses, fearing that doing so would make them 'less hirable'. These environments made them more aware of the Anglo-American value system that their previous employment had been structured upon. An assumption among the interviewees was that the non-Indigenous companies were based on individualistic cultural dimensions and values not only alienated employees who valued family and community fulfilment alongside success in work, but presented additional barriers to success to those whose cultural values placed a greater emphasis on family and community, mirroring the cultural dimensions theory of Hofstede and Bond (1984).

Generally (a word used with caution, though this value identification has been supported by the statements of Indigenous activists, scholars, and community leaders), circumpolar Indigenous peoples maintain cultures that place a higher emphasis on work and work cultures that respect individual commitments to family and community (Galbraith et al. 2006; Melton 1995; Bingham et al. 2014; Haar et al. 2012). This can result in conflict and debate when the assumptions and standards of one sphere are transferred into another. At a 2017 North American/Northern European tourism conference, a small debate erupted over the conduct of one event attendee, who was attending on the behalf of her company. She had brought her baby to the conference, and during a question and answer session, when the baby began fussing, she began to nurse them in front of the panel audience. Some attendees argued that while this may have been acceptable within her own workspace (which was Indigenous-owned), this was not standard behaviour in a 'broader professional' space, which could go on to negatively impact her professional growth.

Meanwhile, companies with Euro-centric value systems and employee bodies seek to hire people whose 'cultures' (whether personal or societal) match their own (Rivera 2012). This creates biased hiring processes based on a misalignment of what employees expect from their employers (e.g. a paycheck, flexible hours for family commitments), what the employers expect of their employees (e.g. regional expertise, a willingness to put in extra hours), and what employment conditions/benefits are considered reasonable or non-negotiable. Southern business models run the risk of being less responsive to community needs and values, thus perpetuating a disproportionately low number of local Indigenous peoples employed in the tourism industry.

Legal Obstacles

The final challenge identified in the case study literature is pervasive and stymying unto itself, while also representing the formalization of systemic soft biases and colonial practices. Indigenous peoples face many legal obstacles in advancing in the Arctic tourism industry. Moreover, the legal situations of Indigenous peoples vary from state to state, making it a problem that is harder to solve through transboundary teaching/learning relationships or tourism success stories. Indigenous tourism employment is impacted by historic political-economic barriers and variance in national legal standings. For example, Norway is party to the International Labor Organization Convention 169 (also known as the Indigenous and Tribal Peoples Convention, or ILO 169), while Sweden and Finland are not. The convention is the only binding international treaty open for ratification that deals exclusively with the rights of Indigenous peoples, forbidding assimilationist policies, asserting the rights of Indigenous peoples to choose to integrate or maintain their cultural and political independence as they desire, and that land and natural resource rights are essential to economic and cultural survival.

This variation is credited as a large influence in why Norway passed the Finnmark Act, an Act that recognizes the ownership and right of Sámi over land and waters in Finnmark, Norway, while such measures have not been taken in the other Fenno-Scandinavian states (Ravna 2015; Graver and Ulfstein Geir 2004). In Finland there are about 6000 Sámi, though they do not have exclusive rights to reindeer herding as Norwegian and Swedish Sámi ('Sami in Finland' 2018). The Finnmark Act attempts to strengthen Sámi rights by giving the entire population of Finnmark influence of the land and water in the county, whereas previously the legal position was that nomadic peoples were unable to exert any formal legal rights over land. While the Act does not cover fishing rights in saltwater, mining rights, or oil rights, it does set a comprehensive and strong legal precedence that empowers Sámi to exercise control over institutions and industries that rely on land its geophysical features. In the case of tourism, the ability to exercise control over economic development is seen as a mechanism to greater self-determination (Joona 2005).

While Norway was largely considered to have good involvement of Sámi in tourism development, resulting in activities that are focused on experience and activity (Lyngnes and Viken 1998), it was not until the establishment of the Finnmark Act that Indigenous communities were able to take advantage of the economic advantages of land-reliant tourism or benefit from its hiring practices. In the 1990s, none of the large hotels in the Norwegian Arctic were owned by Sámi (Pedersen and Viken 1996). After the establishment of the Finnmark Act, Sámi employment in hotels and hospitality sharply rose, particularly in the following two years (Henriksen 2007). Similarly, since Greenland implemented home rule (a referendum granting greater autonomy from Denmark, with sovereignty over education, health, fisheries, and the environment), there have been significant advances in the development of Inuit-led tourism initiatives (Christensen 1992; Johnston and Viken 1997). Tourism is most often seen as an activity that boosts the economic capability of rural and Arctic communities, which benefits communities when combined with a shared sense of responsibility and social cohesion through cultural heritage, ties to land or a region, and interpersonal connections. Weaknesses in state arrangements and land laws hinder its development and lead to the development of 'economic communities' consisting of tourism infrastructure and seasonally employed individuals that are segregated and distinct from historically situated communities (Müller 2015).

Ultimately, there is no comprehensive legislation regulating or protecting the unique nature and culture of Sámi across Sápmi, though there are protected enclaves in the form of nature and culture reserves. The legal field that Sámi navigate is inherently uneven across the countries they reside within. This presents a deeper challenge among communities of how they can preserve characteristics unique to their local identities in their tourism strategies. After all, they are playing on a field where communities face drastically different barriers and political realities as they try to derive economic benefit from tourism in a region that the public largely believes is given value through the work and activity of southern actors.

Discussion

Combined, these barriers weave a complex and comprehensive tapestry of barriers Sámi people must overcome to participate in and benefit from employment in the Arctic tourism industry. Despite the number of challenges, there have been community successes in particular cases that inform later industry battles. Sámi are frequently used in tourism promotion and marketing, particularly because they are portrayed as Europe's only Indigenous peoples living in Europe's last wilderness (Müller and Huuva 2009). While this portrayal causes problems and perpetuates colonial framing, it is true that tourism development is a potential solution to problems in Sámi communities, offering new (and repeated) sources of income and employment. To learn how tourism may contribute to (rather than extract from) native communities, Nordic populations may look to international success stories, such as those of the Maori and Inuit. In the past decade, successes have been seen in the ways in which Sámi communities confront culturally extractive industry practices. Some of the highest Indigenous employment rates are seen within Indigenous-led projects, such as seasonal festivals and circuit events (Müller and Huuva 2009; Müller and Pettersson 2006; Heberlein et al. 2002). These spaces allow for communities to create and propagate their own understandings of 'authentic' culture. At the same time, these success cases point to a truth long-held by native communities underserved by tourism industries: to see greater and sustained employment of Indigenous peoples in the tourism industry, Indigenous peoples must be present as industry leaders. Effective employment of Indigenous peoples in tourism is not merely a matter of overcoming Eurocentric biases in hiring; it is acknowledging and actively working to minimize the colonial structures that pervade the tourism industry.

Thus far, the tourism field benefits from a critical analysis of its racial-ethnic discrepancies. The field, as it addresses Arctic and Northern Scandinavian tourism, has an excellent coverage of contemporary challenges Indigenous peoples face in gaining and maintaining employment in an industry that largely economically benefits players from the south. This chapter, containing a deeper exploration of the historical, social, and political factors that provided a foundation for this extraction, also benefits the field through its understanding of the legacy of imaginations and southern imagery on contemporary challenges. To conclude, I provide a series of considerations drawn from this exploration.

Considerations and Conclusion

As the discussion acknowledges, there have been developments in Indigenous employment within the tourism industry, but existing barriers are persistent and pervasive. As I conclude this chapter, I present five considerations for ongoing and future tourism development which may inform employment decisions. These considerations consider cultural values within the Nordic model of business (Strand et al. 2015; Gjølberg 2010). The purpose of fitting these reflections within the broader Nordic model (given that this chapter's primary alignment is with decolonization literature) is in recognition of the need for pan-Nordic, trans-ethnic action to see meaningful change in tourism employment. Indigenous communities find themselves in a situation where they must appeal to Nordic values to encourage non-native actors to take action (and in a very real sense, mollify them to mitigate the resistance of current players in Arctic tourism). These recommendations thus fit within a broader Nordic model with the intent to strategize success.

Monitor Developments in Land Rights Laws: legal developments like the Finnmark Act set precedents for the further removal of legal obstacles for Indigenous tourism employment. Binding legal developments and normative agreements can occur on three levels: at the local, national, or international levels, or along a combination of two or more levels. The legal developments that occur on multiple levels create a complicated playing field for Indigenous actors. Organizations in the Far North like the Arctic Council seek to consolidate information and influence states to adopt national-level policies that are in agreement with soft international standards; however, discrepancies and a lack of coordination among states and relevant associations make it increasingly important *and challenging* to monitor developments in land right laws.

Maintain Transnational Learning Relationships: Greenlandic Inuit and Sami have relationships with Canadian Inuit, engaging in best-practice sharing when it comes to political activism and rights gains (Ahlness and Gauto 2019). Scholars have identified the strengths of inter-tribal, inter-national teaching/learning relationships (Fall 2017; Wieser 2017; Waterman et al. 2018). Learning relationships can draw from shared or similar histories, ecosystems, and experiences to help guide growing communities on best practices in tourism employment and infrastructure.

Challenge Traditional/Developed Dichotomies: Employing Indigenous individuals in 'non-traditional' tourism activities dismantles social and economic prejudices (Kramvig and Flemmen 2016). While there are certainly problems associated with employing non-Indigenous peoples in activities that have traditional or spiritual connotations, it is seriously problematic to propose that the solution is to employ Indigenous peoples only in Indigenous tourism. A better path forward is for employers to use affirmative hiring practices to encourage growth and entrepreneurship among Indigenous peoples in 'non-traditional' tourism, which will also support lower unemployment rates among rural Indigenous communities (Coates 1988).

Develop Co-management/Tourism Linkages: successes in resource co-management employment may translate to tourism employment. Co-management agreements between Indigenous peoples, other stakeholders, and state agencies offer substantial promise as a way of dealing with natural resource-based conflicts (Castro and Nielsen 2001; Berkes 2009). Less contentious are the economic benefits of tourism, suggesting a potential transferability of co-management agreement structures, given their success in higher-stakes issues.

Conduct Mindful Classification: While the Nordic tourism industry generally considers itself ecological (in an interesting case of self-labelling intended for self-consumption), Indigenous communities are wary of automatically labelling Northern or Indigenous tourism as 'ecological' merely because of its Indigenous components. These worries emerge from concerns of non-Indigenous consumers treating the label like a neo-'noble savage': a label that external consumers suppose is flattering (in a model minority-esque treatment) but instead perpetuates assumptions about minority practices and values.

As Arctic tourism continues to garner interest—especially as the Far North continues to maintain its mystique as a *pristine* and *distant hinterland*—states and their tourism industries feel increased urgency to develop their northern tourism infrastructure and Arctic industries. The Nordic case is revealing for the future of sustainable and inclusive tourism employment across circumpolar states and their own tourism futures.

References

Abram, Simone. 2016. Green ice? Tourism ecologies in the high north. In *Tourism Ecologies in the European High North: Green Ice*, ed. Simone Abram and Katrin Anna Lund, 1–24. London, UK: Palgrave Macmillan.

Ahlness, Ellen. 2019. Incompatible futures: Frontier nostalgia and southern discourses of the Arctic. *Arctic Yearbook* 2019 (1): 1–20.

Ahlness, Ellen, and Maria Gracia Gauto. 2019. Decolonizing gray spaces: Storytelling and art as political activism among Sámi Herders. *Ecologia Politica* 57 (1): 62–67.

Aikio, Eeva-Maarit. 2015. Tourism development in Saami communities in Finland. Septentrio Conference Series.

"Arctic Scandinavia: Climbing, Hiking & Mountaineering." 2018. SummitPost. https://www.summitpost.org/arctic-scandinavia/185383.

Berkes, Fikret. 2009. Evolution of co-management: Role of knowledge generation, bridging organizations and social learning. *Journal of Environmental Management* 90 (5): 1692–1702.

Bingham, Jennie L., Quintina Bearchief Adolpho, Aaron P. Jackson, and Louise R. Alexitch. 2014. Indigenous women college students' perspectives on college, work, and family. *Journal of College Student Development* 55 (6): 615–632.

Byström, Joakim, and Dieter Müller. 2014. Tourism labor market impacts of national parks. *Zeitschrift Für Wirtschaftsgeographie* 58 (1): 115–126.

Castro, Alfonso Peter, and Erik Nielsen. 2001. Indigenous people and co-management: Implications for conflict management. *Environmental Science & Policy* 4 (4–5): 229–239.

Christensen, Torben. 1992. Greenland wants tourism. *Polar Record* 28 (164): 62–63.
Coates, Kenneth. 1988. On the outside in their homeland: Native people and the evolution of the Yukon economy. *Northern Review* 1: 73–89.
Engström, Christina, and Karla Aileen Boluk. 2012. The battlefield of the mountain: Exploring the conflict of tourism development on the Three Peaks in Idre, Sweden. *Tourism Planning and Development* 9 (4): 411–427.
Exner-Pirot, Heather. 2013. What is the Arctic a case of? The Arctic as a regional environmental security complex and the implications for policy. *The Polar Journal* 3 (1): 120–135.
"Explore Sami, Maasai and Hadzabe." 2017. Visit Natives. https://www.visit-natives.com/.
Fall, Juliet. 2017. *Drawing the Line: Nature, Hybridity and Politics in Transboundary Spaces*, 1st ed. London: Routledge.
Fay, Ginny, and Anna Karlsdóttir. 2011. Social indicators for Arctic tourism: Observing trends and assessing data. *Polar Geography* 34 (1–2): 63–86.
Fjellestad, Marthe T. 2016. Picturing the Arctic. *Polar Geography* 39 (4): 228–238.
Ford, James D., and Tristan Pearce. 2010. What we know, do not know, and need to know about climate change vulnerability in the Western Canadian Arctic: A systematic literature review. *Environmental Research Letters* 5 (1): 014008.
Forsyth, Janice. 2017. The illusion of inclusion: Agenda 21 and the commodification of Aboriginal culture in the Vancouver 2010 Olympic Games. *Public* 27 (53): 22–34.
Friedman, Robert Marc. 2010. Making the aurora Norwegian: Science and image in the making of a tradition. *Interdisciplinary Science Reviews* 35 (1): 51–68.
Galbraith, Craig S., Carlos L. Rodriguez, and Curt S. Stiles. 2006. False myths and Indigenous entrepreneurial strategies. *Journal of Small Business & Entrepreneurship* 19 (1): 1–20.
Gjølberg, Maria. 2010. Varieties of corporate social responsibility (CSR): CSR meets the 'Nordic model'. *Regulation & Governance* 4 (2): 203–229.
Graver, Hans Petter, and Ulfstein Geir. 2004. The Sami people's right to land in Norway. *International Journal on Minority and Group Rights* 11 (4): 337–377.
Haar, Jarrod M., Maree Roche, and Daniel Taylor. 2012. Work–family conflict and turnover intentions of Indigenous employees: The importance of the whanau/family for Maori. *The International Journal of Human Resource Management* 23 (12): 2546–2560.

Halpern, Nigel. 2008. Lapland's airports: Facilitating the development of international tourism in a peripheral region. *Scandinavian Journal of Hospitality and Tourism* 8 (1): 25–47.
Hamilton, Clive. 2017. *Defiant Earth*, 1st ed. Malden: Polity Press.
Heberlein, Thomas A., Peter Fredman, and Tuomas Vuorio. 2002. Current tourism patterns in the Swedish mountain region. *Mountain Research and Development* 22 (2): 142–149.
Henriksen, John Bernhard. 2007. Norway. In *Affirmative Action in the Employment of Ethnic Minorities and Persons with Disabilities*, 65–74. Geneva: International Labor Office.
Hofstede, Geert, and Michael H. Bond. 1984. Hofstede's culture dimensions. *Journal of Cross-Cultural Psychology* 15 (4): 417–433.
Hollenhorst, Steven J., S. Houge-Mackenzie, and David M. Ostergren. 2014. The trouble with tourism. *Tourism Recreation Research* 39 (3): 305–319.
Hough, Peter. 2013. *International Politics of the Arctic*. London: Routledge.
Huggan, Graham. 2016. Introduction: Unscrambling the Arctic. In *Postcolonial Perspectives on the European High North*, 1–29. London, UK: Palgrave Macmillan.
Janta, Hania, Peter Lugosi, Lorraine Brown, and Adele Ladkin. 2012. Migrant networks, language learning and tourism employment. *Tourism Management* 33 (2): 431–439.
Johnston, Margaret E., and Arvid Viken. 1997. Tourism development in Greenland. *Annals of Tourism Research* 24 (4): 978–982.
Joona, Tanja. 2005. The political recognition and ratification of ILO convention No. 169 in Finland, with some comparison to Sweden and Norway. *Nordic Journal of Human Rights* 23 (3): 306–321.
Keskitalo, Carina. 2007. International region-building. *Cooperation and Conflict* 42 (2): 187–205.
Kramvig, Britt, and Anne Brit Flemmen. 2016. Kramvig, Britt, and Anne Britt Flemmen: 'What alters when the Sami traditional costume travels? Affective investments in the Sápmi.' In *Sensitive Objects: Affects and Material Culture*, ed. Jonas Frykman and Maja Povrzanovic Frykman, 179–198. Lund: Nordic Academic Press Checkpoint.
Kvist, Roger. 1993. The racist legacy in modern Swedish Sámi policy. In *Presentation, American Folklore Society*, 21–23. Eugene, OR.
Langhelle, Oluf, and Aslaug Mikkelsen. 2008. Framing oil and gas in the Arctic from a sustainable development perspective. In *Arctic Oil and Gas: Sustainability at Risk?* 29–58. New York: Routledge.

Lemelin, Raynald Harvey, Margaret E. Johnston, Jackie Dawson, Emma S. Stewart, and Charles Mattina. 2012. From hunting and fishing to cultural tourism and ecotourism: Examining the transitioning tourism industry in Nunavik. *The Polar Journal* 2 (1): 39–60.

Leu, Traian C., and Dieter K. Müller. 2016. Maintaining inherited occupations in changing times: The role of tourism among reindeer herders in Northern Sweden. *Polar Geography* 39 (1): 40–57.

Lewis-Jones, Huw. 2013. *Imagining the Arctic: Heroism, spectacle and polar exploration*. New York: I. B. Tauris.

Lyngnes, S, and A. Viken. 1998. Samisk Kultur Og Turisme På Nordkalotten. Sandvika.

Martyn-Hemphill, Richard. 2017. In Norway, fighting the culling of reindeer with a Macabre display. *The New York Times*, December 6.

McCannon, John. 2012. *A History of the Arctic: Nature, Exploration and Exploitation*. London: Reaktion Books.

Melton, Ada Pecos. 1995. Indigenous justice systems and tribal society. *Judicature* 79 (3): 126–133.

Müller, Dieter K. 2015. Issues in Arctic tourism. In *The new Arctic*, ed. Birgitta Evengård, Joan Nymand, and Øyvind Paasche Larsen, 147–158. Cham: Springer.

Müller, Dieter K., and Stina Kuoljok Huuva. 2009. Limits to Sami tourism development: The case of Jokkmokk, Sweden. *Journal of Ecotourism* 8 (2): 115–127.

Müller, Dieter K., and Robert Pettersson. 2006. Sámi heritage at the winter festival in Jokkmokk, Sweden. *Scandinavian Journal of Hospitality and Tourism* 6 (1): 54–69.

Olsen, Kjell Ole Kjærland. 2019. Indigenous tourism and the Barents Euro-Arctic Region. *Arctic and North* 34 (1): 35–45.

Pedersen, K., and A. Viken. 1996. From Sami nomadism to global tourism. In *People and Tourism in Fragile Environments*, ed. M.F. Price and V.L. Smith, 69–88. Chichester: Wiley.

Pedersen, Steinar. 2004. Gränslös Näring i Sverige-Norge. In *Samer—Ett Ursprungsfolk i Sverige*, ed. Karin Kvarfordt, Nils-Henrik Sikku, and Michael Teilus, 34–35. Västerås: Aros, Edita Västra.

Penikett, Tony. 2017. *Hunting the Northern Character*. Vancouver: Purich Books.

Pettersson, Robert. 2002. Sami tourism in Northern Sweden: Measuring tourists' opinions using stated preference methodology. *Tourism and Hospitality Research* 3 (4): 357–369.

Pettersson, Robert. 2004. Sami tourism in Northern Sweden: Supply and demand and interaction. Umeå Universitet.

Pettersson, Robert. 2006. Ecotourism and Indigenous people: Positive and negative impacts. In *Ecotourism in Scandinavia: Lessons in Theory and Practice*, ed. Stefan Gössling and Johan Huttman, 166–177. Cambridge: CABI.

Potter, Russell A. 2007. *Arctic Spectacles: The Frozen North in Visual Culture, 1818–1875*. Seattle: University of Washington Press.

Proshansky, Harold M., Abbe K. Fabian, and Robert Kaminoff. 1983. Place-identity: Physical world socialization of the self. *Journal of Environmental Psychology* 3 (1): 57–83.

Ravna, Øyvind. 2015. Recognition of Indigenous lands through the Norwegian 2005 Finnmark act: An important example for other countries with Indigenous people?" In *Indigenous Peoples' Governance of Land and Protected Territories in the Arctic*, ed. T.M. Herman and T. Martin, 189–208. New York: Springer.

Rivera, Lauren A. 2012. Hiring as cultural matching. *American Sociological Review* 77 (6): 999–1022.

Robinson, M. 1999. Cultural conflicts in tourism: Inevitability and inequality. In *Tourism and Cultural Conflicts*, ed. M. Robinson and P. Boniface, 1–32. Wallingford: CAB International.

Ruggie, John. 1993. Territoriality and beyond: Problemetizing modernity in international relations. *International Organization* 47 (1): 139–174.

Ryall, Anka, Johan Schimanski, and Henning Howlid Wærp (eds.). 2010. *Arctic Discourses*. Cambridge: Cambridge Scholars Publishing.

Rybråten, Stine. 2013. Enacting nature in Unjárga-Nesseby, Northern Norway. University of Oslo. https://www.duo.uio.no/bitstream/handle/10852/40947/1/PhD-Rybraaten.pdf.

Saarinen, Jarkko. 2003. The regional economics of tourism in Northern Finland: The socio-economic implications of recent tourism development and future possibilities for regional development. *Scandinavian Journal of Hospitality and Tourism* 3 (2): 91–113.

Sahlberg, Bengt. 2000. *Going North—Peripheral Tourism in Canada and Sweden*. Östersund: European Tourism Research Institute.

"Sami in Finland." 2018. https://www.uoc.edu/euromosaic/web/document/sami/an/i1/i1.html.

"Scandinavian Tours & Nordic Travel." 2019. 50 Degrees North. https://us.fiftydegreesnorth.com/.

Steinberg, Philip, Jeremy Tasch, Hannes Gerhardt, Adam Keul, and Elizabeth Nyman. 2015. *Contesting the Arctic: Politics and Imaginaries in the Circumpolar North*. I. B. Tauris: New York.

Strand, Robert, R. Edward Freeman, and Kai Hockerts. 2015. Corporate social responsibility and sustainability in Scandinavia: An overview. *Journal of Business Ethics* 127 (1): 1–15.

Tervo-Kankare, Kaarina, C. Michael Hall, and Jarkko Saarinen. 2013. Christmas tourists' perceptions to climate change in Rovaniemi, Finland. *Tourism Geographies* 15 (2): 292–317.

"The Arctic Drive." 2019. https://www.thearcticdrive.com/.

Tommasini, Daniela. 2013. Ittoqqortoormiit and the national park of Greenland: A community's option for tourism development. *Polar Record* 49 (3): 237–239.

Veijola, Soile, and Hannah Strauss-Mazzullo. 2019. Tourism at the crossroads of contesting paradigms of Arctic development. In *The Global Arctic Handbook*, 63–81. Cham: Springer.

Viken, Arvid, and Dieter K. Müller. 2006. Introduction: Tourism and the Sámi. *Scandinavian Journal of Hospitality and Tourism* 6 (1): 1–6.

Waterman, Stephanie J., Shelly C. Lowe, and Heather J. Shotton. 2018. *Beyond Access : Indigenizing Programs for Native American Student Success*. Sterling: Stylus Publishing.

Wieser, Kimberly G. 2017. *Back to the blanket: Recovered rhetorics and literacies in American Indian Studies*. Norman: University of Oklahoma Press.

Zhang, Jundan, and Dieter Müller. 2018. Tourism and the Sámi in transition: A discourse analysis of Swedish newspapers, 1982–2015. *Scandinavian Journal of Hospitality and Tourism* 18 (2): 163–182.

19

Hospitality Through Hospitableness: Offering a Welcome to Migrants Through Employment in the Hospitality Industry

Tone Therese Linge, Trude Furunes, Tom Baum and Tara Duncan

Introduction

At the heart of the integration of new migrants into work and the wider community within which they and their families locate is hospitality (Irimiàs and Michalkò 2016). This hospitality is in the form of the welcome they may or may not receive, and the speed at which distinctions

T. T. Linge (✉) · T. Furunes
Norwegian School of Hotel Management,
University of Stavanger, Stavanger, Norway
e-mail: tone.linge@uis.no

T. Furunes
e-mail: trude.furunes@uis.no

T. Baum
Department of WEO, University of Strathclyde, Glasgow, Scotland, UK
e-mail: t.g.baum@strath.ac.uk

T. Duncan
Department of Regional Studies, Dalarna University, Falun, Sweden
e-mail: tdu@du.se

A. Walmsley et al. (eds.), *Tourism Employment in Nordic Countries*,
https://doi.org/10.1007/978-3-030-47813-1_19

between host and guest become blurred. This chapter explores one of the under-conceptualised paradoxes of migration in the contemporary world. Brought to us through the work of Derrida (1999, 2000) and others (for example Pechlaner et al. 2016; Smith 2019), newly settled migrants, still very much guests in their own right, are widely asked to deliver the hospitality that welcomes fellow guests (tourists) to a community and destination. The paradox lies in expectations of this welcome, with its overlays of place, stories, representation and culture.

Put simply, migrant workers come as guests and interact with, at one level, hosts who, in cultural terms, may be colleagues, neighbours in the local community or, indeed, customers. At the same time, there is a workplace expectation that their role will be one of hosts from day one, representing local place and culture through its stories, myths and traditions. Through four narratives showing immigrant experiences of integration processes of migrant workers within the hospitality industry in Norway, we aspire to understand the complexity of this paradox and engage with the role changes from guest to host that take place during the journey to integration.

In structuring this chapter, we explain the background and context of our narrative in focusing on tourism employment and migration in Norway, a country that is highly dependent on migrant workers to deliver frontline service work in the tourism sector. We then explore the extant literature on tourism and migrant work and locate this theoretically within the work of Derrida and others, highlighting the central Derridian paradox. The main questions addressed in this chapter are: (1) How do different migrant experiences influence hospitality employees', or "guests'", possibilities of fulfilling their role as "host" to other tourist guests? (2) How can the use of narratives illustrate challenges related to the Derridian paradox of how newly settled migrants are expected to act as hosts to fellow "guests", or tourists, to a community or destination? The chapter first accounts for the background and context of migrant workers in the Norwegian hospitality industry, before presenting relevant theoretical perspectives for the study. Subsequently, four narratives from migrant workers in the Norwegian hospitality industry are presented and analysed in light of the theoretical perspectives. The conclusion presents managerial implications and challenges for the hospitality industry based on the preceding discussion.

Background and Context

During the last 15–20 years, the Nordic countries have seen an increasing number of new migrants settling with Norway and Sweden receiving the largest numbers of migrants (Calmfors and Gassen 2019). However, labour market participation of new migrants remains below national averages with employment rates for those born outside the EU being 10–20% lower than natives' populations or those born in the EU (Calmfors and Gassen 2019). From 2003 to 2013, the number of immigrants to Norway more than doubled, largely due to labour immigration from new EU-countries in Eastern Europe after 2004 (Østbye and Henriksen 2013). In 2010, the Norwegian hotel industry counted 1476 firms and 26,537 employees (Dokka et al. 2015). Operating income amounted to NOK 22.1 billion with 18.4 million guest nights (Dokka et al. 2015). Over 27,000 people were employed in the accommodation industry in Norway in 2013. Of these, more than 10,000 had an immigrant background (Linge 2015). According to the Norwegian Hospitality Association (Annual Report 2018), the tourism industry is one of the few sectors in Norway with continued growth in employment, in a period generally characterised by downsizing and automatisation.

Of the over 165,000 people working in tourism-related industries (mainly the accommodation and food service industry) in Norway in 2017, 43% had a foreign background, in contrast to 16% in other industries combined (Norwegian Hospitality Association 2018). Furthermore, every third employee in the tourism industry was under 24 years of age. The tourism industry thus points itself out as a central employment engine for both people with immigrant background and young people in Norwegian society. One of the main reasons for this is that the tourism industry is a labour-intensive industry with low demands for language skills due to the many backstage positions, thus making it a low-threshold point of entry into the labour market for many job seekers with immigrant backgrounds.

The hospitality industry is subject to high labour intensity, low margins, strong competition and seasonality. It is also a sector that ranges in scale from micro-operations in the formal, informal and gig economies to affiliates of major international chains. Key characteristics of the hospitality workforce include a high percentage of part-time and precarious workers on non-permanent labour contracts with high labour turnover rates, working in jobs that require little formal education, host language competence or occupational training (Baum 2018). Such working conditions and low status are all factors that may contribute as barriers to integration in a society, because they can hinder possibilities of putting down economic and cultural roots, and lead to social peripherality and isolation. This may, in turn, present central challenges to the Nordic welfare model, which is based on a high degree of participation in work life and public support for a comprehensive welfare system.

The Nordic countries are typically characterised by having regulated labour markets, including a range of low-skill/low-wage jobs that are accessible to newcomers (Grødem 2016). A focus on negotiated compromise, wage equality and a strong trade union movement characterises the Nordic model of industrial relations. This is accompanied by a strong tradition for collective bargaining and employee participation, permanent contracts and a high degree of trust between employers and employees (Underthun and Aasland 2015). Good working conditions and job security are ensured through regulation and effective labour laws (Dokka et al. 2015). The hotel sector, however, shows tendencies towards the deregulation of employment contracts, more use of temporary employment, lower rates of unionisation, higher staff turnover and widespread use of migrant labour (Underthun and Aasland 2015). The question is whether these tendencies may contribute to undermine what Dokka et al. (2015) describe as the spirit of compromise and collaboration between capital and labour that serves as the foundation of the Nordic model of industrial relations.

Nordic Research on Migration and Tourism

Research in the Nordic countries has mainly studied migration as a potential avenue for population growth and entrepreneurial activities among migrants, mainly in rural areas. Workforce migration includes both *permanent* (permanent migrants searching work) and *temporal* migration (individuals coming to work for a shorter period, or repeatedly) (Thulemark et al. 2014). Cohen et al. (2015) challenge these definitions, noting that there is an increasing fluidity between travel, leisure and migration. Population patterns in many rural Nordic regions show an ageing population, depopulation as young people move away and restructuring of employment (i.e. decreasing employment in areas such as manufacturing, forestry and fishing; see e.g. Lundmark 2005). In this context, researchers have looked into tourism employment through migration as a potential avenue for growth (Harbo et al. 2017). Most studies conclude that relatively few are permanent migrants. For instance, Thulemark et al. (2014) and Lundmark (2006) found that the reason to seasonally migrate for tourism employment is lifestyle-related (for example to the ski fields or beach locations), and thus not intended to be permanent from the onset. Similar issues are discussed by Möller et al. (2014) who studied seasonal tourism workers as potential in-migrants in six rural ski resorts in Norway and Sweden. However, Thulemark et al.'s (2014) study suggests that tourism employment can also be a pull factor for in-migration of highly skilled and well-paid workers. In a recent longitudinal study, Möller and Amcoff (2018) found that the net population change among young adults is more positive in tourism-dominated areas than in non-tourism-dominated areas in rural Sweden. Tourism-dominated areas have a younger population and a better gender balance, but population turnover also is higher. In-migration seems to be a driver of positive population change.

Given that some sparsely populated tourism destinations are dependent on mobile tourism workers in order to accommodate tourist flows in the peak season (Tuulentie and Heimtun 2014), studies have looked into entrepreneurship among migrants to understanding the entrepreneurial motivation and activities among migrants (i.e. lifestyle migration and entrepreneurship) (Iversen and Jacobsen 2016). Drawing on tourism–work typologies, Tuulentie and Heimtun (2014) distinguished between five types of workers in relation to their ways of being mobile and explored how their place attachments are based on work-oriented and tourist-related pursuits. Differences in these relationships show that mobility varies from lifestyle mobility to more economic and necessity-based mobility. Place attachment thus varied from perceptions of places where they merely kept their suitcases; to where they lived, worked and were able to undertake leisure activities. In a similar vein, Lundmark et al. (2014), found self-employment to be more common among in-migrants from culturally proximate Western countries, followed by Asia and the Middle East. Length of time in Sweden played a significant role in the incidence of self-employment in tourism, with restaurants dominating as the type of establishment, and it is questioned how much they contribute to rural change and transformation. A recent study by Carson et al. (2018) examined personal migration drivers, lifestyle, and business motivations of international winter tourism entrepreneurs who have moved to a "low-amenity" rural area (i.e. areas with economic decline, outmigration and limited tourism development) in northern Sweden, suggesting that the northern winter and the undeveloped low-amenity character of the place were key factors in migration choices. Consumptive lifestyle interests around counter-urban living and winter outdoor hobbies were prominent, yet there was diversity in terms of business aspirations and considerable seasonal lifestyle–business balancing. However, Carson et al. (2018) question the long-term contributions that migrant entrepreneurs will have to winter tourism development in the low-amenity north, as in line with other studies, they identified a sense of temporariness and expected onward migration among the studied entrepreneurs. A study of migrant tourism entrepreneurs to Norway (Iversen and Jacobsen 2016) found that lifestyle factors (i.e. life stage, migration unit, previous experience with self-employment) also influenced motivations and decision-making.

In Sweden, Eimermann (2016) studied Dutch rural tourism entrepreneurs, their notions of countryside capital, and experiences with Swedish national tourism policy and local populations. He suggests that a tourism–migration nexus occurs where the entrepreneurs are attracted by countryside capital before migration and they use this capital in their firms to attract new tourists after migration. However, rural areas risk losing the entrepreneurs' socio-economic motivation because of frustration among entrepreneurs due to tourism policy issues.

A more fine-grained discussion is Skaptadóttir and Loftsdóttir's (2016) contribution on two kinds of mobility that have played an important role in shaping Iceland's identity, namely the rapid growth of employment-related geographical mobility and the escalation in tourism to Iceland from the 1990s onwards. Their discussion circles around how tourists and migrants are positioned differently in society and culture and thus are perceived differently by the host society. Work-migrants have been important for the economic transformation of Iceland after the crisis in 2008 and are often found in low paid jobs, not recognised for their contribution, whereas the contribution of the privileged tourist is acknowledged. Many recent work migrants to Iceland come to work in the tourism industry. What seems to be lacking is studies on migrant experiences in urban areas, and how they relate to their simultaneous roles as hosts and guests.

Theoretical Approach

To reiterate, our central paradox is that, following the work of Derrida (1999, 2000) and others (Pechlaner et al. 2016; Smith 2019), new migrants, who are often still guests in their own right, are asked to deliver the hospitality that welcomes fellow guests (tourists) to a community and destination. As we have said, the paradox lies in expectations of this welcome. With this paradox comes problems inherent for these hosts, themselves newly arrived guests and the understandings of place, stories, representation and culture. Paradox is inherently embedded in social systems such as tourism and tourism employment and is socially constructed through language and discourse. Paradox theory recognises that there are contradictory yet interrelated elements to many everyday situations (Lewis and Smith 2014). Each dimension may seem logical in isolation, but is absurd when appearing simultaneously and this is certainly the case in our context of guest as host and host as guest.

According to Duncan et al. (2013), the 'social distance' between tourist and hospitality employee becomes less clear due to changes in the travel patterns of people in so-called developed countries. The transitory and seasonal nature of much hospitality work has caused concern in the industry about the degree of authenticity of the service or experience that migrant workers can provide. As they say (ibid. 2013,

p. 6), "Such multiplicity can add spice to the experience of tourism and hospitality but can also challenge or undermine the very reason that visitors seek to consume (or experience) hospitality in a particular location". Thus, employee mobility and the influx of migrant workers in the hospitality industry have challenged conceptions of quality, form and authenticity of interactive service delivery (Duncan et al. 2013, p. 8).

In hospitality, the host–guest relationship traditionally implies that the host is situated in and represents a certain location and its culture, while the guest is a visitor or stranger, seeking to learn about the destination. However, employee mobility and migration challenge these roles of host and guest. Sherlock (2001, quoted in Duncan et al. 2013, p. 10) suggests that "This negotiation of identity acts to destabilize one of the common conceptions behind commercial tourism and hospitality where mobile guests become commercial hosts serving tourist guests". However, migrant hospitality workers have different motives and reasons for travelling and working in another country. It is important to differentiate between hospitality workers who are free, legally, culturally and economically, to travel across borders to work, and those with limited economic, cultural or social capital, who travel due to necessity as refugees or forced economic migrants (Duncan et al. 2013, p. 12). An analysis of hotel workers in Oslo identified different groups of workers based on among other things, ethnic background, occupational mobility and expected time horizon in the hospitality industry (Underthun and Aasland 2015). The study differentiated between career workers who have a certificate of hotel work and are satisfied with their job, and who experience mobility and job permanence; passing workers who are young with short time horizons, and who prefer informal relations with management over trade union membership; and stuck workers who might want other positions in the hotel, but who perceive their chances of occupational mobility as low, and who, because of this, frequently have long often low-level prospects in the hotel job.

Migrants who are mobile due to necessity are often represented in backstage jobs—invisible, low-paid and insecure, cleaning or working in the kitchen—what Underthun and Aasland (2015) characterise as "stuck workers". Duncan et al. (2013) point out that not all bodies fit in all workspaces, and that there exists a "hierarchy of acceptability" around different tourism and hospitality employees and jobs (Duncan et al. 2013, p. 10). Thus, it is important to differentiate between different categories of migrant hospitality workers and their varying motivations and needs to become mobile workers. This further raises central questions concerning "the degree to which this mobile workforce disrupt dichotomies such as 'home' and 'away', 'host' and 'guest' and 'producer' and 'consumer'" (Duncan et al. 2013, p 14).

Final caveats need to be recognised in the context of the Derridian paradox. The extent to which the paradox presents a problem may vary greatly according to hosting context—is the host environment (the hotel or other tourism/hospitality operation) one where guests seek authenticity of experience or are they satisfied with the price, speed and efficiency of the transaction (as might be the case with a fast food outlet)? By the same token, do migrant workers always self-identify as guests within their host communities? This may not be the case with temporary or seasonal migrants (Duncan et al. 2013). Either of these two moderated contexts may, in reality, not represent paradox in the way that we have assumed above.

In the following sections of this chapter, these questions are illuminated and discussed through an examination of the Norwegian context of hospitality migrant workers and through presentation of cases that illustrate different perspectives of the experience of migrant workers in the Norwegian hospitality workforce.

The Case of the Norwegian Hospitality Industry and the Host/Guest Dichotomy

This section presents four cases of hotel workers in Norway: three female room attendants from Poland, the Philippines and Vietnam, and a male Finnish receptionist. Their narratives illustrate different aspects of the Derridian paradox of how migrant hospitality workers, as "guests", experience expectations of delivering hospitality to other "guests", or tourists. Through the narratives, topics such as ethnic background, education, informal networks and status are discussed as they relate to understanding the migrant workers' varying experiences. The four narratives express both similarities and differences in how they address such issues. The cases build on a wider study, consisting of 45 interviews about intercultural organisational communication and inclusion in Norwegian hotel workplaces (Linge 2015). The four narratives are not representative, but illustrate archetypical migrant workplace experiences from the Norwegian hotel industry.

Julia

Julia was 42 years old, from Poland, and worked in a large hotel in a Norwegian city. She had lived in Norway for 15 years, was married and had one child. Before coming to Norway, she had worked in a bank in Poland. After arriving in Norway, she took a cleaning certificate and got a job in the housekeeping department in the hotel. Julia was concerned that there were so few Norwegians working in the housekeeping department, stating that there had been more Norwegians when she started eight years ago. Since language requirements were lower in the housekeeping department than in other departments, most of the employees with non-Norwegian background were gathered there, creating a skewed distribution of Norwegian and non-Norwegian employees in the various departments.

Julia was worried that her colleagues and she would not be able to provide the service expected by the hotel guests, due to their lack of Norwegian proficiency. She was anxious that this would reflect badly on the hotel, since they could not communicate effectively with hotel guests. In this sense, Julia's narrative illustrated the paradox between immigrants being both guests and hosts in a society. As initial guests to the Norwegian society and labour market, the housekeeping employees with immigrant backgrounds lacked central language qualifications that could enable them to fill their roles as hosts in their workplace. Through working with predominantly other foreigners, they experienced few opportunities to practice Norwegian, thus decreasing their opportunities to improve their language skills and, in extension, their host practice.

Julia expressed a concern that she experienced a de-learning of her own Norwegian language skills, since there was no place for her to practice Norwegian at work. She claimed that she had been more proficient in Norwegian when she started working there than she was now. Furthermore, Julia felt that colleagues in other departments looked down on her and her housekeeping colleagues. This she attributed to the fact that there were so many foreigners working in the housekeeping department, which contributed to lower the status of the department. According to Julia, many of her colleagues were afraid of losing their jobs and therefore worked as “machines”. Because of the size of the hotel, the different departments did not sit together for lunch or other breaks, but had their breaks at separate times. Thus, there was not much opportunity for Julia and her colleagues to socialise and communicate with colleagues from other departments during their workday. Julia stated that if she could choose, she would prefer to work with Norwegian colleagues, instead of only working with other colleagues with immigrant backgrounds.

Maria

Maria was 37 years old, from the Philippines, and had come to Norway as an au pair nine years earlier. Her aunt and sister, who also lived in Norway, had helped her when she arrived. In the Philippines she had studied hotel and tourism management, and she had worked in the front office of a hotel. Now she worked in a large hotel in a Norwegian city where she had a permanent work contract in the housekeeping department, at times also substituting as an assistant to the housekeeper. She was married to a Norwegian man and had two children. Maria enjoyed working in a multicultural work environment and socialising with people from different countries. However, she commented negatively on the fact that most of the foreign employees worked in the housekeeping department, while most of the employees in the front desk were Norwegian: “But it’s a little bad, because you see, in the reception, it’s all Norwegian… And in the office here, all are

Norwegian… And then it's only us down here, like in housekeeping or like cleaning…it's foreigners".

Maria's narrative illustrates the importance of family and national networks for immigrants seeking work in another country. One of the reasons that she sought work in Norway, first as au pair, and then in the hotel, was that she already had her sister and aunt living here. Through marrying a Norwegian man and having children growing up in Norway, she also had close ties to the Norwegian society through her Norwegian family. Her strong emphasis on the importance of learning Norwegian and integration revealed a sense of commitment towards being a part of the Norwegian society. However, her comments on the segregated work environment in the various hotel departments reflected a feeling of distance from her Norwegian colleagues in other departments. According to the hotel management, the segregated work environment was largely due to language requirements. Thus, even though Maria had studied hotel and tourism management and had hotel front desk experience from the Philippines, she did not think that she could get a front-office job in Norway.

Maria described her work in the housekeeping department as physically tough and hard work, and she was worried about her health should she continue working there in the future. She planned to pursue further studies in order to qualify for other types of jobs, and eventually apply for these other types of jobs, either in the hotel sector or outside. Thus, although she expressed that it could be difficult to get another job, she did not perceive herself as "stuck" in the sense that she planned on staying a long time in the housekeeping department. Her strong family ties to both her Norwegian and Filipino family living in Norway accentuated her goals of getting another type of job.

Anh

Anh was 45 years old, and worked in the housekeeping department in a medium-sized hotel in a Norwegian city. She came from Vietnam where she had finished secondary high school, and had attended a Norwegian language course in Norway. She was temporarily employed, and would get her schedule 2–3 weeks in advance. Anh said that she preferred to speak Norwegian at work: "Because we live in Norway…so it is important we only speak Norwegian". Concerning guest communication, Anh explained that they had been instructed by the hotel management to greet the guests and to treat them with respect. She described her relationship with colleagues from other departments as good, and explained how they would assist the housekeeping staff if they needed help when communicating with guests. Anh referred to her workplace as a big family. "It's like a big family, very nice, we help each other". She and her colleagues in the housekeeping department would often socialise in their spare time, for example going to the beach or barbequing, and would also bring their families. She enjoyed learning about her colleagues' different cultures and sharing food from their respective countries. Because she was happy with the work environment at the hotel she did not want to change jobs in the future, but hoped to stay in the same workplace.

Anh could in some ways fit the characteristics of what Underthun and Aasland (2015) characterise as "stuck workers" in that she expressed that she thought it would be difficult for her to get another type of job and therefore had long-term prospects of working in her current job. However, her description of her work experience and relations with colleagues at work are quite different from Julia and Maria. Her portrayal of the work environment as a "big family" depicts a sense of belonging and commitment to her workplace. Furthermore, she described a sense of community illustrated through narratives of how her colleagues in other departments would help herself and her housekeeping colleagues when they experienced problems communicating with guests. Because the hotel was not so big, employees from different departments would sit together and talk during lunch, thus giving the foreign employees a chance to practice Norwegian. Anh also had close social relationships with her colleagues in the housekeeping department, socialising in their spare time and sharing knowledge about their respective countries. Thus, although she did express that she thought it would be difficult to get another job, she did not describe a sense of feeling stuck in an unwanted job position. Instead, she expressed a sense of belonging to the multicultural work environment and her colleagues, not only from Vietnam, but also from a multitude of different countries. In many ways, Anh's narrative pointed towards living and being integrated into a heterogeneous and multicultural Norwegian society consisting of a multitude of cultures and backgrounds.

Antti

Antti was 25 years, from Finland, and had lived in Norway for almost a year. He had a Norwegian girlfriend, which was one of the reasons that he moved to Norway where he got a job in the front office of a large hotel in a Norwegian city. Antti was the only employee in the front office who was not Norwegian. He had studied economics and international tourism management in England, and planned to continue with an MBA in Economics in the future. One of the things he enjoyed about his job was that he had a certain autonomy to provide service to guests, such as offering free drinks, providing compensations or upgrading rooms if guests complained, without first having to ask his manager. According to Antti, this was important both for providing good service and securing guest satisfaction, and for building his own work satisfaction as an employee. He seldom experienced that hotel guests or colleagues had difficulties understanding him even though he spoke with a Finnish/Swedish accent, but he tried to adapt to Norwegian words to help colleagues and guests understand him better.

Antti did not experience the hotel environment as very hierarchical. During lunchtime, he would, for instance, sit with colleagues from the front office, security or HR, but not from the housekeeping department. He enjoyed working in a multicultural work environment, but admitted that the front office was probably the least multicultural department (himself being the only foreigner). Antti stated that he thought there was a huge difference working in a big versus a smaller hotel, and that it was easier to get to know colleagues in a smaller hotel since one often sat together at lunch and spoke together. This also made it easier to opt to help colleagues when they needed assistance. However, after a period with job rotation at his current workplace, trying out jobs in other departments, he thought they had gained a better understanding for the work done in, for instance, the housekeeping department, thus increasing the tolerance and understanding for their colleagues' work situation. He was content with his current job, but planned on moving forward, saying that he would like a job with more responsibility: "I have always known that this is an intermediate position, and that I actively want to move forward and apply."

Antti's narrative is fundamentally different from the previous narratives from the housekeeping department. For Antti, the hotel job was a springboard for new job opportunities. His reason for moving to Norway was his Norwegian girlfriend, but he did not necessarily view this as a permanent place to stay. In contrast to Maria, Julia and Anh, Antti was the only non-Norwegian working in his department. Similarly to Maria, Julia and Anh, Antti was also a guest to the Norwegian society and labour market. However, he did not view this as an issue of concern or as an impediment. The only thing he commented on in terms of being a foreigner concerned language issues, where he described how he would adjust his Swedish in order to accommodate hotel guests and colleagues. In many ways, Antti exemplified what Underthun and Aasland (2015) characterise as a career worker with a certificate of hotel work, who is mainly satisfied with his job, and who experiences large opportunities for occupational mobility.

Analysis and Discussion

The four cases share both differences and similarities. In all of the cases, language comes up as a challenge when discussing their dual role as guests and hosts in the hospitality industry. Issues of inclusion and/or exclusion are also central in the narratives. One of the topics that distinguish the first two cases from the latter two concerns a sense of place and belonging in the workplace. Both Julia's and Maria's observations reflect a sense of exclusion from certain job arenas based on national background and immigrant status. Furthermore, Julia and Maria both expressed a sentiment of being "stuck" in certain occupational spaces in the hotel. This was illustrated through their stories of feeling stigmatised and "out of place" when entering other arenas than the housekeeping department in their workplace, which further accentuated their paradoxical position as both guest and host on their workplaces. Julia especially pointed out difficulties of fulfilling her and her colleagues' roles as host when encountering hotel guests, due to lack of communication possibilities and status. This again created a sense of dissatisfaction in that she felt that they were not able to do the job as well as they should. Maria's close networks with both her Norwegian family and her Filipino family living in Norway, gave her a sense of belonging and commitment to improving her Norwegian skills and further integrating into the Norwegian society and work life. This again seemed to strengthen her belief in her own possibilities for occupational mobility through further education and qualifications.

Anh's and Antti's narratives are very different from each other. However, both express a sense of belonging and community in their work environment. Anh's job satisfaction came from a sense of community with her co-workers in all of the hotel departments. While she had the same challenges concerning language and guest communication as Julia, her colleagues helped her overcome this and fulfil her role as host through assisting when she encountered communication problems. In contrast to Julia, Maria and Antti, Anh stated that she was satisfied with her job in the hotel, and that she did not want another job. One might question to what degree this contributed to her satisfaction with her job, since she, unlike the others, was not concerned with possibilities of occupational mobility.

Antti's situation was again quite different, in that he belonged to a more "privileged" group of workers consisting of Nordic, well-educated, internationally mobile and career-oriented young people. Although he could be considered just as much a guest as Julia, Maria and Anh, he did not experience this as strongly as the others in relation to his role as host towards hotel guests. Antti was given a higher degree of autonomy than the housekeeping employees to deal with hotel guests and solve potential problems. This contributed to his positive experience of successfully combining his dual role as "guest" and "host", thus not presenting this as a paradox in the same way that the other three narratives do.

Julia, Maria, Anh and Antti described varying degrees of workplace inclusion and exclusion, touching upon the importance of being part of a workplace community. The narratives illustrate that a multicultural work environment is in itself no guarantee for workplace inclusion. In her work on inclusion, racism and diversity in institutional life, Sara Ahmed comments on how some people are excluded from certain spaces in light of being perceived as "strangers", while others may be seen as "rightful occupants of certain spaces" (Ahmed 2012, p. 2). Thus, some people will feel included in institutions because their physical appearances or ways of looking and behaving are viewed as the norm, while others experience a sense of being out of place or excluded from certain spaces in the workplace. This is also reflected in Duncan et al.'s (2013) observation that there exists a "hierarchy of acceptability" around various tourism and hospitality workers and jobs, where not all bodies fit into all workspaces.

Ahmed (2012) uses the concept "conditional hospitality" to describe how institutionalised diversity works, where the stranger, or foreigner, is welcomed on condition they return that hospitality either by being integrated into a common organisational culture, or by letting the institutions celebrate their diversity. However, according to Ahmed: "This very structural position of being the guest or the stranger, the one who receives hospitality, allows an act of inclusion to maintain the form of exclusion" (Ahmed 2012, p. 43). Both Julia and Maria expressed a sense of being out of place and excluded from certain spaces in their workplace, based on their status as "guests" in the Norwegian society and labour market. This contributed to making it difficult to fulfil their roles as hosts to the hotel guests satisfactorily. However, Maria's family networks, both Norwegian and Filipino, gave her a stronger sense of belonging and inclusion. Julia, on the other hand, described a more negative experience of working in a department with almost no Norwegians. Her statement that she would prefer to work with Norwegian colleagues accentuates her narrative of being a reluctant outsider, or guest, on her workplace. This was also reflected in her concerns that her Norwegian language skills were actually becoming poorer due to lack of interaction with Norwegian colleagues.

Anh expressed a clearer feeling of being included in her work environment, in part due to the fact that she worked in a smaller hotel than Julia and Maria, which made it easier to socialise and communicate with colleagues from other departments. Increased opportunities for socialisation and communication appeared to encourage a community spirit, where colleagues chipped in and helped each other when necessary. Thus, although Anh also experienced language challenges when communicating with hotel guests, she did not experience it as detrimental to guest services. However, Anh's satisfaction could also be related to her lack of wanting to change jobs.

For Antti, it was not difficult to feel included, in that he experienced close cultural proximity to his Norwegian colleagues. In his front office position he had ample guest contact, and especially enjoyed the autonomy he had when providing service to hotel guests and handling guest complaints. His professional position thus seemed to be least impacted by his paradoxical role as both guest and host in Norwegian worklife.

Antti's remark about how job rotation contributed to creating increased understanding between colleagues in different departments, also increasing respect for the work done by other colleagues, is in this sense interesting. Furthermore, his comment about how the size of the hotel was important for creating a sense of community and increasing a sense of "pulling the load together" and helping each other as colleagues, is also relevant when discussing how to create a more inclusive work environment and alleviate experiences of being guests or strangers in a work context.

Conclusion and Implications

The introduction of this chapter pointed to two main questions. The first question addressed how different migrant experiences influence hospitality employees', or "guests'" possibilities of fulfilling their role as "host" to other tourist guests. The cases and discussion illuminate how different migration backgrounds and experiences contribute to varying perceptions of how migrant workers are able to fulfil their role as host in a hospitality work context. Variations in language competency, educational background, networks, work status, and opportunities for occupational mobility are central to how the migrant workers perceived their own possibilities for presenting the expected hospitality towards tourist guests.

The second question concerned how the use of narratives can illustrate challenges related to the Derridian paradox of how newly settled migrants are expected to act as hosts to fellow "guests", or tourists, in a community or destination. This also brings into view one other point. How long is "newly-settled"? As Julia's and Maria's examples highlight, the issue associated with being a "guest" in a host society does not necessarily dissipate over time. While enthusiasm and engagement in jobs and with people may allow migrants (some) moments of inclusion, what Julia and Maria illustrate is that these issues of being a guest-host continue long after migrants have supposedly "settled" in a new place (Etemaddar et al. 2016). In this chapter, the use of narratives contributes to illuminate the complexity inherent in the different migrant workers' situations, and shows the importance of taking into account varying backgrounds and motivations of hospitality migrant workers when discussing their dual role as guest and host. Furthermore, the narratives point out the importance of differentiating between the more economically and culturally privileged migrant workers and those who travel due to necessity, either as refugees or as forced economic migrants.

The narratives illustrate how foreign workers in the Norwegian hospitality industry are often concentrated in cleaning or housekeeping departments where language requirements are not so strict. This inhibits possibilities of improving host language skills, which further restricts their possibilities of presenting the desired hospitality towards tourist guests. Low status and stigma related to being a foreigner or "stranger" in the workplace are accentuated by the fact that many migrant workers are concentrated in typical backstage, low-wage and "low-skilled" jobs, and that there is little structural integration across the workplace. However, integration in both formal and informal networks of friends and family is highlighted as an important factor for further integration, thus enabling migrant hospitality workers to expand their role as host when interacting with tourist guests at work.

The previous discussion raises some managerial implications for the tourism and hospitality industry. Firstly, filling the role as a host in the hospitality industry requires effective communication skills and language competency. For immigrant employees, this implies being able to master host language competency in order to communicate efficiently and appropriately with hotel guests. Today, it is often considered to be the responsibility of each individual employee to provide and maintain adequate language skills. Providing employees with opportunities for language training and practice, for instance in the workplace, could contribute to improving immigrant employees' service towards guests and increase interaction with colleagues, which again could contribute to a more inclusive and effective work environment. Secondly, creating more inclusive work environments could contribute to alleviate the paradox between immigrant workers' status as both guests and hosts. Organising the workforce into smaller departments or teams, inviting employees to participate in job rotation, and creating informal venues for interaction and communication across departments are some actions that might contribute to creating more inclusive workplaces for all. Finally, addressing issues of structural integration (or lack of such) and accumulation of immigrant workers into low-wage and low-status departments may create increased understanding of the mechanisms behind why some employees feel excluded from certain workspaces, thus cementing their status as "guests".

The Norwegian hospitality industry is largely dependent on migrant labour, and is one of the biggest employers of migrant labour in Norway today. Requirements for service quality and hospitality are constantly sharpened to meet the fierce competition in the tourism and hospitality sector. Ensuring that all employees are able to present the expected hospitality to tourists and guests, should therefore be a main concern for all tourism and hospitality organisations. Through addressing these challenges, the tourism and hospitality industry can contribute to improving the service provided by all parts of the workforce; thus gaining a competitive advantage in addition to creating an inclusive workplace for all employees.

References

Ahmed, Sara. 2012. *On Being Included: Racism and Diversity in Institutional Life*. Durham, NC: Duke University Press.

Baum, Tom. 2018. "Sustainable Human Resource Management as a Driver in Tourism Policy and Planning: A Serious Sin of Omission?" *Journal of Sustainable Tourism* 26 (6): 873–889. https://www.tandfonline.com/doi/full/10.1080/09669582.2017.1423318.

Calmfors, Lars, and Nora Sánchez Gassen. 2019. "Integrating Immigrants into the Nordic Labour Markets: Background, Summary and Policy Conclusions." In *Integrating Immigrants into the Nordic Labour Markets*, ed. Lars Calmfors and Nora Sánchez Gassen, 9–36, Copenhagen: Nordic Council of Ministers.

Carson, Doris Anna, Dean Bradley Carson, and Marco Eimermann. 2018. "International Winter Tourism Entrepreneurs in Northern Sweden: Understanding Migration, Lifestyle, and Business Motivations." *Scandinavian Journal of Hospitality and Tourism* 18 (2): 183–198. https://www.tandfonline.com/doi/full/10.1080/15022250.2017.1339503.

Cohen, Scott, Tara Duncan, and Maria Thulemark. 2015. "Lifestyle Mobilities: The Crossroads of Travel, Leisure and Migration." *Mobilities* 10 (1): 155–172. https://rsa.tandfonline.com/doi/full/10.1080/17450101.2013.826481#.XaSoKjMzY2w.

Derrida, Jacques. 1999. *Adieu to Emmanuel Levinas*, trans. P. Brault and M. Naas. Stanford, CA: Stanford University Press.

Derrida, Jacques. 2000. *Of Hospitality: Anne Dufourmantelle Invites Jacques Derrida to Respond*, trans. R. Bowlby. Stanford, CA: Stanford University Press.

Dokka, Synne, Hege M. Knutsen, and Sylvi B. Endresen. 2015. "Outsourcing with a Human Face? Variegated Workplace Regimes in the Norwegian Hotel Industry." In *A Hospitable World? Organising Work and Workers in Hotels and Tourists Resorts*, ed. David Jordhus-Lier and Anders Underthun, 84–100. London: Routledge.

Duncan, Tara, David Scott, and Tom Baum. 2013. "The Mobilities of Hospitality Work." *Annals of Tourism Research* 41: 1–19. https://www.sciencedirect.com/science/article/abs/pii/S0160738312001508.

Eimermann, Marco. 2016. "Two Sides of the Same Coin: Dutch Rural Tourism Entrepreneurs and Countryside Capital in Sweden." *Rural Society* 25 (1): 55–73. https://www.tandfonline.com/doi/full/10.1080/10371656.2016.1152033.

Etemaddar, Mitra, Tara Duncan, and Hazel Tucker. 2016. "Experiencing 'Moments of Home' Through Diaspora Tourism and Travel." *Tourism Geographies* 8 (5): 503–519. https://www.tandfonline.com/doi/full/10.1080/14616688.2016.1220973.

Grødem, Anne Skevik. 2016. "Migration as a Challenge to the Nordic Welfare States." In *The Current Nordic Welfare State Model*, ed. Noralv Veggeland, 149–169. New York: Nova Science Publishers.

Harbo, Lisbeth Greve, Timothy Heleniak, and Åsa Ström Hildestrand. 2017. "From Migrants to Workers: Regional and Local Practices on Integration of Labour Migrants and Refugees in Rural Areas in the Nordic Countries." NORDREGIO Working Paper: 5.

Irimias, Anna, and Gábor Michalkó. 2016. "Hosting While Being Hosted: A Perspective of Hungarian Migrant Hospitality Workers in London, UK." *Tourism and Hospitality Research* 16 (2): 172–183. https://journals.sagepub.com/doi/10.1177/1467358415600211.

Iversen, Ingvild, and Jens Kristian Steen Jacobsen. 2016. "Migrant Tourism Entrepreneurs in Rural Norway." *Scandinavian Journal of Hospitality and Tourism* 16 (4): 484–499. https://www.tandfonline.com/doi/full/10.1080/15022250.2015.1113887.

Lewis, Marianne W., and Wendy K. Smith. 2014. "Paradox as a Metatheoretical Perspective: Sharpening the Focus and Widening the Scope." *Journal of Applied Behavioral Science* 50 (2): 127–149. https://journals.sagepub.com/doi/10.1177/0021886314522322.

Linge, Tone. 2015. "Including or Excluding Workplace Communication? Analyzing Communication Competence in Multicultural Hotel Workplaces." PhD diss., University of Oslo.

Lundmark, Linda. 2005. "Economic Restructuring into Tourism in the Swedish Mountain Range." *Scandinavian Journal of Hospitality and Tourism* 5 (1): 23–45. https://www.tandfonline.com/doi/full/10.1080/15022250510014273.

Lundmark, Linda. 2006. "Mobility, Migration and Seasonal Tourism Employment: Evidence from Swedish Mountain Municipalities." *Scandinavian Journal of Hospitality and Tourism* 6 (3): 197–213. https://www.tandfonline.com/doi/full/10.1080/15022250600866282.

Lundmark, Linda, Marcus Ednarsson, and Svante Karlsson. 2014. "International Migration, Self-Employment and Restructuring Through Tourism in Sparsely Populated Areas." *Scandinavian Journal of Hospitality and Tourism* 14 (4): 422–440. https://www.tandfonline.com/doi/full/10.1080/15022250.2014.967995.

Möller, Peter, and Jan Amcoff. 2018. "Tourism's Localised Population Effect in the Rural Areas of Sweden." *Scandinavian Journal of Hospitality and Tourism* 18 (1): 39–55. https://www.tandfonline.com/doi/full/10.1080/15022250.2016.1259584.

Möller, Cecilia, Birgitta Ericsson, and Kjell Overvåg. 2014. "Seasonal Workers in Swedish and Norwegian Ski Resorts—Potential in-Migrants?" *Scandinavian Journal of Hospitality and Tourism* 14 (4): 385–402. https://www.tandfonline.com/doi/full/10.1080/15022250.2014.968365.

Norwegian Hospitality Association Annual Report. 2018. https://www.nhoreiseliv.no/om-oss/arsrapport/arsrapport-2018/. Accessed on July 17, 2019.

Østbye, Lars, and Kristin Henriksen. 2013. "Innvandrere – Hva Vi Nå Vet og Ikke Vet." *Samfunnsspeilet* 5: 2–10. https://www.ssb.no/befolkning/artikler-og-publikasjoner/_attachment/151663?_ts=143715dbff0. Accessed on July 17, 2019.

Pechlaner, Harald, Christian Nordhorn, and Xenia Poppe. 2016. "Being a Guest—Perspectives of an Extended Hospitality Approach." *International Journal of Culture, Tourism and Hospitality Research* 10 (4): 424–439. https://www.emerald.com/insight/content/doi/10.1108/IJCTHR-02-2016-0016/full/html.

Skaptadóttir, Unnur Dis, and Kristín Loftsdóttir. 2016. "The Tourist and the Migrant Worker: Different Perceptions of Mobility in Iceland." In *Mobility to the Edges of Europe: The Case of Iceland and Poland*, ed. Dorota Rancew-Sikora and Unnur Dis Skaptadóttir, 17–37. Warsaw: Scholar Publishing House.

Smith, Rachel. 2019. "Be Our Guest/Worker: Reciprocal Dependency and Expressions of Hospitality in Ni-Vanuatu Overseas Labour Migration." *Journal of the Royal Anthropological Institute* 25: 349–367. https://rai.onlinelibrary.wiley.com/doi/full/10.1111/1467-9655.13029.

Thulemark, Maria, Mats Lundmark, and Susanna Heldt-Cassel. 2014. "Tourism Employment and Creative in-Migrants." *Scandinavian Journal of Hospitality and Tourism* 14 (4): 403–421. https://www.tandfonline.com/doi/full/10.1080/15022250.2014.968000.

Tuulentie, Seija, and Bente Heimtun. 2014. "New Rural Residents or Working Tourists? Place Attachment of Mobile Tourism Workers in Finnish Lapland and Northern Norway." *Scandinavian Journal of Hospitality and Tourism* 14 (4): 367–384. https://www.tandfonline.com/doi/full/10.1080/15022250.2014.967998.

Underthun, Anders, and Aadne Aasland. 2015. "Diverging Work Experiences and Time Horizons: Mapping Hotel Workers in the Oslo Metropolitan Area." In *A Hospitable World? Organising Work and Workers in Hotels and Tourist Resorts*, ed. David Jordhus-Lier and Anders Underthun, 101–121. London: Routledge.

20

Tourism Employment in Nordic Countries: Trends, Practices and Opportunities

Andreas Walmsley, Kajsa Åberg, Petra Blinnikka and Gunnar Thór Jóhannesson

This text set out to provide an overview of issues relating to tourism employment in Nordic countries. We did not begin the journey with a specific agenda in mind, save to offer contributors a platform to present their work and to explore themes that were being researched in relation

A. Walmsley (✉)
International Centre for Transformational Entrepreneurship,
Coventry University, Coventry, UK
e-mail: ad3412@coventry.ac.uk

K. Åberg
Region Västerbotten, Umeå, Sweden

P. Blinnikka
JAMK University of Applied Sciences, Jyväskylä, Finland
e-mail: petra.blinnikka@jamk.fi

G. T. Jóhannesson
Faculty of Life and Environmental Sciences,
University of Iceland, Reykjavík, Iceland
e-mail: gtj@hi.is

A. Walmsley et al. (eds.), *Tourism Employment in Nordic Countries*,
https://doi.org/10.1007/978-3-030-47813-1_20

to tourism employment in Nordic countries. Essentially, we wanted to know what the issues were that researchers in this area were dealing with: what were the trends, practices and what the resultant opportunities? We were delighted with the wide range of themes that were proposed, documenting the richness and underlying complexities of tourism employment, and in equal measure acknowledge the resurfacing of some common themes. Thus, the contributions point to trends, practices and opportunities that those with a background in the area may recognise, and yet equally offering material that should stimulate further interest and reveal new perspectives. In this chapter we review and comment on some of the common themes that have threaded their way throughout the text, culminating in an agenda for future research, with implications for policy and practice also.

The Nordic countries are, of course, unique in their own right, and yet we argue share much common ground to merit a publication looking at employment issues as they relate to tourism through a 'Nordic lens' (see Chapter 1). This lens is in part economic, and yet also climatic and cultural and more than just of academic interest: Ahlness picks up on the need for this common Nordic focus in Chapter 18 where she argues for pan-Nordic action to see meaningful change in tourism employment practices. Adopting and extending this logic we would also argue that while the text was designed to appeal to those interested in a Nordic perspective, the issues explored in this volume arguably find parallels across the globe. As such the contributions here should appeal to those interested in employment in tourism more generally, including the many areas touched by tourism's extended reach, such as regional and rural economic development, migration, entrepreneurship, human resources management, cultural studies and sustainability/sustainable development.

In addition, those approaching the chapters from a general employment rather than tourism specific view will certainly find issues here that are relevant to the changing structure of employment and the nature of work in the twenty-first century. The development of tourism may from such a macro perspective be regarded as one out of many sectors that face new challenges, but also opportunities relating to issues such as technological developments culminating in the Fourth Industrial Revolution (World Economic Forum 2016), the rise in artificial intelligence more generally, and broader socio-political developments with implications for the meaning and nature of work in the twenty-first century.

Many of the issues presented in the individual chapters inevitably mirror issues commonly associated with tourism work, such as high levels of youth, female and migrant employment, high rates of labour turnover brought about in part by high levels of seasonality, informal recruitment and selection practices, informal HRM practices more generally, and lack of structured career paths. The manifestation of tourism-typical employment patterns varies by country. We can see this most clearly perhaps in the dramatic growth of international tourism to Iceland and consequently the chapters on Iceland all, more or less, grapple with accommodating this growth that has created various challenges, not least relating to employment. Thus, approximately one-third of Iceland's tourism workforce are migrants. While not as stark as in Iceland, all Nordic countries have witnessed growth in tourist arrivals in recent years and thus much of the book has dealt with reactions to and anticipation of further growth in tourism.

With anticipation of further growth, one theme that received much attention is workforce needs and how policy-makers plan to address these within the wider context of tourism employment (e.g. seasonality, attitudes towards tourism work) (see Bohn and Bernardi's chapter, also Harju-Myllyaho et al., Duncan et al. and Visbech). This also relates to workforce segmentation and labour mobility. As Bohn and Bernardi argue, however, a segmentation of the workforce could run against the grain of an egalitarian society such as is promoted by Nordic countries (see introductory chapter). Linge et al. also pick up on this issue when they review migrant work in Norway pointing out that a lack of integration in employment has implications for a lack of integration into society. Here Linge et al. also demonstrate how within the tourism migrant workforce stratification exists, reminiscent of Adler and Adler's (2004) review of resort workers in Hawaii. Here then we find tensions that are not unique to Nordic countries but that are intensified in societies that not only promote egalitarianism, but that seek via policy initiatives to enact it. How Nordic countries continue to manage the contradictions inherent in tourism employment within a society that, using Hofstede's (1980) terminology, avoids power distance will remain an interesting subject of study for academics and of practical concern to policy-makers.

Many of the (seeming) contradictions in tourism employment, what Harju-Myllyaho et al. with reference to Ladkin's (2011) work more generally refer to as 'the complex nature of tourism work and all the presumptions that affect the circumstances in the industry', are increasingly brought to the fore by the move towards precarious employment, generally, but also specifically in tourism (Robinson et al. 2019). One form of precarious employment, not regularly regarded as such perhaps but critical in its implications for sustainable employment practices, is voluntourism. This is picked up very directly by Einarsdóttir and Rafnsdóttir who explore the situation of voluntourists in Iceland, also commented on by other Icelandic authors. One can immediately see the attraction for many voluntourists of spending time working in some of the most dramatic landscapes Europe has to offer, and to experience 'the Icelandic way of life'. However, it is clear that the relationship between the host organisation and the volunteer is based on an economic exchange, and even where hosting businesses are matter-of-fact about the nature of the work that is required, this does not detract from the fact that much of this work falls outside the scope of labour market regulations. In fact, Einarsdóttir and Rafnsdóttir draw attention to the potential breach of labour regulations, specifically where for-profit organisations recruit individuals for unpaid labour. Regulations that are easily flouted lose their credibility and ultimately purpose. Júlíusdóttir and Halldórsdóttir are equally critical of many employment practices as they relate to the use of volunteers in tourism in the Icelandic economy. This seems to be, for Iceland at least according to the contributions here, a serious issue.

What the text has also done in many chapters is bring together micro and macro perspectives. While the majority of studies of tourism (and hospitality) employment tend to focus on individual firms (or small groups of firms), with discussions of sustainability considerations surrounding employment only infrequently making their way to policy-making, i.e. macro levels (Baum 2018), many of our contributors do make these connections. For example, Gjerald and Furunes focus on co-creation at the level of the individual organisation but draw conclusions related to the wider discussions surrounding the meaning of decent work. Here they then refer specifically to the notion of Nordic Leadership which revolves around notions of democracy, human dignity, responsibility, obligations, rights and the individual's role in relation to the community (Andreasson and Lundqvist 2018). Again, we encounter the intersection of broader policy ideals, arguably Nordic in flavour, with organisational priorities.

Duncan et al.'s chapter demonstrates that the Swedish government is very aware of the need for a sustainable approach to policy-making in tourism and specifically to tourism employment, which does not make the issue any easier to tackle. The issue of wanting tourism growth, but at the same time trying to ensure this growth is sustainable, leads to the creation of decent jobs, not just 'any kind of job', remains one of the key challenges for those charged with tourism policy and planning. Indeed, the issue of decent work and economic growth has broader societal relevance, featuring as one of the UN Sustainable Development Goals. There is a constant danger that policy-makers but particularly employers only pay lip service to the ideals of decent work, and that the (immediate) economic needs ultimately take precedence (Duncan et al.). Indeed, it could be argued, that with the changing nature of work (e.g. increased labour market flexibility and the use of short-term, temporary or even zero-hour contracts) the challenge for sectors such as tourism to become more sustainable in relation to its employment practices becomes even more challenging. Ironically, where human resources management is said to have superseded personnel management with the former's focus on commitment rather than compliance, or a long-term rather than short-term perspective (e.g. Guest 1987), one wonders whether we are now reverting to a 'new' form of employment relationship. Precarity is the sign of the times in many sectors (International Labour Office 2018), including tourism (Cañada 2018; Robinson et al. 2019)—this presents particularly Nordic countries with their developed social welfare systems and strong social contracts with a significant challenge. Ultimately, as Duncan et al. argue, picking up also on other recent contributions (Baum 2018; Robinson et al. 2019), decent work in tourism cannot be divorced from wider societal issues, not least what counts as legitimate business practices. Here, what happens in tourism needs to be tied to broader political discourse, including the relationship between business and society, between capital and labour, where, despite affirmations to the contrary, maximisation of shareholder value still seems to hold much sway when considering the relationship between tourism employer and tourism worker (Walmsley and Partington 2014). Aligning the economic growth with decent work agenda will continue to challenge policy-makers.

Given the rise in precarious employment this is still an area that could do with more research in the area of tourism employment such as that offered here by Meged. Her investigation of tour guides in Copenhagen is instructive as it provides a clear demonstration of how the traditional organisation of work and workers' rights via trade unions is being undermined by advances in what Meged, with reference to Johannessen's (2019) work, refers to as the aforementioned Fourth Industrial Revolution. In her chapter, Meged describes how certified guides are facing increasing competition from 'guide lights', frequently migrants with little or no training and education in tour guiding who undercut the traditional guides, by for example accepting 'tips-only' work. This is another example of how the loosening of labour market regulations, including the free movement of labour, is affecting tourism work. There are clear tensions here as Meged recognises between Denmark's 'flexicurity' system (see also Introduction) and changing dynamics in the labour market. The shift in working practices is for some moving away from the notion of decent work.

Labour migration is another key theme running throughout the book, again, reflecting broader global issues. Nonetheless, immigration within and to Nordic countries presents its own characteristics such as highly seasonal demand patterns, highly concentrated tourism development, generous welfare systems, high levels of union representation and consequently working conditions largely the result of negotiation rather than legislation, not forgetting cultural and natural settings that attract neither strictly economic migrants nor migrants fleeing persecution (refugees). As Eimermann et al.'s chapter illustrates, northern Europeans, here Dutch nationals, emigrate to Nordic countries not for economic gain but for lifestyle reasons. Labour migrants are a highly diverse group; their work experiences may differ substantially even within the same business as Wendt et al. in their comparison of receptionists and housekeepers illustrate. Of course, functionally these are two different roles, but it was not so much the work function that led to varying experiences but the ability for social interaction with the native population that resulted in quite different experiences for both groups of migrant workers. From a more utilitarian perspective, interaction with 'the locals' also assisted in the acquisition of language skills which in turn promoted labour mobility within the country, something also identified by Baum (2012) previously, as well as being highlighted in Einarsdottir and Rafnsdottir's chapter.

Immigration to Nordic countries occurs for a number of reasons. We can see in Eimermann et al.'s chapter that immigrants forego economic gain, choosing to move to a Nordic country (in this case Sweden) for a slower pace of life more aligned with personal values. At the same time, while much immigration to politically stable Nordic countries undertaken by those fleeing conflict and political turmoil, there is arguably an economic dimension to much immigration from other European countries: political push and economic pull factors exist side by side. There is also the temporary immigration mentioned in numerous chapters, undertaken by voluntourists (the notions of tourist and migrant converge) or seasonal tourism workers, neither economically nor politically motivated.

As a final note to Eimermann et al., it is worth highlighting their finding that as a collective, the Dutch lifestyle entrepreneurs did little for the Swedish economy (either by avoiding taxes, or paying few taxes, or by being so far at the economic margins that their businesses hardly had any impact, e.g. on employment or innovation, on their communities)—something recognised more generally in the entrepreneurship literature (e.g. Sautet 2014). However, taking just an economic perspective would be, one feels given Eimermann et al.'s conclusion, ignoring the bigger picture. Thus, 'new residents' provide an important social function to communities, which is, so Eimermann et al. argue, in the interest of the Swedish welfare state.

Seasonality has featured strongly in the text and is a facet of much Nordic tourism. Though a regularly discussed phenomenon of much tourism and tourism employment, because of its major role in shaping the lives of those in seasonal destinations it continues to attract interest from policy-makers and academics. But, as Ericsson et al. maintain, there are still new perspectives to be taken on seasonality. Thus, in their chapter, Ericsson et al. argue that high rates of labour turnover and the lifestyle orientation of many of those who assist in Norwegian and Swedish ski resorts could in fact be contributing to innovation and competitiveness. The movement of labour between destinations can increase levels of innovation, which is particularly interesting to consider given that this may reduce the learning and absorptive capacity within the individual tourism organisation itself (Thomas and Wood 2015). This aspect of seasonality and labour churn are rarely discussed in the tourism literature, though interestingly Ateljevic and Doorne (2000) in a New Zealand context, not that dissimilar to the Nordic countries discussed here, also highlight how a lifestyle orientation can be a means of achieving competitiveness via proximity to the consumer/tourist.

The background to Kugapi et al.'s chapter is also the desire to tackle some of the challenges raised by seasonality where craft-based activities have been introduced in the setting of Finnish Lapland. Here a programme aimed at supporting craft-based entrepreneurs, moving them from being a 'crafter' to a 'service provider' was described. Although one frequently considers capacity-building in the context of emerging and developing economies, there is evidence in Kugapi et al.'s work that capacity-building among Finnish tourism entrepreneurs in Lapland can also play a role in developing the industry. However, not all crafters want to grow beyond their micro-status and concern was expressed by some crafters that a focus on the business side of their craft would in fact take time away from the artistic side. This 'growth-reluctance' mirrors Eimermann et al.'s lifestyle entrepreneurs' attitudes towards business development. Without wanting to be too judgemental, it could be argued that both Kugapi et al.'s 'crafters' and Eimermann et al.'s 'lifestyle entrepreneurs' share perhaps an element of business naivety. Similar to Kugapi et al.'s chapter, Visbech describes an initiative 'Digital Growth Development Project' in southern Denmark aimed at enhancing the skills of owner/managers of tourism SMEs in line with the Danish government's goal of improving quality in tourism. With a continued issue of retaining talent in the sector (see discussion below), the issue of upskilling owners/managers of tourism firms themselves may partially addresses the need for tackling skills gaps in the sector.

Further insights into underlying complexity inherent in tourism work are provided by Lundmark who uses labour force data in Sweden to examine the causes of labour mobility. Of particular note here is the comparison with other sectors, something rarely done which highlights that tourism is in many respects not (that) unique (other sectors, notably retail, share many characteristics of tourism employment such as low pay, long hours and general precarity, after all), at least not with regard to the impact of structural factors. What Lundmark then points to is the possibility of cultural drivers within the industry that at least in part explain high levels of labour turnover and mobility. He writes therefore of a 'culture of mobility' which could be something worth exploring further in both a Nordic and non-Nordic context. It has long been noted that an aspect of modernity is mobility, including looser ties to home, family and traditions (Bauman 2000; Rees 2017) that then also spill over into the workplace, and yet the extent to which mobility is desirable (or at least accepted as part of modern life) will vary across nations: Chandler (2016) notes that in the United States workers typically move around a lot more than in Europe, although both Finland and Norway also demonstrate high levels of geographical labour mobility. If mobility between employers and occupations is the new normal, then tourism employment has long been normal.

Chapters that take a macro, policy-making perspective such as that by Harju-Myllyaho et al. who review the relationship between tourism, policy-making and employment in Finland demonstrate how tourism's impact is felt differently in different regions within the country. Overall though, despite (modest) tourism growth in Finland the authors (Harju-Myllyaho et al.), drawing on Vesterinen and Hakkarainen (2013), suggest we are witnessing a decline in study places in tourism in secondary and tertiary level educational institutions. This is in part attributed to a persistent weak perception of tourism work.

Unsurprisingly perhaps given the recognition that for many youth their first experience of work is in the tourism (and hospitality) industry and that the sector is known for high levels of youth employment

(Robinson et al. 2019; Walmsley 2015) the relationship between youth and tourism employment is addressed in a number of chapters (Harju-Myllyaho et al.; Einarsdottir and Haraldsdottir, Visbech). This relationship, so Dagsland et al. who look at the situation of youth employment in Norway specifically through the role of apprenticeships, is problematic. Apprentices, while offering a keen and 'cheap' workforce are often looked down upon, seen by the industry as a burden, and treated accordingly. This is hardly an atmosphere that will encourage the industry's future talent to stay in the sector, leading, as Dagsland et al. argue, to the likelihood of a 'premature exit'. Precisely this issue is also commented upon by Harju-Myllyaho et al. in a Finnish context and Visbech in Denmark. Where Finland is facing labour shortages for an expanding tourism sector, attempting to find ways of attracting new talent, Harju-Myllyaho et al. pithily comment: 'In addition to attracting fresh workforce to the industry, it would be useful to consider how to ensure that the existing workforce would stay'. Similarly, Visbech focuses on the Danish tourism strategy, recognising its aim for improved quality while at the same time she highlights the problems with attracting and retaining qualified staff (her focus is particularly on youth and migrants), especially given the issue of seasonality.

The youth employment theme also features as the focus of Einarsdottir and Haraldsdottir's chapter. Here they focus specifically on teenage girls (16–17 years old) whose first work experience is in tourism. The dramatic growth of tourism in Iceland, already commented upon, has seen the industry faced with labour shortages. These appear to be filled by migrants but in Iceland's case also Icelandic youth. The experience of youth employment again is multifaceted, even contradictory as expressed so frequently in their chapter: 'The interviewees are in general satisfied with their wages, provided that they get paid on time…'; 'The girls enjoy having real work and overall seem to like what they are doing. However, the working days can be long and the number of employees on a shift insufficient'. 'In certain instances, the girls had responsibilities that seemed beyond their maturity, while in others their jobs were repetitious and "boring" and not likely to provoke interest or ambition'. The chapter overall does strike the same tone as the other chapters with a focus on youth employment in that if the sector is to remain attractive as youth grow into adults, then it needs to offer meaningful career paths. Whether tourism will remain a stopgap, and employer of first and last resort for youth is yet to be seen; it certainly remains one of the challenges of tourism employment.

There are parallels between apprenticeships and voluntourism: the employer obtains an educated workforce at low cost (apprenticeship) or no direct cost (voluntourist). While employers may argue they have to invest in young or inexperienced workers and so it is only right that they are paid less, this claim is somewhat suspect in light of the general view of tourism being a low-skills sector, supported by the low wages it is notorious for. However, it also underlines the need for a rhetoric where tourism skills are valued in order for educational institutions to recruit students and the educated to be employed by the tourism sector (Åberg 2017). Hypothetically, such a change could well affect the attraction of the sector on lifestyle migrants as cases presented by Eimermann et al. show that few consider their own lack of experience in the tourism sector to be a hindrance to venturing into it.

Despite frequent well-meaning calls to improve the lives of tourism workers, it is somewhat ironic that the voices of these workers are often ignored, or at least not heard directly. This text has also offered tourism workers a voice, for example in Linge et al.'s review of migrant workers in Norway or Asgeirsson et al.'s review of hotel receptionists' work in Iceland. In the latter case, the voices of employees and their managers are heard, set within a human resource management context of improving employee motivation and satisfaction. Another vehicle for accessing and then ensuring the tourism worker's voice is heard is via trade unions. Although as Harju-Myllyaho et al. argue there may be less unionisation in Nordic countries in tourism compared to other industries such as agriculture (fishing), Nordic countries do set themselves apart from other countries in terms of high trade union membership and collective bargaining. As Júlíusdóttir and Halldórsdóttir very forcefully argue, trade unions in Iceland are caught between rapid tourism growth and the legacy of a neoliberal economic policy. Interviews with trade union representatives here very effectively capture some of the problems tourism workers, and particularly migrant workers face, not least where breaches in employment legislation are commonplace. Again, we encounter the ongoing struggle between Nordic models of industrial relations and economic value systems.

A final word perhaps to the differences and gaps. Ahlness begins her chapter by bringing to our attention the error of postulating 'the' Arctic experience as often promoted by industry, thereby denying the 'diverse landscapes and rich mosaics of cultures' that constitute 'the' Arctic. What we have also attempted in this volume then is not to excessively interfere with the views and interpretations of the contributors for fear of overlaying their views with our own. As all four editors are based in different countries, one of them not a Nordic country it should be noted, this has made the danger of viewing Nordic countries through a single lens less of a risk. Nonetheless, we are aware of the danger and though offering comments to each chapter during the editorial phase these tended to be more to aid clarification than to challenge points of view.

Future Research Agenda

The diverse range of issues discussed in this volume offers several avenues for further research, with implications also for practice. It demonstrates a healthy amount of research being undertaken on tourism employment in Nordic countries; in fact, claims of any aspect of tourism being under-researched depend on what benchmark is used. Thus, compared to other aspects of research in tourism, for example destination marketing or consumer behaviour, research on tourism employment may be regarded as undeveloped. Our text demonstrates that claims such as this miss the mark: there is much work on tourism employment that is highly developed, itself drawing on the intellectual foundations of extant research in this area, as well as drawing on theoretical frameworks outside of the tourism-specific literature.

Opportunities for original research derive from two sources: further investigation into existing phenomena with the aim of bringing hitherto undiscovered aspects of these to light, or research on new phenomena where researchers have not yet had the opportunity to explore these. Both approaches can be employed in relation to tourism employment in Nordic countries. Aspects of tourism employment that have received a considerable research focus for some time include labour turnover, working conditions and skills needs, for example. This does not mean these issues have been exhausted in their potential of offering new insights. The role of seasonal workers and their contribution to innovation levels is a case in point (see Ericsson et al.). Even the issue of working conditions, frequently acknowledged and yet arguably under-researched in terms of actually hearing voices of the tourism worker, offers further scope for empirical investigation. Here, gaining access to workers can prove an almost insurmountable problem which is where trade unions can represent the voice of the tourism worker, not just in negotiations with employers, but also to wider society. This is especially the case where many tourism workers come from marginalised strata of society, compounding the issue of not being heard. For those studying tourism employment in a Nordic context, the high levels of union membership in these countries offer an opportunity not necessarily as accessible in other locations (examples of this happening can be seen in the Icelandic Tourism Research Centre's recent work with trade unions, for example, see Júlíusdóttir and Halldórsdóttir's chapter).

Staying with the voice of the worker, but also those of the employers, we are quite used to seeing surveys used to test or refine theories, with so-called resulting 'hard' data appealing to a sense of what counts as valid knowledge (especially among policy-makers). We should not however ignore the value of exploring the meaning of work in a Weberian *Verstehen* tradition (Weber 1949). We need both, of course, the quantitative to provide broad-based, generalisable results, but to affect change we also need to appeal to the 'human' element in its diversity and the values underlying the rationalities of work. It is this that could provide a bridge between the regularly encountered theory-practice gap that is highlighted by Asgeirsson et al.'s study. As per Kurt Lewin's maxim, 'there is nothing as practical as good theory' we suggest that interpretive research is required not just to explore the unknown in an inductive sense, but interpretive research can be used to further refine broad-based claims and to help explain how they apply in particular circumstances. The same goes for case studies which may be criticised for only giving us a view of a specific locale (in essence many of the chapters here are case studies)—however, that view is detailed and rich and can provide insights that transfer to other situations (Kennedy 1979). The same issue of generalisability (transferability) from the sample population to other contexts applies to quantitative studies too of course (two-span inference as per Cornfield and Tuckey 1956) but is rarely, in our experience, questioned in quantitative research.

Concerning 'new phenomena' the twenty-first-century world is already providing developments that are changing tourism and changing the world of work. Regarding the latter, the pace of technological change is rapidly altering the nature of many occupations, notably developments in artificial intelligence and automatisation. Many tourism workers may be regarded as being at less risk of being replaced by AI where there are limited routine and much customer-specific focus, the kind of jobs less prone to replacement by AI (Frey and Osborne 2017). This could give policy-makers added impetus to promote tourism as a means of employment generation. The relationship between technological developments and tourism work certainly offers scope for further research, both at the level of the individual tourism worker and at a macro-level for policy-makers.

A further aspect that will continue to provide research opportunities are ongoing tensions between different politico-economic models, or the specific manifestation of those models, and their impacts on labour markets. We began the book by outlining how the Nordic model is witnessing ongoing challenges and changes, with recent decades seeing a move towards less security for workers and more flexibility for employers (Knudsen and Lind 2018; Kvist and Greve 2011) challenging the flexicurity system Nordic countries (particularly Sweden and Denmark) are known for. Precarious work is on the rise, one could even argue the nature of work itself is changing. Evidence of these challenges abounds in this text. Further research can seek to make sense of how growing pressures for precarious employment and the resulting pressures to drive inequality (e.g. dual labour market theory, hour-glass economy) tally with the Nordic ideal of egalitarianism. This also offers research opportunities from perspectives such as power dynamics between worker and employer, the experience of work itself and not least the meaning of work to the individual, how to manage workers that are not tied to the organisation, especially performance and reward mechanisms. The meaning of work and careers at a society level, what counts as a respectable or desirable job, will also have ramifications for tourism firms' recruitment challenges and the composition of the tourism workforce.

An issue that has not directly been addressed in this text, but we feel deserves some mention, if admittedly speculative in terms of its implications on employment is the issue of climate change. We hypothesise this will have implications for Nordic countries at several levels. It is recognised that climate change will hit the world's poorest hardest; climate change is likely to lead to increased migration pressures, whether driven by poverty or by conflict. How developed economies should respond to these increased pressures is something governments in the 'affluent North' will be confronted with. Tourism as a sector that disproportionately employs migrants, evidenced in numerous chapters in this text, is more susceptible to policy towards migrants (witness employer concerns about labour shortages subsequent to the UK Brexit (KPMG 2017)).

Furthermore, climatic changes in individual locations are masked by global averages that prima facie may not seem that dramatic. The natural environment which constitutes a key draw of much tourism to Nordic countries is already seeing worrying changes. Added to this is a growing awareness of the impacts of individual behaviour on climate change, especially flying. Flight shaming is growing, especially in Sweden (Abend 2019). With pressure on governments to decarbonise their economies also growing, new legislation is likely to be introduced that targets the most polluting sectors. Many Nordic destinations that are very reliant on international arrivals by air will suffer most. The consequences will extend beyond, but also include tourism employment. Perhaps there will be a shift to more domestic away from international tourism.

Finally, a further macro-level change not unrelated to the climate change issue is the question surrounding the notion of (economic) growth. The shortfalls of focussing solely on the univariate measures of per capita GDP or GNP as proxies for development are now widely accepted, and alternatives continue to engender debate. We have started to encounter the idea of 'de-growth', in tourism too, but frequently this is like 'closing the stable door after the horse has bolted', i.e. after irreparable damage may have been done. We can see that conflicting pressures, for untrammelled growth on the one hand, and also for sustainability on the other, will continue to be played out in the employment sphere. Encouragingly, we recognise a growing (or re-awakened) interest in the employment dimension of sustainability considerations, and real steps being taken to embed sustainability at a macro, policy-making level (e.g. Sweden), and yet the devil is in the detail and implementation of a sustainability agenda is less clear cut. As in the corporate world, so it is in public policy, the problem is often not coming up with a strategy, the real problem lies in its implementation.

References

Abend, L. 2019. The season of 'flight shame' takes off. *Time Magazine* 194: 10.

Åberg, K. 2017. Anyone could do that: Nordic perspectives on competence in tourism. Doctoral diss., Department of Geography and Economic History, Umeå University, Umeå.

Adler, P., and P. Adler. 2004. *Paradise Laborers: Hotel Work in the Global Economy*. New York: Cornell University Press.

Andreasson, U, and M. Lundqvist. 2018. *Nordic Leadership*. Copenhagen: Nordic Council of Ministers.

Ateljevic, I., and S. Doorne. 2000. Staying within the fence: Lifestyle entrepreneurship in tourism. *Journal of Sustainable Tourism* 8 (5): 378–392.

Baum, T. 2012. Migrant workers in the international hotel industry. International Migration Papers. International Labour Office, Geneva.

Baum, T. 2018. Sustainable human resource management as a driver in tourism policy and planning: A serious sin of omission? *Journal of Sustainable Tourism* 26 (6): 873–889.

Bauman, Z. 2000. *Liquid Modernity*. Cambridge: Polity Press.

Cañada, E. 2018. Too precarious to be inclusive? Hotel maid employment in Spain. *Tourism Geographies* 20 (4): 653–674.

Chandler, A. 2016. Why do Americans move so much more than Europeans? How the national mythos and U.S. labor laws influence geographic mobility. *The Atlantic*, October 21.

Cornfield, J., and J. Tuckey. 1956. Average values of mean squares in factorials. *The Annals of Mathematical Statistics* 27 (4): 907–949.

Frey, C.B., and M.A. Osborne. 2017. The future of employment: How susceptible are jobs to computerisation? *Techological Forecasting and Social Change* 114: 254–280.
Guest, D. 1987. Human resource management and industrial relations. *Journal of Management Studies* 24 (5): 503–521.
Hofstede, G. 1980. *Culture's Consequences: International Differences in Work-Related Values*. Beverley Hills, CA: Sage.
International Labour Office. 2018. *World Employment Social Outlook: 82*. Geneva: International Labour Office.
Johannessen, J.-A. 2019. *The Works Place of the Future: The Fourth Industrial Revolution, the Precariat, and the Death of Hierarchies*. New York: Routledge.
Kennedy, M.M. 1979. Generalizing from single case studies. *Evaluation Quarterly* 3 (4): 661–678.
Knudsen, H., and J. Lind. 2018. The "danish models" of labour market regulations and their status after recent changes. *New Zealand Journal of Employment Relations* 43 (1): 83–98.
KPMG. 2017. Labour Migration in the Hospitality Sector. A KPMG report for the British Hospitality Association. KPMG and the British Hospitality Association.
Kvist, J., and B. Greve. 2011. Has the Nordic welfare model been transformed? *Social Policy & Administration* 45 (2): 146–160.
Ladkin, A. 2011. Exploring tourism labour. *Annals of Tourism Research* 38 (3): 1135–1155.
Rees, A. 2017. Reading Australian modernity: Unsettled settlers and cultures of mobility. *History Compass* 15 (11): 1–13.
Robinson, R., A. Martins, D. Solnet, and T. Baum. 2019. Sustaining precarity: Critically examining tourism and employment. *Journal of Sustainable Tourism* 27 (7): 1008–1025.
Sautet, F. 2014. Local and systemic entrepreneurship: Solving the puzzle of entrepreneurship and economic development. *Entrepreneurship Theory and Practice* 37 (2): 387–402.
Thomas, R., and E. Wood. 2015. The absorptive capacity of tourism organisations. *Annals of Tourism Research* 54: 84–99.
Vesterinen, N., and M. Hakkarainen. 2013. Matkailualan koulutuksen jatutkimuksen ohjaus Suomessa. In *Oppiva matkailu. Puheenvuoroja matkailualan koulutuksen kehittämisestä*, ed. Minni Haanpää, Maria Hakkarainen and Heidi Kaihua, 19–31. Rovaniemi: Lapland University Press.
Walmsley, A. 2015. *Youth Employment in Tourism: A Critical Review*. Oxford: Goodfellow Publishers.

Walmsley, A., and S.N. Partington. 2014. A stakeholder approach to working conditions in the tourism and hospitality sector. Paper presented at the 2nd International Hospitality & Tourism Conference 2014 Penang, Malaysia.
Weber, M. 1949. *The Methodology of the Social Sciences*. New York: Free Press.
World Economic Forum. 2016. *The Future of Jobs: Employment, Skills and Workforce Strategy for the Fourth Industrial Revolution*. Geneva: World Economic Forum.

Index

A

academia/academic/university/higher education/college 23, 31, 32, 41, 47, 60, 80, 102, 244, 259, 269, 291, 295, 313, 330, 334, 335, 341, 365, 381, 426, 428, 432
accommodation/hotel 38, 43, 52, 59, 106, 108, 109, 114, 115, 124, 172, 176, 258, 306, 332, 355, 403
adolescent/youth/teenagers 10, 13, 26, 52, 101, 144–148, 153, 159–161, 258, 259, 261, 269–272, 315, 343, 427, 433, 434
ads/adverts/advertisements 50, 105–110, 112–118, 384
aesthetic(s) 217, 222, 262, 269, 272
agency (human agency) 216
ambition 160, 162, 296, 306, 315, 320, 321
appraisal 198, 202, 204, 207, 225
apprentice/apprenticeship 259–272, 434, 435
arctic 354, 355, 375–379, 382, 386, 390, 391, 436
arrivals (tourist) 1, 9–12, 39, 124, 171, 187, 196, 427, 439
attraction (tourist) 7, 65, 137
authentic/authenticity 79, 91, 92, 228, 375, 376, 385, 391, 407–409
autonomy 24, 25, 151, 218, 221, 223, 283, 390, 414, 416, 418

B

bargaining 10, 20, 26, 27, 33, 41–45, 52, 170, 175, 435

A. Walmsley et al. (eds.), *Tourism Employment in Nordic Countries*,
https://doi.org/10.1007/978-3-030-47813-1

barrier (to work/employment/self-employment) 127, 226, 258, 354, 367, 386, 388–390, 392, 404
belonging/ness (sense of) 152, 267, 413, 415–417
benefits (in terms of welfare payments) 6, 13, 31, 44
boss (manager/superior) 135, 148, 150, 161, 196–203, 205–207, 214, 215, 219–228, 252, 304, 313, 340, 414, 433, 435
boundary-spanners 241, 242, 244, 252

C

capital 7, 9, 39, 113, 143, 170, 172, 177, 188, 222, 297, 310, 331, 355, 368, 404, 406, 408, 430
capitalism/capitalist 3, 4, 303, 354
career 22, 25, 46, 47, 52, 60, 66, 70, 72, 125, 126, 135–137, 159, 162, 199, 200, 204, 207, 208, 219, 236, 238, 266, 272, 280, 283, 284, 296, 297, 299, 304, 307, 308, 319–321, 329, 333, 342, 349, 350, 354, 367, 408, 415, 427, 435, 439
catering 59, 172, 176, 220, 319, 366
chef 47, 60, 69, 260, 364
child/children/childcare/childhood 4, 6, 7, 66, 89, 107, 110, 112, 114, 117, 145–147, 153, 158–161, 293, 295, 308, 309, 313, 315, 316, 319–321, 387, 410–412
climate change 305, 322, 439
co-creation 83, 214–219, 221–228, 247, 428
collaboration 25, 30, 42, 51, 83, 89, 91, 92, 150, 183, 184, 224, 245, 312, 404
colleague/co-worker 109, 128, 130, 132–136, 138–140, 149, 150, 152, 154, 156, 158, 160, 203, 216, 218, 221, 224, 226, 227, 262, 263, 266, 267, 272, 284, 297, 402, 410–418, 420
colonial/colonialism/decolonisation 103, 377, 379, 381–383, 389, 391
commitment 83, 148, 150, 151, 156, 198, 243, 251, 263, 282, 284, 333, 387, 388, 412, 413, 416, 429
commodification 353, 385
communication 22, 48, 72, 103, 110, 125, 128, 129, 132, 133, 135, 140, 148, 156, 159, 179, 205–208, 226, 243, 247, 355, 386, 410, 413, 416, 418, 420
community/communities/community-based 21, 24, 58, 63, 71, 81, 88, 92, 103, 109, 188, 228, 235–237, 239, 242, 243, 317, 318, 320, 321, 328, 331–333, 343, 344, 350, 377, 378, 380, 382, 384–388, 390–393, 401, 402, 407, 413, 416–419, 429, 431
commute/commuting 280, 282, 296, 297
compensation (pay/wages) 13, 42, 106, 309, 414

competence/competencies/capability/talent 37, 48, 50, 51, 72, 86, 87, 115, 128, 159, 161, 208, 223, 224, 228, 242, 243, 250–252, 265, 268, 281, 334, 364, 365, 390, 404, 433, 434
competition/competitive 9, 13, 20–22, 24, 27, 28, 30, 32, 33, 42, 47, 86, 148, 196, 223, 236, 251, 320, 350, 366, 403, 420, 430
complexity 14, 102, 185, 229, 283, 331, 338, 402, 419, 426, 433
compliance/non-compliance 176, 178, 179, 186, 333, 429
conflict 41, 48, 205, 218, 226, 332, 388, 393, 431, 439
consumer/consumption 9, 10, 48, 82, 216, 217, 224, 239, 307, 308, 320, 351, 368, 385, 393, 409, 432, 436
contracts (employment) 153, 180, 203, 280, 284, 298, 404
cooperation 61, 68, 86, 203, 206, 379
corporate social responsibility (CSR) 183, 184, 339
craft/crafter/handi-crafts 22, 33, 79, 80, 82, 83, 85–87, 89–91, 228, 260, 385, 387, 432
crisis 9, 40, 46, 144, 181, 291, 407

D

Danish/Denmark 3, 6, 7, 10, 12, 13, 26, 28, 31–33, 37–46, 48, 49, 51–53, 171, 184, 221, 285, 390, 430, 432, 434, 438
decent work 3, 13, 31, 185, 186, 328–330, 333–336, 340, 343, 344, 429, 430
decision-makers/decision-making 61, 63, 151, 202, 204, 207, 376, 379, 406
degree/diploma/qualification 10, 24, 25, 30–32, 43, 60, 115, 137, 138, 149, 153, 220, 223, 263, 264, 266, 280, 283, 284, 291, 294, 295, 297, 352, 365, 366, 376, 381, 404, 407, 410, 416
demand (tourist/tourism) 4, 11, 26, 29, 45, 47, 59, 88, 89, 386, 403
dependence/independence 81, 156, 244, 249, 252, 389
destination/destination management/destination marketing/destination marketing-management organisation (DMO) 7, 9–12, 20, 37, 38, 51, 58, 63, 65, 79, 82, 89, 115, 118, 124, 129, 140, 171, 172, 237–240, 243, 248–252, 281, 290, 317, 328, 331, 333, 350, 356, 357, 362, 375, 380, 402, 405, 407, 408, 419, 432, 436
development (professional/tourism) 2, 7, 9, 22, 39, 52, 58, 61, 63, 67, 78, 79, 81, 84, 91, 235, 266, 270, 342, 349, 367, 378, 379, 390–392, 406, 426, 430
digital/digitalisation 21, 37, 40, 47–53
discourse 9, 103, 332–334, 376, 377, 379, 407, 430
dismiss/dismissal 42, 44, 187, 342

dissatisfied/dissatisfaction (with work/job) 266, 284, 296
downshifter/downshifting 303–308, 312, 313, 316, 318–322
dumping (social) 32, 174, 355, 366

E

economy/economics 2, 8, 9, 12, 20, 21, 27–29, 31, 37, 42, 43, 46, 61, 67, 173, 224, 294, 322, 327, 339, 350, 351, 353, 354, 403, 428, 432, 439
education/training 22, 24, 25, 29, 30, 33, 37, 38, 42–45, 47, 51, 52, 65, 70, 79, 80, 83, 84, 92, 93, 102, 127, 187, 217, 222–224, 228, 236, 240, 250, 259, 260, 269, 280, 283, 285, 291, 293, 296, 298, 308, 320, 321, 361–367, 390, 404, 409, 416, 430
efficiency 21, 59, 69, 150, 269, 409
emotions 129, 150, 224, 228, 241, 247, 311
employability 102
empowerment 151, 198, 339
entrepreneur/entrepreneurship 21, 24, 25, 27–30, 32, 33, 61, 63, 71, 72, 78, 83, 86, 88, 89, 92, 304, 308, 311, 313, 315, 319, 320, 322, 357, 363, 393, 405, 406, 426, 432
environment (natural) 439
equality/inequality 6, 10, 32, 170, 185, 186, 304, 305, 309, 365, 366, 404, 438
ethics/ethical 27, 102, 198, 228
ethnic/ethnicity 45, 90, 110, 187, 383, 408, 409
EU 12, 13, 48, 61, 62, 64, 67, 72, 101, 104, 174, 175, 403
Europe 3, 7, 8, 26, 27, 32, 39, 44, 101, 104, 170, 187, 342, 376, 391, 428, 433
events (festivals) 216, 391
exclusion 128, 271, 376, 415, 417
expectation(s) 90, 92, 262, 402
expenditure 13, 38, 39, 46, 102
experience (work experience) 21, 73, 103, 108, 124, 127, 131, 132, 147, 151, 152, 160, 187, 224, 246–248, 250, 258, 281, 413, 417, 433, 434, 438
exploit/exploitation 90, 93, 103, 124, 272

F

farms/farming 103, 107, 108, 111, 112, 114, 117, 183, 315
feedback 115, 151, 199, 202, 205, 219, 225, 247, 264, 268
finance 11, 308, 365
Finland/Finnish 3, 6, 7, 10–12, 57, 59, 60, 62–67, 69–71, 103, 171, 221, 351, 354, 356, 358, 361, 362, 366, 380, 381, 383, 385, 389, 409, 414, 432–434
flexibility 13, 22, 26, 29, 31, 42, 43, 104, 179, 187, 222, 223, 228, 280, 429, 438
flexicurity 13, 37, 41–45, 52, 430, 438
foreign/foreigner 9, 11, 24, 38, 39, 107, 137, 144, 149, 171, 172,

174, 176, 177, 180, 184, 187, 355, 361, 362, 364, 366, 380, 403, 410, 411, 413–415, 417, 419
freedom 80, 124, 175, 202, 204, 207, 306, 316, 329, 366
frustration 179, 184, 225, 380, 406
full-time/part-time employment 10, 45, 46, 59, 61, 80, 87, 236, 258, 281, 297, 299, 305, 321, 333, 351, 355, 367, 404

G

gender 6, 60, 65, 110, 112, 291, 305, 308, 309, 315, 319, 321, 405
generation 2, 30, 83, 223, 319, 355, 358, 365, 378, 438
globalisation 4, 21, 24, 40, 169, 351, 353
governance 58, 186, 354, 366
government 4, 10, 11, 39, 40, 42, 46, 58, 60–63, 67, 153, 170, 174, 176, 182, 186, 334, 341, 343, 354, 356, 363, 365, 383, 439
graduates 23, 364
Greenland 4, 381, 382, 390

H

heritage (cultural, natural) 7, 77, 79, 81, 89, 91, 93, 390
hierarchy 21, 24, 31, 129, 170, 187, 197, 206, 220, 221
holiday/vacation 82, 102, 178, 281, 315, 319
host/host-guest 25, 30, 33, 91, 103–117, 124, 127–129, 138, 331, 362, 402, 404, 407–410, 416–420, 428
(hotel) chain 196, 197, 200–205, 207
housekeeper/house-keeping 110, 124, 125, 131, 132, 134–140, 152, 158, 342, 410–413, 415, 416, 419, 431
human resource management (HRM) 57, 70, 148, 161, 195–198, 200–202, 229, 282, 331, 332, 427, 435

I

Iceland/Icelandic 3, 6–10, 101, 103–107, 109, 110, 112–118, 124, 130–140, 143–148, 152, 158, 159, 161, 169–171, 173–180, 182–188, 196, 200, 221, 376, 407, 427, 428, 434, 435
identity 60, 81, 137, 152, 169, 185, 188, 201, 242, 243, 332, 390, 406
image (of tourism work/employment) 70, 71, 366
immigrant/immigration/migrant 6, 12, 13, 20, 24, 28, 33, 40, 44–46, 52, 69, 116, 123–140, 169–171, 175–188, 226, 228, 239–241, 244, 249, 252, 281, 296, 304–306, 311–313, 317, 319, 330, 334, 339, 342, 354, 355, 364–366, 401–412, 415, 418–420, 427, 430, 431, 434, 435, 439

incentive 148, 151, 198, 202, 204, 207, 284, 383–385
inclusive/inclusivity/inclusiveness 11, 31, 170, 185, 330, 343, 394, 418, 420
income (pay/wage/salary) 20, 23, 25, 27, 46, 58, 59, 65, 80, 88, 170, 188, 236, 280, 286, 293, 296, 298, 304, 305, 307, 309, 312, 313, 318–321, 350, 391, 403
indigenous 82, 90, 91, 93, 376–378, 380, 381, 384–393
infrastructure 62, 67, 90, 92, 159, 315, 344, 390, 393, 394
innovation 10, 29–33, 40, 217, 218, 236, 237, 240, 241, 243–246, 248, 249, 251, 252, 363, 368, 431, 437
institutional setting/structures 187
integration 28, 128, 218, 329, 342, 401, 402, 404, 412, 419, 427
intern/internship 51
inuit 376, 381, 391, 393

J

(job) security 148, 218, 227, 404

L

labour-intensive 2, 19, 58, 124, 126, 196, 197, 219, 236, 403
labour shortage 71, 124, 161, 361, 434, 439
Lapland 11, 59, 63–65, 67, 78, 79, 82–84, 86, 87, 89–93, 103, 115, 432
leader/leadership 70, 198, 215, 220, 221, 224, 226, 228, 240, 247, 263, 264, 266, 272, 328, 386, 388, 391
legislation 12, 13, 79, 145, 153, 158, 283, 297, 352, 368, 390, 430, 436, 439
leisure 64, 65, 70, 183, 237–239, 241, 243, 248, 249, 251, 252, 284, 290, 304, 305, 307–309, 317, 320, 332, 354, 362, 405, 406
lifestyle (entrepreneur/business) 81, 92, 237, 354, 405, 406, 431, 432
linguistic skills/language skills 107, 128, 129, 133, 137, 223, 236, 403, 410, 411, 417, 419, 420, 431
livelihood 78–80, 92, 350, 351, 363, 376, 380, 381, 383
low-paid 24, 101, 148, 281, 282, 298, 354, 365, 368, 407, 408
low-skills/skilled 2, 125, 131, 144, 147, 361, 366, 419, 435

M

macro (focus/level/economic) 13, 270, 350, 428, 438, 439
marketing 27, 40, 49, 68, 69, 71, 86, 90, 92, 104, 105, 108, 114, 117, 118, 199, 216, 240, 247, 311, 322, 333, 357, 379, 391
minority 171, 187, 342, 383, 394
mobility 101, 114, 123, 125, 126, 136, 137, 139, 140, 170, 187, 188, 236–240, 249, 251, 270,

279–284, 286, 294–299, 342, 351, 354, 361, 362, 364, 405, 406, 408, 415, 416, 419, 427, 431, 433

N

narrative 160, 170, 269, 271, 335, 378, 382, 402, 409, 410, 412–417, 419
negotiation 41, 174, 182, 336, 430, 437
(neo) liberal 3, 103, 169, 170, 174–176, 332, 353, 366, 367, 435
network(s) 13, 21, 23, 86, 87, 92, 125, 130, 131, 136, 138–140, 182, 183, 186, 236, 240, 241, 243, 247, 251, 280, 318, 385, 386, 409, 412, 416, 417, 419
Norway/Norwegian 3, 6, 7, 9, 10, 38, 171, 176, 214, 220, 236, 237, 244, 245, 258–261, 264, 269, 272, 342, 354–356, 358, 362, 363, 367, 368, 380, 381, 383, 384, 387, 389, 402, 403, 405, 406, 409–420, 427, 433–435

O

occupation 10, 19, 28, 124, 125, 131, 132, 214, 218, 220, 223, 224, 226, 259, 262, 263, 272, 293, 349, 433, 438
OECD 4, 6, 40, 61, 78, 80

P

pension(s) 155, 319, 320
performance 48, 69, 109, 124, 126, 128, 148–151, 157, 160, 197, 199, 200, 203–208, 218, 219, 262, 309, 387, 439
periphery/peripheral 2, 7, 60, 67, 143, 236, 251, 311, 333, 341, 343, 354
personnel 31, 196, 198, 200, 203, 204, 208, 240, 241, 247, 269, 281, 365, 429
planning 47, 63, 81, 83, 84, 135, 197, 318, 319, 322, 330, 333, 350, 353, 354, 356, 363, 366, 367, 429
policy/policymakers 3, 9, 13, 42, 52, 61, 62, 186, 197, 203, 330, 333, 350–354, 356–358, 361, 363–368, 379, 383, 389, 392, 406, 426, 427, 429, 432, 435, 437–440
politics 3, 65, 71, 352, 379
power 4, 33, 42, 47, 61, 103, 170, 187, 223, 294, 298, 309, 321, 332, 382, 428, 438
practitioner 208, 241, 242, 248, 251, 335, 341
precarity/precarious/precariat/precarious work 12, 21, 24, 25, 29, 32, 59, 124, 126, 174, 181, 186, 187, 228, 239, 329, 332, 343, 350, 351, 355, 365, 366, 368, 404, 428, 429, 433, 438
productivity 43, 45, 150, 159, 160, 198, 281

profession/professional 22, 25, 26, 28, 30, 47, 60, 125, 127, 131, 136, 138, 159, 236, 238, 240, 242, 259, 261, 263, 267, 293, 296, 340, 342, 349, 351, 362, 364, 388, 418

R

reception/receptionist 125, 131–140, 154, 200, 204, 260, 267, 409, 411
recruitment 57, 64, 65, 68, 71, 103, 104, 159, 196, 201, 280, 361, 427, 439
refugee 171, 187, 306, 408, 419, 430
regulation 43, 103, 220, 316, 350, 355, 404, 428, 430
relationship (work) 70, 205, 207
residents 46, 59, 238, 321, 376, 383
resort 249, 355, 427, 435
responsibility 6, 86, 126, 146, 148, 149, 152, 156, 160, 186, 188, 201, 202, 204, 205, 217, 223, 224, 228, 247, 252, 268, 271, 272, 307, 356, 358, 367, 368, 390, 420, 429, 434
restaurant 38, 43, 47, 49, 52, 64, 72, 78, 108, 115, 124, 131, 145, 146, 152, 203, 245, 247, 271, 279, 285, 290, 291, 293–296, 315, 317, 319, 332, 342, 406
retail/retailer 38, 43, 49, 59, 88, 90, 239, 295, 298, 332, 433
retired/retirement 289, 307, 319, 368
revolution (Fourth Industrial) 2, 20, 21, 24, 29, 31, 426, 430
reward 22, 156, 199, 200, 203, 207, 208, 331, 365, 439
rights (labour/employment/worker) 7, 22, 26, 28, 43, 47, 52, 157, 318, 343, 383, 389, 402, 426, 435
risk 32, 45, 181, 187, 198, 214, 215, 218, 219, 221, 226, 272, 280, 289, 290, 318, 365, 388, 406, 436, 438
rural 2, 11, 51, 58, 60, 124, 130, 139, 146, 153, 159, 236, 251, 304, 311, 313, 316, 317, 319, 321, 322, 341, 343, 362, 363, 380, 390, 393, 404–406, 426

S

sámi 90, 91, 376, 379, 380, 383–385, 387, 389–391
satisfaction (job) 144, 148, 150, 151, 156, 160, 196, 198–200, 207, 218, 263, 264, 282–285, 416
Scandinavia 31, 249, 376, 377
school 47, 107, 118, 146, 159, 259, 316, 319, 412
seasonal/seasonality 9, 10, 19, 25, 45, 46, 52, 59, 63–65, 69, 72, 78, 88, 117, 126, 143, 146, 159, 235–241, 243–252, 284, 290, 298, 329, 350, 354, 361–363, 366–368, 391, 403, 406, 407, 427, 430–432, 434, 437
segregation 64–66, 132
self-employed/self-employment 13, 21, 290, 316, 321, 322, 406
service work 65, 70, 225, 353, 402
skiing 236, 243, 249, 250, 252, 291

skills 2, 22, 29, 40, 43, 47, 48, 50–53, 63, 64, 67, 70, 79, 80, 87, 92, 102, 104–106, 108, 114, 115, 125–130, 136, 137, 143, 150, 152, 153, 160, 161, 215, 223, 226, 240, 242–244, 247, 248, 252, 258, 262, 264, 268, 283, 316, 333, 365, 433, 435
SMEs 46–49, 51–53, 432
socialisation 130, 261–264, 271, 418
stakeholder 61, 83, 84, 92, 213, 228, 229, 333, 356, 357, 368, 393
strategy 11, 12, 40, 44, 46, 49, 51, 52, 58, 61, 63, 86, 182, 188, 195, 197–199, 201–204, 206–208, 217, 307, 318, 319, 329, 342, 351, 356, 358, 361, 362, 364, 366–368, 386, 390, 434, 440
stress 151, 159, 215, 218, 219, 226, 244, 263, 284, 306, 315
student 29, 30, 41, 259, 435
sustainability 30, 32, 79–82, 90, 183, 208, 320, 327, 328, 331–336, 340, 342–344, 350, 353, 363, 364, 367, 368, 426, 428, 440
sustainable (development/tourism) 9, 81, 82, 89, 92, 170, 185, 188, 328, 330–334, 340, 341, 343, 344, 350, 356, 363, 367, 426
Sustainable Development Goals (SDGs) 3, 31, 185, 329, 330, 335, 336, 342–344, 429
Sweden/Swedish 2, 3, 6, 7, 10, 31, 38, 40, 44, 78, 171, 176, 184, 236, 237, 245, 281, 282, 285, 286, 296, 304, 309–311, 313, 315–317, 328, 329, 331–335, 341–344, 350, 354–356, 358, 364, 366, 380, 381, 383–385, 389, 403, 405, 406, 431, 433, 438–440

T

tariffs 19, 20, 23, 25, 26, 32
teams/teamwork 87, 201, 205, 218, 420
technology/artificial intelligence 2, 20, 49, 224, 283, 352, 427, 438
traditions 91, 93, 132, 220, 384, 402, 433
trends 10, 37, 182, 258, 320, 336, 376, 377, 426

U

unemployment 6, 10, 11, 13, 26, 42, 44, 64, 101, 102, 104, 137, 187, 297, 308, 332, 363, 393
union (trade)/unionisation 6, 12, 13, 25, 27, 41, 42, 45, 174, 329, 341, 355, 404, 408, 430, 435, 437
urban/urbanisation 2, 6, 7, 30, 145, 235, 296, 316, 321, 342, 407

V

values (work) 263, 307, 437
visitor 9, 10, 81, 236, 337, 341, 375, 408
voluntary (work)/voluntary/involuntary 104, 109, 115, 116, 260, 289, 290, 306, 333

volunteer/voluntourism 102–118, 179, 187, 320, 336, 355, 428

W

wage/low wage/low-wage 2, 10, 12, 22, 41, 64, 66, 116, 126, 127, 148, 155, 159, 174, 175, 177–181, 186, 239, 269, 270, 308, 309, 349, 355, 363, 366, 367, 404, 419, 434, 435
welfare 4, 10, 11, 31, 40, 42, 52, 59, 160, 186, 308, 309, 322, 332, 344, 352, 353, 356, 358, 361, 366–368, 404, 429, 430
wellbeing/well-being 67, 71, 72, 125, 146, 150, 186, 199, 208, 218, 262, 312, 313, 320, 358, 363, 366, 367

Z

zero-hour (contracts) 65, 332, 351, 355, 429

The manufacturer's authorised representative in the EU is Springer Nature Customer Service Centre GmbH, Europaplatz 3, 69115 Heidelberg, Germany. If you have any concerns regarding our products, please contact ProductSafety@springernature.com

Printed and bound by CPI Group (UK) Ltd, Croydon, CR0 4YY
30/09/2026
02225649-0017